SELFORGANIZOLOGY

The Science of Self-Organization

SELFORGANIZOLOGY

The Science of Self-Organization

WenJun Zhang
Sun Yat-sen University, China

World Scientific

NEW JERSEY • LONDON • SINGAPORE • BEIJING • SHANGHAI • HONG KONG • TAIPEI • CHENNAI • TOKYO

Published by

World Scientific Publishing Co. Pte. Ltd.
5 Toh Tuck Link, Singapore 596224
USA office: 27 Warren Street, Suite 401-402, Hackensack, NJ 07601
UK office: 57 Shelton Street, Covent Garden, London WC2H 9HE

Library of Congress Cataloging-in-Publication Data
Zhang, Wen-Jun.
Selforganizology : the science of self-organization / WenJun Zhang, Sun Yat-Sen University, China.
pages cm
Includes bibliographical references and index.
ISBN 978-9814699488 (hardcover : alk. paper)
1. System theory. 2. Management. 3. Organizational behavior. I. Title. II. Title: Self organizology.
Q295.Z43 2015
003--dc23

2015026523

British Library Cataloguing-in-Publication Data
A catalogue record for this book is available from the British Library.

Typeset by Stallion Press
Email: enquiries@stallionpress.com

Printed in Singapore

In memory of my father, GuoXiang Zhang, and mother, GuiFang Niu

Preface

This book is the first monograph on selforganizology in the world. It covers contents such as organization and organizational theory, selforganizology: the science of self-organization; agent-based modeling, intelligence principles, catastrophe theory and methods; self-adaptation and control systems, celluar automata and spatial diffusion models, artificial neural networks, ant colony optimization, fish and particle swarm optimization, synergy, coevolution and evolutionary algorithms, correlation analysis, community succession and assembly, and mathematical foundations. From this unique book, researchers, teachers, and students will have an in-depth and complete insight on selforganizology and its applications.

I am so grateful to the teachers and scientists who gave me instructions in the past years, Profs. ShiZe Wang, HonSheng Shang, ZhenQi Li, DeXiang Gu and KG Schoenly. I am also indebted to the people who have contributed valuable suggestions, Profs. Yi Pang, GuangHua Liu, and ZhiGuo Zhang, *et al.* Special thanks are given to the anonymous reviewers for their comments and suggestions on this book. I thank my postgraduates and undergraduates LiQin Jiang, LiMin Luo, Yue Zhao, and other students for their help and cooperation in field investigation, manuscript preparation and paper publication. Also, special thanks are given to my family members, Yanhong Qi and Qi Zhang. I thank Ms Darilyn Yap and Ms Yugarani Thanabalasingam for their patience and suggestions on editing this book. This book is supported in part by International Academy of Ecology and Environmental Sciences (IAEES) and Discovery and Crucial Node Analysis of Important Biological and

Social Networks (2015.6-2020.6), from Yangling Institute of Modern Agricultural Standardization, China.

I hope this book will help readers be aware of theories and methods of selforganizology. To keep track of the new theories and methods in selforganizology, readers can browse articles published in the journal, *Selforganizology*, by the journal page: http://www.iaees.org/publications/journals/selforganizology/online-version.asp.

WenJun Zhang
Sun Yat-sen University, China
International Academy of Ecology and Environmental Sciences, Hong Kong
E-mail: wjzhang@iaees.org, zhwj@mail.sysu.edu.cn

Contents

Chapter 1

Organization and Organizational Theory

To better understand self-organization, it is necessary to have a basic knowledge on organization and organizational theory. Some details can be found in Zhao and Zhang (2013).

1.1 Definitions

1.1.1 *Organization*

The definition of organization has been updated and improved over the past hundred years.

Daft and Armstrong (2007) treated organization as a goal-directed social entity that is designed as a deliberately structured and coordinated dynamic system connecting with the external environment. An organization cannot thrive without successful and powerful traits. Faced with threats and chances, it should be sensitive to external changes and keep adapting and learning (Hannah and Lester, 2009). Learning is not limited to the scope of knowledge per se but a "problem-oriented action" or "knowing" (Kuhn and Jackson, 2008). Roberts (2007) discussed contemporary organization by summarizing several influential books and has managed to address relevant problems. Rashman *et al.* (2009) reviewed the literature on organizational learning and knowledge relevant with public organizations particularly, and maintained their uniqueness by using the dynamic model. The external situations in the environment are also vital issues. Analyzing the community context will revitalize the research on organizations (Freeman and Audia, 2006), since organizations function with other social units interdependently. King *et al.* (2010) noted that we should locate the organization in a wider social landscape and then explore its uniqueness as a social actor.

In addition to external conditions, internal components are important. An organization cannot survive or exist without rational structure and design of the system. Rank (2008) argued that although considerable researches aim at unveiling the complicated function of organizational systems, little attention has been given to the "structural interdependencies between formal organizations and informal networks."

Santos and Eisenhardt (2005) stressed organizational boundaries, which may facilitate the understanding of organizations. Kulic and Baker (2008) also held that it was difficult to draw boundaries clearly under real-world situations. As a response, they proposed another method to cover various views of organizations in a simulative environment using computational organizational theory.

Organization can be classified into two basic categories, i.e., self-organization and external-organization (Zhang, 2013). Major difference between the two categories of organizations is whether the organizational instructions/forces come from outside the system or from inside the system. The organization with organizational instructions/forces from inside the system is called self-organization.

1.1.2 *Organizational theory*

Organizational theory is the sociological study of formal social organizations, such as businesses and bureaucracies, and their interrelationship with the environment in which they operate (Wikepedia: http://en.wikipedia.org/wiki/Organizational_theory). Tompkins (2005) argued that organizational theory was the study of how and why complicated organizations behave the way they are. A complex organization is always too enormous and structurally differentiated to be effectively represented by a single individual. Organizational theory is neither a single of theory nor an integrated body of information but a field of studies which cover various scientific fields.

Donaldson (2003) viewed organizational theory as a positive science. Driven by the environment, scientific methods validate and testify these positive but normative theories. So far, organizational science has made a huge progress by using the positivist approach. Organizational theory has proved to have strong potentiality in the future with regards to the positivist approach.

Hatch and Yanow (2003) called organizational theory an interpretive science. Many interpretive researchers held that social world and natural world ought to be ascertained in different ways.

Willmott (2003) viewed organizational theory as a critical science. While Chia (2003) thought organizational theory as a postmodern science and drew our attention to the requirement for managers and policymakers. Obviously they were more aware of the basic information and situation of our society and industries.

Ocassionally, some peculiar things could boost our understanding on organizational theory. Jones and Munro (2005) examined the works of eighteen researchers on modern organizational theory in the last twenty years. Many topics and debates were discussed including some basic concepts and postmodernism. Warner (2007) mentioned modern literary Guru Franz Kafka. His works shed light on the deep examination of organizations, and were further compared with Max Weber. In addition, some researchers explored the organizational theory in terms of its logics with novel insights and methods (Hannan, 2007; Kamps, 2009; Durand, 2008), which was mainly shaped in the book, *Logics of Organizational Theory: Audiences, Codes, and Ecologies* (Hannan, 2007). However, this approach focuses on the entire process of theory-shaping, which deviates from the traditional way of organizational ecology (Kamps, 2009). Audia *et al.* (2006) connected the theories of organizational ecology and social network and dug into the variations "in rates of foundings over geographic locales" affected by the structure of relations in various populations.

Organizational design is a major field in organizational theory. It is becoming more important due to a series of development trends, such as advances in information technology, attempts by large global economies to rapidly transform the organizational infrastructure of their administration, the professionalization of the non-governmental organization (NGO) and charity sectors, and increasing attempts of multinational corporations to exploit globally distributed intellectual resources, etc. (Puranam, 2012).

1.2 History of Organizational Theory

Organizational theory has appeared for a very long time in history, which aims to pursue scientism, managerialism, and enhanced efficiency and effectiveness (Üsdiken and Leblebici, 2001). It is an ancient but also modern science. Organization research has occupied its status in science since Aristotle's era (Rosvall and Bergstrom, 2011). The research of organizations began its journey along with the human civilization. However, organizational theory was not recognized as a science until the 1960s (Cunliffe, 2008). Since the 19th century, in particular the era of industrial revolution, organization studies have quickly developed, especially in such areas as socio-political questions (Wolin, 1961).

According to Cunliffe (2008), the development of organizational theory can be divided into four stages: (1) classical and scientific management/modernism; (2) systems and contingency theories [contingency theory is a class of behavioral theory that claims that there is no best way to organize a corporation, to lead a company, or to make decisions. Instead, the optimal course of action is contingent (dependent) upon the internal and external situation. See Wikipedia: http://en.wikipedia.org/wiki/Contingency_theory]; (3) social constructionism (social constructionism, or the social construction of reality, is a theory of knowledge in sociology and communication theory that examines the development of jointly constructed understandings of the world. See Wikipedia: http://en.wikipedia.org/wiki/Social_constructionism); and (4) postmodernism (postmodernism is a late 20th century movement in the arts, architecture, and criticism that was a departure from modernism. Postmodernism includes skeptical interpretations of culture, literature, art, philosophy, history, economics, architecture, fiction, and literary criticism. See Wikipedia: http://en.wikipedia.org/wiki/Postmodernism). The first stage is classical and scientific management stage, and Adam Smith, Carl Marx, Taylor, Weber, *et al.*, were representative researchers during this period. These researchers have drawn and distilled various theories from routine and social activities, and built fundamental concepts of organizations. The second stage, system and contingency theories — i.e. modernism — was led by such researchers as Parsons, Woodward, etc. They emphasized the optimization of production efficiency and stressed the need of treating organization as a sophisticated system (Barzilai, 2010). The third stage, social constructionism, was mainly contributed by Berger, Goffman, Weick, etc. They held the shared belief that organizations were important because they were actually communities interacting with each other. In the last stage — postmodernism, more researchers appeared, such as Harvey, Cooper, and Burrell. During this period, various thoughts on organizations formed and evolved.

A little different from the classification described above, Docherty (2001) classified the development of organizational theory into three big stages: (1) classical theory; (2) neoclassical theory; and (3) contemporary theory. In the first stage, the mass production facilitated the overall formation of organizations and relative theories. Focuses in this stage were the studies of some contents about laborers, labor division, and scientific management — such as hierarchy, span of control, the degree of centralization and the specialization of work. Unlike the first one, in the second stage — neoclassical theory, organizational theory puts its main focus on the individuals and their mutual relationship (or interactions). In the last stage, contemporary theory, various theories appeared and organizational theory stepped into a new era.

1.3 Major Theories

Organizational theory is made up of different theories. These theories are summarized as follows (Zhao and Zhang, 2013).

1.3.1 *Classical organizational theory*

Overall, it consists of three sub-theories — scientific management, Weber's bureaucratic theory, and administrative theory.

(a) Scientific management. Frederick W. Taylor is recognized as the pioneer of scientific management. It is thus called Taylorism. He proposed the theory with the observation, analysis, and synthesis of workflows. The focus is how to minimize cost, maximize economic efficiency, and achieve specialization and standardization. However, this theory is criticized for the reason that this system overlooks the human's perception and senses.

(b) Weber's theory of bureaucracy. Max Weber's theory considered that an organization is governed by top-down rules and regulations. Employees work on strictly defined responsibility and own limited power: Bureaucratic officials need expert training; Rules are implemented by neutral officials; Career advancement depends on technical qualifications judged by the organization, not individuals (See Wikipedia: http://en.wikipedia.org/wiki/Bureaucracy#Max_ Weber).

(c) Administrative theory. Administrative theory was pioneered mainly by Henri Fayol. It takes the form of hierarchical pyramid as its structure. It was concerned principally with achieving the "most rational" organization for coordinating the various tasks specified within a complex division of labour (See: http://www.encyclopedia.com/doc/1O88-administrativetheory.html). Fayol put forward 14 principles to advise managers on how to mandate and fulfill their responsibility. In addition, he outlined five basic elements of management: planning, organizing, command, coordination, and control.

1.3.2 *Neoclassical organizational theory*

The Neoclassical theory began with the Hawthorne studies in the 1920s. Significantly different from the early thoughts and approaches, this theory particularly emphasized "affective and socio-psychological aspects of human behaviours in organizations." (See Wikipedia: http://en.wikipedia.org/wiki/Organizational_theory#Neoclassical_perspective). Personnel relationships among workers, employees, and managers were stressed, reflecting the growing need of humane and emotional care of workers. Laborers with high concentration and volition contribute positively and meticulously, so the company and factory would benefit more and function better. A lot of studies were thus conducted by sociologists and psychologists, among which Elton Mayo contributed the most.

1.3.3 *Contemporary theories*

Dozens of new theories have appeared in modern times. Modern theories evolve quickly into various shapes and structures in rapidly-changing environments. Several major theories or methods are described as follows.

(a) The systematic theory. According to this theory, an organization is viewed as a system which composes of many mutually connected components, aiming benefit both internally and externally. Overall an organization consists of three parts: components, linking processes, and goals of organization (Bakke, 1959).
(b) Contingency theory. Contingency theory, proposed by Lawrence and Lorsch, suggests that there is no best way to direct an enterprise, because the requirements for corporations vary enormously in different environments and conditions. A series of factors may work as variables, including environmental uncertainty, technology, size, strategy, resource dependence, and public accountability (Tompkins, 2005).
(c) Other theories. There are some other theories that play a better role in the modern organizational theory, such as quality management theory, organizational culture, leadership theory, and so forth.

After sharing Sedov's statement that moving matter has only two characteristics: the intensity of movement and the organization of movement, Aleinikov proposed organizology — the science on organization of movement — following this binary logic (Aleinikov and Smarsh, 2010). It is true that one cannot find any entry about this terminology in Webster's dictionary and even by searching it on the Internet. Actually, this terminology was crafted by Aleinikov in 2003. Science per se needs to be ordered and organized properly. Moreover, considering the development of science and organization a science of organization should have been founded. But this did not happen until recent years due to the absence of measurement or organization. The proposition of organizology is a beneficial attempt in this aspect. Organizology was founded on one basic measurement to address complex problems with one simple explanation (Aleinikov, 2004). But the defining and refining of this measurement proved to be a challenge and a new unit of organization expressed by the formula T/L (time divided by space) was offered. Also, a few cases were provided to explain its nature in different kinds of organizations and this unit of organization was named "aleandr" (Aleinikov, 2005). Furthermore, an accurate prediction based on mathematics was given to elucidate this science (Aleinikov and Gera, 2006). It could be expected that any nascent theory would confront controversy and criticism and organizology is no exception. In any case, a breakthrough thought to the classic theory ought to be encouraged.

1.4 Prospect of Organizational Theory

Organizational theory is an unavoidable outcome of historical development, under the impact of multiple forces: industrial revolutions, technological revolution, digital revolution, and the third industrial revolution (Rifkin, 2013). A series of revolutions have produced novel thoughts and minds, industries and sciences, craftsmanship and technologies, which endowed humans with new lifestyles and jobs, new ideas to understand the world, and new management methods. Since the late 1980s, the trajectory of organizational theory has changed from "paradigm-driven work to problem-driven work" (Davis and Marquis, 2005). Huge changes of organization theory come as a result of discontinuous and fast-changing environment (Marshak, 2004). Accordingly, a new model of reconfigurable organization has been given (Stefanovic *et al.*, 2011). Organizational theory absorbs nutrients from other scientific fields and industries unprecedentedly, and eventually evolves into various shapes and colors. No one can give an exact answer to the question: what is the future of organizational theory? Every researcher elucidates his own ideas and imaginations and as a consequence, various thoughts emerge and evolve.

Walsh *et al.* (2006) have posed three fundamental questions about the future of organizational theory. "How can we understand, live in and live with today's organizations?" He held that the trend is hardly traceable and predictable, considering this fast-changing world and the influences of globalization. But as difficult as it may be, we can work hard to collect data and empirical evidences and try to handle it. Burrell (2003) proposed another two questions concerning the subsequent research areas to deal with and methodology and epistemology approaches to use. Davis and Marquis (2005) argued that the central aim in the 21th century would be to explain the economic institutions.

Andriai and McKelvey (2009) introduced 15 kinds of scale or scale-free scientific theories, among which eight theories can be used in organizational theory:

(1) Phase transition.
(2) Creation of spontaneous orders (McKelvey and Lichtenstein, 2007).
(3) Preferential attachment (Albert and Barabási, 2002). The theory states that the larger the node is, the more capacitity it attracts a new agent, thus the node gets bigger and bigger (Zhang, 2012a).
(4) Power-law distribution theory (Barabasi, 2002; Newman, 2006). It states that multiple-exponential distribution, or logarithmic-normal distribution, or their combinations will lead to power-law distribution as complexity increases.
(5) The principle of least effort (Zipf, 1949). It postulates that animals, people, even well-designed machines will naturally choose the path of least resistance or "effort" (Wikipedia: http://en.wikipedia.org/wiki/Principle_of_least_effort).
(6) Square-Cubic rate. It states that body surface area for absorbing energy increases at the rate of square-level, but life grows at the rate of cubic-level, which leads to an imbalance and thus, fractal will emerge to offset the decline in the ratio of surface area *vs.* volume.
(7) Connection cost (Simon, 1962). It states that the number of cell divisions (n) increases at the rate of square-level, but the number of possible connections is $n(n-1)/2$, which leads to an imbalance

between the benefit of cell divisions and the cost for sustaining connections and thus, life must generate some modules to reduce the connection cost.

(8) Critical points of self-organization (Bak, 1996).

1.5 Applications of Organizational Theory

Phillips (1992) held that some methods used in natural science had been partially accepted by social science and vice versa (Burrell and Morgan, 1979). As a social science, organizational theory has been used in such natural sciences as biology and ecology (Zhao and Zhang, 2013).

Biological systems like social systems represent hierarchical organizations with sub-modules that cover multiple scales (Rosvall and Bergstrom, 2011). Organizational ecology has been an attractive research in the last decade mainly in the USA. It aims to explain "how social, economic and political conditions affect the relative abundance and diversity of organizations and to account for their changing composition over time" (Baum and Amburgey, 2005), and to emphasize the "evolutionary dynamics of processes" behind them (Singh and Lumsden, 1990). On the other hand, to understand the organizational diversity means to answer a question: "Why are there so many (or so few) kinds of organizations?" (Hannan and Freeman, 1993). This theory uses biological and ecological models to analyze businesses and issues about organizations (Clegg and Hardy, 1999). Dobrev *et al.* (2006) argued this theory as a "research paradigm," and by this theory, "multivariate models" are used for various potential reasons. Hannan and Freeman (1977) proposed a perspective of population ecology to analyze organization-environment relationship. Carroll (1984) reviewed research on organizational ecology and especially three different levels of analysis and methods: organizational, population, and community, followed by developmental, selection, and macro-evolutionary method, respectively. The development on organizational taxonomies was also recommended. Within more than thirty years, organizational ecology has taken a huge step. Innovations of knowledge and theory are required to revive it. Particularly, ecological theory is of vital importance and most connected with evolution and it can be given more attention (Amburgey and Rao, 1996).

Baum and Amburgey (2005) maintained that ecological approaches are radically different from the traditional ones. The former methods focus on the contextual factors. However, organizational ecology that has built mainly on the population models indeed confuses some sociologists (Hannan, 2005). Using biological theories or metaphors to explain organizational changes has been critically misunderstood (Singh and Lumsden, 1990).

In recent years, Reuter *et al.* (2010) proposed new approaches to explain ecological interactions across scales. They stressed multiple organizational hierarchies and their mutual effects. Cross-scale interactions are among the most prominent concerns in ecology and biodiversity problems, invasive species and long-term effects of habitat change (Kerr *et al.*, 2007). Lidicker (2007) came up with the fourth level — "ecospace" of levels of ecological organizations, apart from the other three — organism, population and community — given the hierarchical arrangements being increasingly favored by many ecologists.

Actually, organization assembles organisms to some extent (Morgan, 2006). They function as a whole. Accordingly, organizational theory may be rationally used to fit with the natural world. Organizations interact with communities they dwell in (Freeman and Audia, 2006). The population ecologists hold the view that the ability of seizing a resource niche and defeating its rivals are what really matter. The concept of ecological niche has been successfully adopted by organizational theory researchers. In contrast with the niche in the natural world, the organizational niche reveals itself in the social and economic world (Boone *et al.*, 2002). Therefore, sociologists have made their minds to discover the appropriate niche by which an organization can develop and prosper (Hannan, 2005). In fact, this view does not satisfy many organization theorists, since they emphasize the role of managers and decision makers, not merely the environments, in choosing organizations (Morgan, 2006). For instance, some critics refuted the contingency theory that an organization can self-adjust to fully utilize its environment (Pfeffer and Salancik, 2003).

Great theories pave the way for the dawn of typical practices. Couzin (2006) has paid attention to the social organization in one of the most complex "fission-fusion" systems in nature — elephant populations. However fully elucidating and understanding this system by using various technologies and constructing mathematical models proves a huge challenge still. Co-evolutionary method plays an important role in helping researchers who study the natural environment and organizations to revise their organizational theory (Porter, 2006). It works as a propeller (Lewin and Volberda, 2003). However, it should be noted that organizational co-evolution involves some aspects that are absent from its biological counterpart, knowledge, learning, demand, actor traits and behavior, strategy and tactics (Malerba, 2006; Zhang, 2012a). Organizations not only hold a position in large-scale environments but also make them the fundamental concepts in cell biology: self-organization has emerged to explain and understand the components and compartments of the cells (Karsenti, 2008).

Chapter 2

Selforganizology: The Science of Self-organization

The scientific terminology, selforganizology, originated from Zhang (2013a). The journal, *Selforganizology*, can be referenced at http://www.iaees.org/publications/journals/selforganizology/selforganizology.asp, for some of the most recent advances. In this chapter, I will discuss selforganizology with more details.

2.1 Self-organization

2.1.1 *Theories and principles of self-organization*

As described earlier, the organization with organizational instructions/forces from inside the system is called self-organization. Self-organizing systems are those systems which can evolve and improve the organization's behaviors or structure by themselves. In a self-organizing system, the system evolves spontaneously to form an ordered structure based on some compatible rules. A system is called a self-organizing system if there is no specific intervention from the outside during the process of evolution. The stronger a system's self-organization capacity is, the stronger the system's ability to generate and maintain new functions.

Self-organization is a process that some form of global order or coordination arises out of the local interactions between the components of an initially disordered system. This process is spontaneous, i.e., it is not directed or controlled by any agent or subsystem inside or outside of the system; however, the laws followed by the process and its initial conditions may have been chosen or caused by an agent. It is often triggered by random fluctuations that are amplified by positive feedback. The resulting organization is wholly decentralized or distributed over all components of the system. As such it is typically very robust and able to survive and self-repair substantial damage or perturbations (Zhang, 2013a, b, 2014; Wikipedia, 2014).

Unlike other organizations, the self-organizing system arises only from the interactions between the basic components of system, without external instructions and forces. During the process of self-organization, some structural components can interact and cooperate to display behaviors that only a group will have. The dynamic interactions between low-level components typically include attraction and repulsion, that is, positive and negative feedbacks.

Generally, self-organization arises from the increase in complexity or information. According to the thermodynamic laws, this situation will only occur in the open systems far from thermodynamic equilibrium. For most systems, this means the energy supply to the system is needed for generating and maintaining a certain mode. In an abstract sense, self-organization is a dynamic process that makes an open system change from the disorder to order states and thus reduces system's entropy by absorbing "negative entropy" from outside the system (Glansdorff and Prigogine, 1971; Nicolis and Prigogine, 1977).

From the perspective of systematic theory, self-organization is an irreversible dynamic process. Each component in the system will spontaneously aggregate to form an organic entity without external instructions. From the view of mathematics and physics, self-organization means the dimensional reduction of state space or the reduction of degrees of freedom, i.e., the system converges spontaneously to one or more steady states — attractors. In such a system, the local interactions between the basic components of the system can change the modes of the system's organization, and the global behaviors of the system cannot be understood intuitively. They cannot be understood by simply observing existing laws and behaviors of between-component interactions (Zhang, 2012a, 2013a, 2014). In other words, the global properties of self-organizing systems are not predictable.

Self-organization are usually based on three conditions (Bonabeau *et al.*, 1999): (1) strong nonlinear dynamic interactions, even though they do not necessarily correlate to the positive or negative feedbacks; (2) an equlibrium between development and exploration; and (3) complex and diverse interactions.

Prigogine believed that conditions for self-organization include: (1) the system must be an open and dissipative system, where a dissipative system is a thermodynamically open system that operates out of and often far from thermodynamic equilibrium in an environment with which it exchanges matter/energy (Wikipedia: http://en.wikipedia.org/wiki/Dissipative_system); (2) the system is far from thermodynamic equilibrium, which allows for entering the non-linear zone; (3) nonlinear interactions exist between components; and (4) some parameters fluctuate and if fluctuations reach some thresholds, the system will change to unstable from steady state, catastrophe will occur and the system may thus exhibit a highly ordered state. For example, we often try to put sand dune higher, but to a certain height, the addition of a little amount of sand will cause landslide of a large amount of sand in the sand dune, and the sand dune cannot be piled further. In fact, before landslide occurs, sand dune has reached a critical state (threshold), and thus small perturbations can lead to instability.

2.1.1.1 *Thermodynamic basis of self-organization*

The spontaneous formation of new structures, for example, crystallization process, Bénard phenomenon, Belouzov-Zhabotinsky reaction (Sun and Lin, 2004), etc., are all self-organization processes, i.e., the formation of a structure or a mode does not need to be imposed by any external force. It seems that the components of these systems are arranged into a more ordered pattern by themselves. It appears that self-organization violates the second law of thermodynamics. This law holds that the entropy of a closed system can only increase rather than decrease. In other words, the second law of thermodynamics means that an isolated system should evolve in a uniform, simple, difference-eliminating way, which is in fact an evolution to a low-level organization.

In the example of crystallization process, the randomly moving molecules, which have been bonded into a crystal structure and thus have been fixed inside it, will transmit their kinetic energy to the liquid that they are dissolved in. Thus, the reduction in the entropy of crystallization process is offset by the increase in the entropy of liquid. The entropy of the whole system has actually increased, and therefore it is consistent with the second law of thermodynamics.

For self-organizing systems, which are not in their equilibrium states, it is hard to determine whether the second law of thermodynamics is true or not. Prigogine started to study systems far from thermodynamic equilibrium states since the 1950s and proposed the dissipative structure theory (Glansdorff and Prigogine, 1971; Nicolis and Prigogine, 1977), in which the most used models to explain the dynamic self-organization process are Bénard cells and the Brusselator. The dissipative structure theory tries to address such problems as: under what conditions a system will be able to evolve from the disordered to the ordered state, and form a new, stable, and internally-dynamic structure. The theory concludes that such a structure must be an open system, i.e., there are energy/matter flows in the system; the system will continuously generate entropy, but at the same time the entropy will be actively dissipated from the system or output from the system. Thus, at the cost of environmental disorder, the system will be able to increase order of its own. The system will be able to follow the second law of thermodynamics simply by removing its excess entropy. This dissipation can be mostly found in life systems. Plants and animals obtain energy and matter by absorbing light or food with low entropy, and then output energy and matter by draining the metabolic waste of high entropy. This will reduce its internal entropy to offset the degradation process required by the second law of thermodynamics.

The output of entropy cannot explain why and how self-organization happens. Prigogine believed that self-organization would mostly occur in nonlinear systems far from thermodynamic equilibrium.

2.1.1.2 *Principles of self-organization*

In the first symposium on self-organization held in 1959 in Chicago, British cybernetic expert, Ashby, proposed "principle of the self-organizing dynamic system" (Ashby, 1947). It states that any deterministic dynamic system, ignoring its classification or composition, will automatically evolve towards a equilibrium state, i.e., an attractor. It will leave behind all non-attractor states, i.e., the attractor's basin, and select the attractor out of all others (Wikipedia, 2014). The further evolution of the system is thus constrained to remain in the attractor. This theory reduces our uncertainty on the system's state and solves the problem of entropy in systematic science. It is equivalent to self-organization, which finally reaches equilibrium and the final equilibrium can be considered to be a state of mutually adaptation of all components in the system. Another cybernetic expert, Heinz von Forster, proposed the principle of order from noise (Foerster, 1960, 1996). It states that self-organization is facilitated by random perturbations (i.e., noise) which allow the system to explore a variety of states in the state space. It increases the chance that the system would arrive in the basin of a "strong" or "deep" attractor, from which it would then quickly enter the attractor itself. This idea is very simple: The larger the state space that a system moves through is, the faster it will reach the attractor. If the system stays at its initial state, it will not reach the attractors, and self-organization will thus not occur (Foerster, 1996). A similar principle was formulated by the thermodynamicist Prigogine as "order through fluctuations" (Wikipedia, 2012).

Generally, there are multiple attractors in a nonlinear system. The attractor theory holds that the behavioral trajectories of a complex system in the state space can be represented by the dynamic equations. These dynamic equations are always determined by a set of "attractors." The attractor the system will move towards depends on the attraction domain that the initial state falls into. The attractor the system eventually reaches is uncertain. To find attractors of the system is the key to understanding the self-organization of the system. Small fluctuation of some parametrical values will cause the system to change. Prigogine thus proposed a related principle, i.e., order through fluctuations.

Both fluctuation and dissipation are necessary for self-organization of a system. They correspond to attraction and repulsion in the system respectively. Dissipation is the dominant factor for self-conservation of system, and fluctuation is the dominant factor for self-innovation of system. The system with fluctuation and without dissipation will lose its stability; the system with dissipation and without fluctuation will quickly terminate its evolution. The system will evolve steadily once the dissipation is dominant. The system will be governed by random process and which final state is reached depends on which fluctuation has been initiated initially and what environmental selection was made.

2.1.1.3 *Major theories on self-organization*

So far, self-organization theory is generally believed to mainly consist of three parts, dissipative structure theory (Glansdorff and Prigogine, 1971; Nicolis and Prigogine, 1977; Brogliato *et al.*, 2007), synergetics (Haken, 1978, 2004), and catastrophe theory (Zeeman, 1976, 1977; Woodcock and Davis, 1978; Saunders, 1980; Thompson and Michael, 1982). Some basic thoughts and theoretical kernel of self-organization can be derived from dissipative structure theory and synergetics.

(1) Dissipative structure theory. Prigogine officially proposed the theory of dissipative structures in a symposium on theoretical physics and biology in 1969 (Nicolis and Prigogine, 1977). The theory is proposed mainly to explain the exchange of matter/energy between the system and the environment and its effect on the self-organizing system. The structure, established on the basis of the exchange of matter/energy between the system and the environment, is a dissipative structure, such as living organisms, convection, cyclones, hurricanes, etc. A dissipative structure has a reproducible steady state. This steady state may be reached by systematic evolution (or by artifice). It is characterized by the spontaneous appearance of symmetry breaking and the formation of complex or chaotic structures where interacting particles exhibit long range correlations (Wikpedia: http://

en.wikipedia.org/wiki/Dissipative_system). Far from thermodynamic equilibrium, the openness of system, and nonlinear interactions between different components in the system, are three conditions for the formation of a dissipative structure. Far from thermodynamic equilibrium refers to the concept that the distribution of matter/energy in different areas of a system is extremely uneven. The dissipative structure theory is mostly used to treat the evolution of complex systems. It uses two levels of approaches, i.e., deterministic and stochastic approaches in the analysis of systematic evolution. Deterministic approach uses macroscopic physical variables to describe dynamics and features of the sytem. Stochastic approach treats macroscopic physical variables as the average of corresponding random variables. Analyzing random variables will not only produce averaged values but also help understand fluctuation characteristics of the system.

(2) Synergetics. Inspired by the laser theory, Haken proposed the concept of "Synergy" in 1976, and another science on self-organization, synergetics, was thus established (Haken, 1978, 1983, 2004; Graham and Wunderlin, 1987). It is a science that explains the formation and self-organization of patterns and structures in open systems far from thermodynamic equilibrium. Synergetics mainly discusses the coordination (synergy) mechanism between internal components of the system studied. It holds that the coordination between various components in the system is the basis of self-organization process. In synergetics, Haken held that on one hand, many subsystems interact to produce the structure and function at the macroscopic scale; on the other hand, there are many different scientific disciplines cooperating to explore the general principles for governing self-organizing systems. The order parameters generate and govern subsystems by competition and cooperation between various subsystems. Here, the order parameters can be treated as the amplitudes of unstable modes that determine the macroscopic pattern of system. Serving of various subsystems to order parameters reinforces order parameters themselves and further promotes the serving

of subsystems to order parameters, so that the system can spontaneously organize by itself (Haken, 1978, 2004). Competition and cooperation between order variables will result in different forms of evolution of self-organization. Competition and cooperation between order parameters of the system are direct forces for the formation of new structures. Because of independent evolution of components in the system, various local and collaborative evolution, as well as random interferences by environmental factors, the actual state of system always deviates from the average. The magnitude of such a deviation is called fluctuation. When the system is in its transition from one steady state to another one, and if the independence evolution and collaborative evolution between system components move into a balance, any small fluctuation will be amplified, and quickly spread to the whole system. The resultant giant fluctuation will promote the system to move into an ordered state. In addition, Harken proposed the concept of "functional structure," i.e., the function and structure are dependent on each other. If the energy or matter flow is cut off, the physical and chemical system will lose their structure, though a biological system is mostly able to maintain its structure for a fairly long time. Such biological systems seem to combine non-dissipative and dissipative structures together (Haken, 1978, 2004).

(3) Catastrophe theory. Catastrophe theory was first proposed by the French mathematician, Thom R, in 1969. Since the 1970s, Zeeman and other scientists have further enriched catastrophe theory by considering a special case where the long-run stable equilibrium can be identified with the minimum of a smooth, well-defined potential function (i.e., Lyapunov function), and applied to various aspects of physics, biology, ecology, medical science, economics and sociology, and produce significant impacts (Zeeman, 1976). Catastrophe theory is in essence a branch of bifurcation theory in the study of dynamical systems and is a special case of singularity theory in geometry. Bifurcation theory studies phenomena characterized by sudden shifts in behavior arising from small changes in circumstances, and analyze

> how the qualitative nature of equation solutions depends on the parameters that appear in the equation (Wikipedia: http://en.wikipedia.org/wiki/Catastrophe_theory). This theory was thus built on the basis of the stability theory.

It considers a catastrophe process as a process that transit towards a new steady state through an unsteady state from an original steady state. Small changes in certain parameters of a nonlinear system can cause equilibria to appear or disappear, or to change from attracting to repelling and vice versa, leading to large and sudden changes of behavior of the system (Wikipedia: http://en.wikipedia.org/wiki/Catastrophe_theory). In mathematical view, this means the changes of values of a set of parameters and mathematical functions that denote the states of the system. Thus it is a theory to describe the phenomena that continuous change of parameters leads to the discontinuous change of the states of system.

It treats with systems that have structural stability in the state space but some structural instability on some point sets of measure 0. The basic characteristics of catastrophe systems include: multiple steady states, reachability, jumping, lagging, and divergence (which reflects the sensitivity of evolutionary trajectory to the path of control parameters). Catastrophe theory holds that different outcomes may occur even if it is the same process that corresponds to the same controlling factors and critical values; different new steady states may be achieved at different probabilities (Saunders, 1980). Generally, catastrophe theory itself does not reveal the mechanism to produce catastrophe phenomenon. It just provides a reasonable mathematical model to describe the phenomenon of catastrophe in the real world, and classifies various catastrophe types.

In addition, there are also other theories on self-organization, for example, Eigen's super circle theory. Super circle theory is a self-organization theory of molecular evolution. However, it is a scientific hypothesis and the impact is still limited.

2.1.1.4 *Properties of self-organization*

Interacting with the environment, a self-organizing system can evolve to form new structures and functions. Unlike conventional mechanical systems, it owns its peculiar properties. Those properties can be used as part of the definition of self-organization and self-organizing systems, for example, no centralized control, and continuous adaptation to the changing environment, etc.

(1) Local interactions generate global order. The most obvious change of the self-organizing system is the formation of global order. Local interactions follow basic physical processes; any impact from a region to another one must first explore all transition regions. When it explores transition regions, all the processes will be disturbed by the turmoils occurred in the transition regions. First assume the system is disordered and all components of the system evolve randomly. The impact of any passed will be quickly dispersed and ultimately destroyed by these random turbulence. The result is thus that, starting at the chaotic state, the distant parts of the system are actually independent: they do not affect each other. During the self-organization process, all components are closely connected. Understanding the structure of a local component will be valuable in understanding the structure of components of its adjacent regions.
(2) Distributed control. We tend to consider that a highly organized system is commanded and controlled by external or internal forces. This control is called centralized control. In a self-organizing system, the control is distributed throughout the system. All components contribute to the final arrangement of the states of system. Despite some of the advantages of centralized control with respect to distributed control, at some levels centralized control must be based on distributed control. For example, the function of human brain disperses over the network of interacting neurons. Different brain regions perform specific functions, but a neuron or a group of neurons do not have the ability to control the brain. This is a result of self-organization.
(3) Nonlinearity and feedbacks. Nonlinearity means the whole is not equal to the simple sum of its parts, i.e., superposition principle is not met. Suppose a system is represented by a function: $y = f(x)$. If the following condition is satisfied: $f(\alpha x_1 + \beta x_2) =$

$\alpha f(x_1) + \beta f(x_2)$, where $\alpha, \beta \in R$, then it is a linear system, otherwise it is a nonlinear system (Zhang, 2010, 2012). Judging from the mechanical movement, a linear phenomenon is generally manifested as smooth motion in time and space; it can be described by functions with good performance, and continuous and differentiable functions. The nonlinear phenomenon is a movement from regular motion to irregular motion, with obvious jumping and intermitting features. From the view of perturbation and parameter theory, the response of a linear system is smooth and proportional to changes, but a nonlinear system will exhibit substantial changes in some key points because of the small changes in parameters and form, and maintain spatially regular and order structures. Linear relationship is independent of each other, while the nonlinear relationship is an interactive one, which makes it violate superposition principle and produce gain or loss. In nonlinear systems, there are feedbacks between system components; each component affects the others, and other components in turn affect it. The positive feedback plays a role similar to the input so that if the system's deviation increases, and the system's oscillation is thus amplified. The negative feedback causes a reverse outcome as compared to the input's role, the system's output error can thus be reduced and the system is thus stabilized.

In complex self-organizing systems, there are often several chains of positive and negative feedbacks, so a change can be amplified in a certain direction but suppressed in the other directions. This will result in very complex behaviors that are difficult to be predicted.

(4) Far from thermodynamic equilibrium. Equilibrium is a special state of a system. At this state, the measurable macroscopic physical properties of the system are uniform throughout the system (so that there is no macroscopic irreversible process inside the system). At the equilibrium state, the system follows the first law of thermodynamics: $d_E = d_Q - pd_V$, i.e., the increment of energy inside system is equal to the absorbed heat subtracted by outward work done by the system. It also coincides with the

second law of thermodynamics: $d_S/d_t \geq 0$, that is, the spontaneous evolution of system is always towards the direction of entropy's increase. For the system in thermodynamic equilibrium state, it must remove its extra energy; it will remain in the minimum energy state without the input of external energy. A system will likely move across a nonlinear region if the system is far from thermodynamic equilibrium. The system far from thermodynamic equilibrium is more sensitive and more vulnerable to environmental changes due to its dependency on external energy input. But it is more powerful in its response to changes. On the other hand, the surplus of external input energy allows the system to amplify the self-organization process, for example, offsetting small perturbation or maintaining positive feedbacks longer in the aid of strong interactions. This makes the system more vigorous and more adaptive to external changes.

(5) Systematic termination and organizational hierarchy. The interactions between individual components of the self-organizing system can be, to some extent, defined as an ordered structure. However, the order does not mean organization. Organization is an ordered structure and can achieve a particular function. In a self-organizing system, this function is to maintain a particular structure under various disturbances. The general characteristics of self-sufficiency can be understood as a closure. A process with causal relationship can be represented as a chain or a sequence: A $\rightarrow$ B $\rightarrow$ C $\rightarrow$ D $\rightarrow \ldots$, where A initiates B, B initiates C, C initiates D, and so on. Overall, this will lead to a continuous change. However, there may be a termination of the sequence such that a loop system result, for example, O returns J, so the cycle of the system becomes J, K, L, M, N, O, J, K, L. Thus, the corresponding arrangement of system will always be maintained or recycled. In addition, if the loop is placed in a negative feedback region, it is relatively unaffected by the impact of external interference (Foerster, 1960).

A self-organizing system may contain a lot of autonomous and organization-closed subsystems. Those subsystems will interact in a more indirect way. They will also adapt to the structure for termination and determine subsystems at a higher level. New generated subsystems will contain the original subsystems as their components. Each self-organizing system constitutes a series of subsystems. A self-organizing system thus forms a layered structure. Each self-organizing system belongs to a high-level self-organizing system and contains low-level self-organizing systems. It interacts with other self-organizing systems at the same level. Therefore, the hierarchization of the system's structure and function is also a characteristic of self-organizing systems.

2.1.2 *Existing algorithms of self-organization*

Because it is hard to predict the complex behaviors of self-organizing systems, sometimes we use mathematical modeling and computer simulation to describe these systems. They also help people understand how these systems work. A mathematical modeling method for self-organization is to use differential equations, and another method is to use cellular automata (Wolfram, 2002; Ballestores and Qiu, 2012; Zhang and Gu, 2001; Zhang *et al.*, 2011; Zhang, 2011, 2012, 2013a, 2014).

Many optimization algorithms can be considered as a self-organization system because optimization aims to find the optimal solution to a problem. If the solution is considered as a state of the iterative system, the optimal solution is essentially the selected, converged state or structure of the system, driven by the algorithm based on the system landscape (Yang *et al.*, 2013; Yang, 2014). In fact, we can view an optimization algorithm as a self-organization system.

In the sense of optimization, existing algorithms of self-organization can be classified into four hierarchies, Monte Carlo method, heuristic method, meta-heuristic method, and hyper-heuristic method (Mirjalili *et al.*, 2014, Zhao, 2014).

2.1.2.1 *Monte Carlo method*

Monte Carlo method tries to obtain deterministic solution by using random numbers. Monte Carlo method is used to test the statistical characteristics, approximate the distribution of statistic with asymptotic approximation, estimate the variance and test statistical significance, and compute the expectation of function of random variables (Manly, 1997; Zhang and Shoenly, 1999a, b; Zhang and Schoenly, 2001; Ferrarini, 2011; Zhang, 2010, 2012; Zhang, 2011a, b, c; Zhang and Zheng, 2012).

2.1.2.2 *Heuristic methods*

Heuristic methods use the heuristic information contained in the problem itself to guide search process. This information is usually localized, limited, and incomplete. Heuristic methods use some simple heuristic rules to search for solution in a limited search space, which can greatly reduce the attempts and quickly reach the solution even though the search process occasionally failed. The key of the methods is to design simple and effective heuristic rules.

The greedy algorithm is a typical heuristic method. A greedy algorithm is an algorithm that follows the problem solving heuristic of making the locally optimal choice at each stage with the hope of finding a global optimum (Cormen *et al.*, 1990). A greedy strategy does not in general produce an optimal solution, but may yield locally optimal solutions that approximate a global optimal solution in a reasonable time (Wikipedia: http://en.wikipedia.org/wiki/Greedy_algorithm).

In addition, gradient-based algorithms such as Newton's method, conjugate gradient method, etc., can be called heuristic methods using different levels of heuristic information.

Generally, the efficiency of a heuristic method depends on the available amount of heuristic information that a problem can provide. For example, Newton's method uses the heuristic information in Hessian matrix to solve problem in the local area to achieve quadratic convergence rate. Thus, Newton's method is highly efficient and specialized. However, a general greedy algorithm with limited heuristic information is inefficient but may be widely applicable.

2.1.2.3 *Meta-heuristic methods*

Roughly speaking, meta-heuristic methods can be viewed as the population-based heuristic methods with some stochastic characteristic. It tries to have both speciality and high-efficiency of heuristic methods and simplicity of Monte Carlo method. Nevertheless, the two requirements are contradictive for each other. How to avoid its disadvantages is extremely important.

Meta-heuristic methods can be generally divided into three categories: evolutionary, physics-based, and swarm intelligence algorithms. Sometimes meta-heuristic methods are also equivalent to population-based optimization algorithm, natural computation, computational intelligence, intelligence computation (the latter three methods can be viewed as evolutionary algorithms), etc. Many complex problems are hard to be addressed by conventional artificial intelligence technologies. Intelligence computation is a powerful technique in addressing more complex problems. In the intelligence computation, computation is intelligent. It can automatically adjust parameters during the process of computation, and thus achieve optimal results (Koza, 1992). Evolutionary computation searches for the optimal solution by simulating biological evolution in nature, for example, genetic algorithms, etc. Swarm intelligence algorithms are a kind of new evolutionary algorithms (swarm intelligence can be viewed as the population-based optimization with population size 1), which are closely related to evolutionary strategies and genetic algorithms.

2.1.2.3.1 Evolution- and population-based method

The fundamental rule for the development of these algorithms is Darwinian natural selection, survival of the fittest (Chen *et al.*, 2014). Main operations of algorithms include recombination/crossover, mutation, selection, and other operations. Driving evolution of agents is achieved by using these operations. They are also called evolutionary algorithms, evolutionary computation, bio-inspired computing, etc. In a narrow sense, these algorithms include genetic algorithm (GA), evolution strategy (ES), evolutionary programming (EP), and genetic programming (GP). Broadly speaking, in addition to these algorithms (GA, ES, EP, and GP), further expansion include immune optimization algorithm (IOA), differential evolution (DE), biogeography-based optimizer (BBO), and memetic algorithm, etc.

Among these IOA also include four algorithms, clonal selection algorithm (CSA), artificial immune network (AIN), and negative selection algorithm (NSA).

Genetic algorithm is a kind of stochastic search algorithms that simulate the evolution of organisms (survival of the fittest, genetic mechanism). It was first proposed by Holland (1975). It aims to explain the adaptive processes of natural and artificial systems. Main characteristics of genetic algorithm include: (1) directly operating structural objects; (2) there is no assumptions on derivative and function continuity; (3) implicit parallelism and better search performance on global optimization; (4) using probabilistic optimization-searching method which can automatically obtain and guide optimized search space; and (5) adaptively adjust the search direction without determinant rules. These properties make genetic algorithm to be widely used in combinatorial optimization, machine learning, signal processing, adaptive control, and artificial life. Genetic algorithm is considered the key technology that will significantly impact the future of computing technology, along with adaptive systems, cellular automata, chaos theory, and artificial intelligence, etc.

2.1.2.3.2 Physics-based method

The mechanism of these algorithms is different from the genetic mechanism. They construct population-based optimization algorithm according to some physical laws. By using some rules inspired by physical laws, agents can mutually exchange information and move in the search space, and finally the population solution is achieved. They include simulated annealing (SA), gravitational search algorithm (GSA), chemical reaction optimization algorithm (CRO), gravitational local search (GLSA), Big-Bang Big-Crunch (BBBC), Central Force Optimization (CFO), charged system search (CSS), black hole (BH) algorithm, ray optimization (RO) algorithm, small-world optimization algorithm (SWOA), galaxy-based search algorithm (GbSA), curved space optimization (CSO), Tabu search algorithm (Tabu Search, TS), etc.

2.1.2.3.3 Swarm Intelligence (SI)-based method

The concept, swarm intelligence, was first proposed by Hackwood and Beni (1992) in their cellular automata system. Swarm intelligence refers to a group of unintelligent entities that can cooperate to solve problems in a distributed way. They can directly or indirectly communicate by changing the local environment. These unintelligent entities behave intelligently through cooperation (Bonabeau *et al.*, 1999; Hu and of Li, 2008; Zhang, *et al.*, 2008). A significant feature of swarm intelligence is that though the behaviors of an individual are simple, when they work together, the system exhibits very complex behaviors. Without centralized control and global model, swarm intelligence provides a solution for distributed problems.

Swarm intelligence simulates population search, collaborative behavior, and emergency phenomenon of biological population to achieve population-based intelligence search behavior that cannot be achieved by a single individual. Through group collaboration, information exchange and social intelligence, the optimal solution can be achieved. Swarm intelligence computation includes particle swarm optimization (PSO), ant colony optimization (ACO), artificial bee colony optimization (ABC), bat-inspired algorithm (BA), artificial fish-swarm algorithm (AFSA) (Li, 2003; Grosenick *et al.*, 2007; Chen *et al.*, 2009), grey wolf optimizer (GWO), weed optimization algorithm (WOA), firefly algorithm (FA), fruit fly optimization algorithm (FOA), etc.

Ant colony optimization is a method for finding the optimal path in the graph. It is a probabilistic algorithm (Colomi and Maniezzo, 1991; Dorigo *et al.*, 1996). It is proposed by Dorigo in 1992 in his doctoral thesis, who was inspired by the behaviors of ants found in the process of looking for the food path (Colomi and Maniezzo, 1991). Ant colony in nature can cooperate to find the shortest path from the nest to the food, and can change strategy as circumstances change and quickly re-find the shortest path.

Ant colony optimization is a self-organization algorithm. At the start of the algorithm, a single artificial ant searches for a solution in a disorder way. After a period of algorithm evolution, the artificial ants spontaneously tend to find some solutions close to the optimal solution, which is a process from the disorder to the order.

Particle swarm optimization (PSO) is an evolutionary computation based on iteration, which was proposed by Kennedy and Eberhart (1995). PSO was originally a graphical simulation of preying behaviors of a flock of birds. The basic idea is inspired by their early findings on group behaviors of birds, and they thereafter used and improved the biological population model. In the particle swarm algorithm, each particle in the particle swarm is equivalent to a bird in the bird flock. They all track the currently optimal particle (which is equivalent to the bird closest to the food) in the solution space, and they constantly update their position and velocity. Through continuous iteration, the algorithm reaches the optimal solution (similar to bird finding food) (Shi and Eberhart, 1998; Eberhart and Shi, 2000; Krink and Løvbjerg, 2002; Clerc, 2004, 2006; Zhang *et al.*, 2007; Niknam and Amiri, 2010).

Stochastic diffusion search was proposed by Bishop (1989) to solve the problem of incentive equivalence in pattern recognition. Stochastic diffusion search is one of the swarm intelligence optimization algorithms. Unlike most swarm intelligence optimization algorithms, stochastic diffusion search uses direct communication between entities (Beattie and Bishop, 1998; Nasuto *et al.*, 1998; Myatt *et al.*, 2004; Meyer, 2004; Meyer *et al.*, 2006). In stochastic diffusion search, each of the entity holders holds an assumed solution about the problem to be solved, and assesses the solution partially. The successful entity directly communicates with unsuccessful entities to repeatedly test its assumption. Thus a positive feedback mechanism is established, so that the group can quickly converge to the optimal solution in the solution space. In the solution space, the regions largely aggregated by entities are considered as candidate solutions. Through the cooperation between the locally-run simple entities, the global solution can be reached in the region with most aggregated entities. The stochastic diffusion search is a truly adaptive algorithm, because even if the optimal solution is found, there are still some entities to explore the solution space, which makes the algorithm adapt to changes in the environment (Nasuto *et al.*, 1998).

2.1.2.3.4 Hyper-heuristic method

Hyper-heuristic method can be roughly understood as the heuristic method to find the heuristic algorithms (Jiang, 2011). Hyper-heuristic method provides a high-level heuristic method, which produces a new heuristic algorithm by managing or manipulating a series of low-level heuristic algorithms. Hyper-heuristic method runs in the search space consisting of heuristic algorithms. Each vertex in the search space represents the combination of a series of low-level heuristic algorithms. It runs to achieve one or more optimal heuristic algorithms. Roughly speaking, hyper-heuristic method is to find an optimal heuristic algorithm in the search space of (meta-) heuristic algorithms.

2.1.2.4 *Further explanation of computation/algorithm*

Denning (2003, 2007, 2010) have discussed the definition of computing/computation/algorithm in detail in a series of his papers. Representation is defined as a pattern of symbols that stands for something. The association between a representation and what it stands for can be treated as a link in a table/database, or as a memory in human brain. A representation has two important aspects, syntax and stuff. Syntax is the rules for constructing patterns. It allows us to distinguish patterns that stand for something from patterns that do not. Stuff is the measurable physical states of the world that hold representations (e.g., in media or signals) (Denning, 2010). A machine can be built to detect when a valid pattern is present. Denning stated that a representation which stands for a method of evaluating a function is called an algorithm, and a representation that stands for values is called data. An algorithm controls the transformation of an input data representation to an output data representation when it is implemented by a machine. The algorithm representation controls the transformation of data representations. According to these definitions, however, there is no algorithm for finding the shortest possible representation of something (Chaitin, 2006).

In the sense of a machine, an implementation is a computation. An information process is a sequence of (changing) representations. A computation is an information process in which the transitions from one element of the sequence to the next are controlled by a representation (Denining, 2010). Strictly, computations are logical orderings of strings in abstract languages. Denning held that implementable representations are the basis of scientific approach to computation.

Computation and its implementation schemes were first defined and discussed as early as in the 1930s by Kurt Godel, Alonzo Church, Emil Post, and Alan Turing. In their definition, computation means the mechanical steps followed to evaluate mathematical functions (Denning, 2003, 2007, 2010). By the 1980s, computing included a series of related fields, computer science, computational science, computer engineering, software engineering, information technology, etc. By 1990, computing had become the standard for referring to these disciplines. Since then, computing was treated as not only a tool for science, but also a new method of thought and discovery in science (Hazen, 2007). So far there is no recognized and uniform definition on computing. Some treat it as a branch of applied mathematics and some refer to it as a branch of computational oriented science. Since the 1990s, people wondered whether all natural information processes are produced by algorithms (Hazen, 2007). If it is true, the traditional view that algorithms are at the heart of computing will be challenged. Under such a situation, information processes may be more fundamental than algorithms (Denining, 2010). Actually, Wolfram (2002) has argued that information processes underlie every natural process in the universe. This leads to a conclusion that computing is the study of natural and artificial information processes, just as stated by Denining (2010): "To think computationally is to interpret a problem as an information process and then seek to discover an algorithmic solution." As a consequence, Denining (2003, 2007, 2010), and Denning and Freeman (2009) developed a set of principles computing framework, which fall into seven categories: computation, communication, coordination, recollection, automation, evaluation, and design.

Self-organization can be treated as an information process and a computation. How to find algorithmic solution is a focus in self-organization studies.

2.1.3 *Case examples of self-organization*

Self-organization is popular in nature and human society, covering many fields such as physics, chemistry, biology, economics, and society.

2.1.3.1 *Physics*

Some physical processes can be treated as self-organization (Glansdorff and Prigogine, 1971). Such examples include structure formation in astrophysics and cosmology (formation of stars, formation of planetary systems, formation of the Milky Way, etc.), phase transition of structures, self-similar expansion, diffusion-limited aggregation, infiltration, reaction-diffusion systems, crystallization, spontaneously magnetization, laser, superconductivity, Einstein-Bose condensation, and spontaneous symmetry breaking, etc.

A phase transition is the transformation of a thermodynamic system from one phase of matter to another phase. The measurement of the external conditions that leads to the transformation is called phase transition. During a phase transition of a given system, induced by certain external condition, some properties of the system change (often abruptly). For example, a liquid may become gas when it has been heated to the boiling point, which results in an abrupt change in volume (Wikipedia: http://en.wikipedia.org/wiki/Phase_transition).

Diffusion-limited aggregation (e.g., electrodeposition, Hele-Shaw flow, mineral deposits, and dielectric breakdown) is a process in which the random walking particles due to Brownian motion aggregate together to form clusters of these particles (Witten and Sander, 1981).

Dendrite growth is a complex nonlinear self-organization process: (1) non-linearity. The partial differential equation used to simulate dendrite growth is non-linear. In particular, both the phase-field parameters and the second-order derivative terms in the phase-field equations are non-linear. (2) Self-organization. Dendrite growth is a self-organization process, that is, given initial and boundary conditions and assume no more imposition of certain external conditions, it will spontaneuously form an ordered structure. This is just a dissipative structure. (3) Complexity. Dendritic growth is a self-organization, so it is much more sensitive to the initial and boundary conditions.

Laser is a time-ordered self-organization. In the Helium-Neon laser-generating mechanism, the laser generator is an open system. Energy is supplied to the laser generator from the external environment by a pump. When the power supply is little, the laser generator emits random and weak natural light only because the frequency, phase, and vibration direction of light emitted by every Neon atoms are different. When the power supply increases to a certain value, the system mutates and self-organization occurs for large number of atoms; and they will emit highly coherent light beam, i.e., laser, with the same frequency, phase, and direction.

Spontaneous symmetry breaking is the realization of symmetry breaking in a physical system (Weinberg, 2011). In spontaneous symmetry breaking, the basic laws are invariant under a symmetry transformation, but the system as a whole changes under such transformations — which is in contrast to explicit symmetry breaking. Spontaneous symmetry breaking is a spontaneous process by which a system in a symmetrical state ends up in an asymmetrical state (Wikipedia: http://en.wikipedia.org/wiki/Spontaneous_symmetry_breaking).

2.1.3.2 *Chemistry*

Self-organization is widely found in chemical processes, for example, molecular self-assembly, self-assembled monolayer film, Langmuir-Blodgett film, B-Z reaction, self-organization of nanomaterials, macroscopic self-assembly under molecular recognition, oscillatory chemical reactions, and autocatalytic networks, etc. (Kim *et al.*, 2006; Pokroy *et al.*, 2009; Coleman *et al.*, 2011; Harada *et al.*, 2011).

Molecular self-assembly (intramolecular self-assembly and intermolecular self-assembly) is the process by which molecules follow a pre-defined arrangement without external commands. Formation of micelles, vesicles, liquid crystal phases, and Langmuir monolayers by surfactant molecules fall under this category. Assembly of molecules in such systems is directed through noncovalent interactions (e.g., hydrogen bonding, metal coordination, hydrophobic forces, van der Waals forces, $\pi-\pi$ interactions, and/or electrostatic) and electromagnetic interactions (Lehn, 1988, 1990; Wikipedia: http://en.wikipedia.org/wiki/ Molecular_self-assembly). It was proved that molecular self-assembly can produce different shapes and sizes (Katsuhiko *et al.*, 2008).

2.1.3.3 *Life sciences*

Self-organization is very popular in biological systems, whether at sub-cellular level or at ecosystem level (Hess and Mikhailov, 1994; Misteli, 2001; Camazine, 2003; Clyde *et al.*, 2003; Motegi *et al.*, 2011).

Self-organization of ecosystems is a fundamental theory in ecology. The essential difference between ecosystems and non-biological systems is its ability in self-organization. The evaluation of self-organizing capacity of ecosystems has become one of the most important methods in revealing the complexity and uncertainty of ecosystems.

In the field of life sciences, there is a rapidly growing emphasis on the phenomena of self-organization *in vivo*. In biological systems, self-organization is a process at the global level. The system is generated only from the interactions between components at the lower levels. Implementing the rules of between-component interactions only requires local information rather than global information (Camazine, 2003). Increasing evidences are proving that many biological systems are close to what is called a critical point: they sit on a knife-edge, precariously poised between order and disorder. This strategy is believed to increase the flexibility in dealing with a complex and unpredictable environment (Ball, 2014).

Almost all biological systems are self-organizing systems (Hess and Mikhailov, 1994; Misteli, 2001; Clyde *et al.*, 2003; Motegi *et al.*,

2011), for example, (1) the self-assembly of proteins, as well as the formation of other biological macromolecules and lipid bilayers; (2) homeostasis, which is a self-organization from cell to tissue; (3) pattern formation and morphogenesis, i.e., the growth and differentiation of living organisms; the interface between two different types of cell will trigger the formation of a third kind of cell at their boundary; an embryo can construct complex tissues this way, with different cell types in all the right places (Davies, 2014); (4) human motion; (5) creation of structures by gregarious animals, such as social insects, bees, ants, etc.; (6) group behaviors (the most typical examples can be found in birds and fish); and (7) in the super cycle theory and autocatalytic theory, life itself is originated from the self-organizing chemical systems.

Phase transitions often occur in biological systems. For example, the lipid bilayer formation, the coil-globule transition in protein folding and DNA melting, liquid crystal-like transitions in DNA condensation, and cooperative ligand binding to DNA and proteins with the character of phase transition (Lando and Teif, 2000).

Gel to liquid crystalline phase transitions is critical in physiological role of biomembranes. Due to low fluidity of membrane lipid fatty-acyl chains, in gel phase, membrane proteins have restricted movement and are restrained. Plants depend on photosynthesis by chloroplast thylakoid membranes when exposed to cold environmental temperatures. Thylakoid membranes retain innate fluidity even at relatively low temperatures, due to high degree of fatty-acyl disorder allowed by their high content of linolenic acid (YashRoy, 1987).

Molecular self-assembly is fundamental for constructing macromolecules in cells of the living organism, this includes the self-assembly of lipids to form the membrane, the formation of double helical DNA through hydrogen bonding of the individual strands, and the self-assembly of proteins to form quaternary structures. Molecular self-assembly of nanoscale structures is important in the growth of β-keratin lamellae/setae/spatulae structures which are used to endow geckos the ability to climb walls and adhere to ceilings and rock overhangs (Daniel *et al.*, 2007; Min *et al.*, 2008; Wikipedia: http://en.wikipedia.org/wiki/Molecular_self-assembly).

In biology, Miller experiment proved that the thundering and lightning in the primitive Earth's atmosphere produced organic compounds (especially amino acids), which demonstrated the chemical evolution of the origin of life. In this experiment, the mixture of hydrogen gas (H_2), helium (He), methane (CH_4), ammonia (NH_3), and other inorganic composition can generate 17 kinds of amino acids after implementation the spark discharge. This process (simple inorganic matters become complex organic compounds when high energy is supplied) is a typical phenomenon of self-organization. In the self-organization, the supply of high energy leads to more ordered matters, and the amino acids of high energy are maintained at the excitation state, and a new state of equilibrium is thus achieved.

In recent years, some scientists have attempted to interpret the origin of life from the view of self-organization. Scientists use a set of biomolecules to show a way in which life might have started. They hold that different chemicals come together due to many forces that act on them and become a molecular machine capable of even more complex tasks. Each living cell is full of these molecular machines. These molecular machines have not done much on their own. When they add fatty chemicals, which form a primitive cell membrane, the chemicals become close enough to react in a highly specific manner. Molecules and cells interact according to simple rules, creating a whole that is greater than the sum of its parts (Davies, 2014). This form of self-organization may be popular for the origin of life on both earth and other planets.

To interpret the origin of life, Stano and his colleagues chose an assembly that consists of 83 different molecules including DNA, which was programmed to produce a special green fluorescent protein (GFP) that could be observed under a confocal microscope (Lehn, 1988, 1990). The assembly can only produce proteins when its molecules are close enough to react with each other. When the assembly is diluted with water, they can no longer react. This can explain why the cell is so compact: to allow the chemicals to work.

In order to recreate this molecular crowding, Stano added a fatty molecule, POPC, to the dilute water solution and these molecules then automatically form liposomes that have a very similar structure to the membranes of living cells. They found that many of these liposomes trapped some molecules of the assembly. Five in every 1000 liposomes had all 83 of the molecules needed to produce a protein. These liposomes produced a large amount of GFP and glowed green under a microscope. Surprisingly, computer calculations reveal that even by chance, five liposomes in 1000 could not have trapped all 83 molecules of the assembly. The calculated probability for even one such liposome to form is essentially zero. This means that some unique mechanism is behind it, and self-organization is one of the reasons.

Davies (2014) even showed how from these interactions we can deduce "rules" of embryo development. For example, cells communicate with each other and tweak their behavior in response to changes in their environments. This is what puts the "adaptive" into adaptive self-organisation, ensuring that development can cope with noise or disruption. An example is the way tiny blood vessels called capillaries cater to different kinds of tissue, even while these tissues are moving and growing (Davies, 2014). A feedback loop exists between oxygen and a cell protein called HIF-1-alpha. Oxygen normally causes HIF-1-alpha to be destroyed. If a tissue lacks oxygen, HIF-1-alpha levels rise, triggering a cellular signal encouraging capillaries to grow. This brings in oxygen, which shuts down HIF-1-alpha and halts capillary development. Should the tissue then grow, oxygen levels will fall again, and the loop is set in motion once more.

2.1.3.4 *Self-organization in sociology*

Self-organization is also called spontaneous order theory in sociology. Complex behaviors, such as herd behavior, groupthink, critical mass, etc., are found to follow some mathematical laws, e.g., Zipf's law, power law, and Pareto principle, which are self-organizing behaviors (Wikipedia, 2014).

Self-referentiality is a social self-organization that can describe the evolution of society and its subsytems (Luhmann, 1991). In a social system, all elements are self-producing communications, that is, a communication will produce more communications and the system can thus reproduce itself given there is dynamic communication (Luhmann, 1991; Wikipedia, 2014).

Self-organization may result in a decentralized, distributed, self-healing system in the human network. By limiting each individual's scope of knowledge, it can protect the individual's security (Wikipedia, 2014).

2.1.3.5 *Self-organization in cybernetics*

As early as in the 1960s, Machol and Gray held that the automatic and continuous identification of the black box problems and subsequent replication fitted the properties of self-organization. In a sense, cybernetics deals with some of the self-organization problems.

2.1.3.6 *Self-organization in networks*

Self-organization is an important mechanism to establish networks (Wikipedia, 2014). Such mechanisms are also referred to as self-organizing networks. It should be noted that only certain kinds of networks are self-organizing. They are known as small-world networks, or scale-free networks. These networks emerge from bottom-up interactions, and appear to be limitless in size. In contrast, there are top-down hierarchical networks, which are not self-organizing. These are typical of organizations, and have severe size limits (Wikipedia, 2014).

2.1.3.7 *Self-organization in mathematics and computer science*

As mentioned above, phenomena from mathematics and computer science such as cellular automata, random graphs, and some instances of evolutionary computation and artificial life exhibit features of self-organization. In swarm robotics, self-organization is used to produce emergent behaviors. In particular, the theory of random graphs has been used as a justification for self-organization as a general principle of complex systems. In the field of multi-agent systems, understanding how to engineer systems that are capable of presenting self-organized behavior is a very active research area.

2.2 Selforganizology

2.2.1 *Problems for self-organization research*

Although a lot of theories and methods were established to describe self-organization, there are still many problems in this area. Self-organization is a universe phenomenon. Nevertheless, we still lack unified theories and thoughts on self-organization. We lack universal basis of methodology in the modeling and simulation of self-organization. Self-organization is categorized as a research area in complexity science. So far it is not an independent science (Zhang, 2013a, b, 2014).

2.2.2 *Selforganizology: A science on self-organization*

For the reasons mentioned above, I proposed a fundamental science — selforganizology (Zhang, 2013a). It was proposed for finding and creating theories and methods from self-organization phenomena in nature, simulating and reconstructing self-organization phenomena, exploring and synthesing mechanisms behind numerous self-organization phenomena, and promoting applications of self-organization theories and methods in science and industry. Selforganizology is a science that deals with self-organization. Many properties, principles, theories and methods on self-organization hold in this scientific field. The theory of dissipative structures, stability theory (e.g., bifurcation theory, singularity theory and catastrophe theory), topology, etc., are fundamental theories in selforganizology. Some theories and methods should be futher improved.

Selforganizology is an interdisplinary science based on systematic theory, computational science, artificial intelligence, mathematics, physics and some other sciences. Evolution-, interaction-, behavior-, organization-, intelligence- and feedback-based theories, such as coevolution theory, coextinction theory, community succession theory, correlation analysis, parrondo's paradox (Harmer and Abbott, 1999a, b; Toral, 2001, 2002), game theory, neural networks, artificial intelligence, behavioral theory, organization theory, automation, and control theory in various scientific disciplines can be reviewed, revised, and introduced to selforganizology.

2.2.3 *Some thoughts on methodological basis of selforganizology*

In selforganizology, the self-organization is considered as a universal mechanism in nature. In a sense, all things, from atom to universe, are the results of various self-organization processes. Without external forces and instructions, a dissipative system far from thermodynamic equilibrium may spontaneuously evolve towards one or more steady states through between-component interactions at different hierarchies in self-organization process. It is thus a self-organizing system. In the self-organizing system, the interactions between components produce different functions and properties and behaviors from that of components, which lead to a system with certain functional characteristics and purposeful behaviors different from the nature of components. A self-organizing system is an aggregation of interactive components, and it has a hierarchical structure. A component is an autonomous and organization-closed subsystem. Some components at a hierarchical level will interact and aggregate to form a component at higher hierarchical level, with or without these components in this component. The most basic and inseparable component is the individual (i.e., a person, bird, plant). Different components at the same hierarchical level or at different hierarchical levels will most likely have different behaviors. Self-organization is a dynamic and spontaneous process from the low-level to the high-level, from the local to the global and from the micro-level to the macro-level.

Following Macal and North (2005), we define a component which satisfies these following criteria (Zhang, 2012a, 2013a, 2014):

(1) A component is an independent and identifiable individual which possesses a set of attributes and rules that forge its behaviors. A component is self-contained and independent. It has a boundary

through which people can easily discern between outside the component and inside the component or shared characteristic.

(2) Each component locates in a certain position and interacts with its adjacent components. A component has a set of protocols that govern its interactions with other components, such as communication protocol, the capability to affect its environment, etc. The component is able to identify and discern the characteristics of other components.

(3) The component is goal-directed. The component behaves to realize some goals.

(4) The component is independent, autonomous, and self-guided. At least within a finite range, the component can independently operate in its environment.

(5) The component is flexible. It is capable of adapting the environment and adjusting its behaviors. The component possesses some high-level rules to adjust its low-level behavior rules. However, the component with fixed behavior rules is also available.

The behaviors of a self-organizing system cannot be described by using deduction, induction, or other formalization methods. However, the behaviors of a component (aggregation behaviors) can be derived from the interactions between components at low hierarchical level. A behavior of an independent component might be a primitive response and decision, or even a complex intelligence. The behavior rules of a component include basic rules and high-leveled rules that govern the basic rules (rule-changing rules) (Casti, 1997; Zhang, 2012a). Basic rules define necessary responses to the environment, and rule-changing rules define adaptation. In a specific study, it is necessary to determine a theory on behaviors. A component may use various behavioral models, including if-then rules, threshold rules, repeat rules, and other equation/model based rules. Knowledge engineering and participative simulation can be used in defining behaviors. Knowledge engineering includes a series of techniques collected for organizing experts' knowledge (Zhang, 2012a).

In a self-organizing system, the basic structure of behavior rules includes: IF-THEN-ELSE rule; (2) GO TO rule; (3) DO WHILE rule; (4) SWITCH CASE DO rule; (5) LET rule; (6) AND/OR rules; (7)

RANDOMIZE rule; (8) other equation/model/algorithm/statement based rules, etc. I think that using these rules for all components at all hierarchical levels will probably produce any complex behaviors of the self-organizing system. Mathematical equations and models (e.g., differential equations) can be used in the simulation and modeling of self-organization phenomena. An algorithm, a problem-solving procedure, etc., can be represented by a structure of hierarchical rules in which each rule has both parent and son rules with exception of the top rule and bottom rules. It is expected that many existing self-organization algorithms, such as Swarm Intelligence algorithms, can be resolved into simple rules based algorithms. We should maximally resolve complex rules into simple rules. Differential equations, for example, can be rewrited into difference equations, and if it is possible, further resolved into many simple rules, until no further simpler rules can be found. Compared to complex rules, simple rules are more useful in exploiting the most fundamental mechanism of self-organization. In general, the simpler rules mean the more effectiveness of simulation method used.

In the sense of systematic simulation, selforganizology may be considered as the science based on self-organization, components, hierarchies, interactions, feedbacks, behaviors and rules, etc.

Some methods, in particular agent-based modeling (Topping *et al.*, 2003; Griebeler, 2011; Zhang, 2012a) can be considered as the methodological basis of simulation and modeling of self-organization. These methods will not only help propose hypothesis on behaviors and mechanism of a self-organizing system but also help propose management strategies on the self-organizing system.

It should be noted that the concepts, definitions, principles, and methodology in classic theory of agent-based modeling can be further revised and improved to reasonably describe and model self-organizing systems. For example, when used in self-organization, agents in agent-based modeling are components.

In selforganizology, we can follow some standard protocol, for example the standard protocol proposed by Grimma *et al.* (2006), to describe the simulation and modeling of self-organization. The core of the protocol is to structure the information about self-organization simulation and modeling in a sequence. This sequence consists of seven elements, which can be grouped in three blocks: overview, design concepts, and details (Grimma *et al.*, 2006):

(1) The overview consists of three elements including purpose, state variables and scales, process overview and scheduling. It provides an overview of the overall purpose and structure of the model. It includes the declaration of all objects (classes) describing the models entities (different types of components or environments) and the scheduling of the model's processes.
(2) The design concepts describe the general concepts underlying the design of the model. The purpose of this element is to link model design to general concepts identified in the field of self-organizing systems. These concepts include the interaction types between components, whether the components consider predictions about future conditions, or why and how stochasticity is considered.
(3) The details include three elements, i.e., initialization, input, and submodels, which present the details that have been omitted in the overview. The sub-models implementing the model's processes are particularly described in detail. All information required to completely re-implement the model and run the baseline simulations should be provided.

The logic behind the protocol sequence is that the context and general information is provided first (overview), followed by more strategic considerations (design concepts), and finally supplied with more technical details (details).

Main procedures of self-organization simulation and modeling include (Zhang, 2012a): (1) determine various types of components and define behaviors of components; (2) identify relations between components, and construct interaction types between components; (3) determine the platforms and environments for self-organization simulation and modeling, and set the strategies for simulation and modeling; (4) acquire necessary data for simulation and modeling; (5) validate the patterns of components' behaviors and system's behaviors; and (6) run the model and analyze the output from the standpoint of linking the micro-scale behaviors of the components to the macro-scale behaviors of the self-organizing system.

I have proposed and presented some ideas for the establishment and development of selforganizology (Zhang, 2013a). Theories and methods of selforganizology should be continuously revised and improved in the future. Further research is needed to promote this fundamental science.

Chapter 3

Agent-based Modeling

Agent-based modeling is a relatively new methodology and has been used in various fields. More details on agent-based modeling can be found in Zhang (2012).

3.1 Complex Systems

3.1.1 *Properties of complex systems*

Generally, complex systems exhibit such properties as follows (Li *et al.*, 2007; Zhang, 2012):

(1) Nonlinearity. Complex systems usually behave nonlinearly. Global behaviors of a complex system cannot be directly derived from the behaviors of individuals.
(2) Emergency. Emergency refers to a dynamic process from low-level to high-level, from local to global and from micro-level to macro-level. Emergency stresses that the interactions between individuals lead to different functions and properties and behaviors from that of individuals. It leads to a system with certain functional characteristics and purposeful behaviors that are different from the nature of individuals.
(3) Dynamics. A complex system will always change with time.
(4) Modularity and hierarchy. A complex system is the aggregation of interactive and correlative individuals/objects, i.e., modularity. A complex system shows a hierarchical structure (i.e., hierarchy). The objects at a hierarchical level will aggregate to form the objects at its parent hierarchical level.

(5) Information flow and associativity. In a complex system, there will be large amounts of information/matter interactions among individuals/objects. Changes of any individual/object may affect other individuals/objects or the entire system (Harman *et al.*, 2009).
(6) Incomputability. Incomputability of complex systems refers to: (A) the behaviors of a complex system that cannot be described by using deduction, induction, or other formalization methods; and (B) the process of a complex system that can be approximated by the inference system based on some rules.

Obviously, self-organizing systems are, in general, complex systems.

3.1.2 *Modeling complex systems*

To study and describe complex systems, we can choose to use differential equations, artificial neural networks, agent-based modeling, network analysis, etc.

3.1.2.1 *Modeling with differential equations*

Differential equations can be used to model complex systems. Given the motion equation of a system:

$$dx/dt = f(x,t)$$

where

$$x(t) = (x_1(t), x_2(t), \ldots, x_n(t))^{\mathrm{T}}$$
$$f(x,t) = (f_1(x,t), f_2(x,t), \ldots, f_n(x,t))^T$$

The system is classified by different forms of $f(x,t) : f(x,t) = f(x,t)$, non-linear and non-stationary system; $f(x,t) = f(x)$, non-linear stationary system; $f(x,t) = Ax$, linear stationary system; $f(x,t) = A(t)x$, non-stationary linear system.

Differential equations are developed in a top-down way (Hraber and Milne, 1997). These models are always differential or finite difference equations with one or more dependent and independent variables (Zhang and Gu, 2001). In differential equations, the types of between-variable interactions should be predefined. Pairwise interaction coefficients in the equations represent the interactions between variables. All variables are assumed to interact with equal probability, which is a property of the interactive system with fully mixed or mean field (Zhang, 2012).

In a sense, a differential equations model is the agent-based model with uniform field. For example, in differential equations, we usually reflect spatial heterogeneity of interactions by defining adjacent interactions and including some factors. These models can be used to predict which variable will dominate the system as time changes. The parameters of differential equations may approximate the apparent properties of a system. However, the parameters in differential equations approximate the phenomenological properties of a system without identifying mechanisms or considering between-individual variation and therefore it is difficult for them to show complex interactions and include various simulated behavior. In addition, the complexity and incomputability of differential equations will exponentially increase with increase of the system's complexity. Thus they can only be used to describe systems with lower complexity.

3.1.2.2 *Modeling with artificial neural networks*

Artificial neural networks (ANNs) are used to simulate, for example, complex and non-linear ecological problems (Zhang and Barrion, 2006; Zhang, 2007a, b; Zhang *et al.*, 2007, 2008a, b; Zhang and Zhang, 2008; Zhang and Wei, 2009; Zhang, 2010). ANNs are learning models that need enough sample data to train and learn from. Learning from samples, ANNs store the intrinsic mechanism of a system or data sets as the connected weights of network. ANNs are models that lie between empirical models and mechanistic models.

There are many types of ANNs. Their capability and effectiveness depend on three factors: (1) complexity of neuron models; (2) velocity and efficiency of leaning rules; and (3) topological structure of network.

The first ANN of importance is Multi-layer Perceptron (MLP). MLP is a feed-forward ANN that neurons distribute in every layer. The neurons at each layer only connect to the neurons in the posterior layer and loops are not allowed. A standard setting for feed-forward ANNs is that there are three types of layers, i.e., input layer, hidden layer, and output layer. There are currently a lot of software packages on ANNs among which most tend to establish feed-forward ANNs that are used in classification, optimization, and regression. ANNS, like BP, radial basis function networks, self-organizing networks, etc., have been widely used in theoretical and technological studies including ecological studies (Zhang and Barrion, 2006; Zhang *et al.*, 2007, 2008a, b; Zhang, 2010).

In a sense, ANNs have the characteristic of fully mixed or mean field systems.

3.1.2.3 *Modeling with network models*

Network analysis is based on graph theory, etc. It treats individuals/objects as nodes in the network, which allows people to analyze the large scale structure of a complex system (Zhang, 2012). This methodology makes it easy to examine the effects of interactions, characterize important elements, and describe global structure of the system. Nevertheless, this methodology is in general insufficient to describe system dynamics.

3.2 Agent-based Modeling

Agent-based modeling (ABM) is rooted in the modeling of complex adaptive systems (CAS). A complex adaptive system may self-organize spontaneously, and reconstruct its components dynamically, in order to survive in the environment. Holland (1995) argued that a complex adaptive system should possess these properties:

(1) Nonlinearity. Simple and linear extrapolation is invalid in the prediction of system's behaviors.
(2) Aggregation. It allows for the formation of a population.
(3) Flows. Flows allow for the transition and transformation of resources and information between components.
(4) Diversity. Different agents may have different behaviors, which are usually conducive to the robustness of the system.

The mechanisms of complex adaptive systems include, (1) labels: it allows agents to be defined and identified; (2) internal pattern: it allows agents to make inference to their world; and (3) establishment of blocks: it allows components and the entire system to be constructed from simple components at various hierarchies.

The earlier ABM originated from cellular automata (CA) (Gardner, 1970). To answer the question of whether a machine can be programmed to reproduce itself, that was proposed by Von Neumann, the physicist Stanislaw Ulam proved this to be true by using CA. A typical CA is a two-dimensional mesh or lattice or dot matrix that composes of cells (grids). At any time and point, each cell has finite number of states. A set of simple rules are used on previous states for determining current value of a cell. The next value of a cell is dependent upon its current value and the values of its eight adjacent cells. Updating rules of each cell is the same.

Wolfram (2002) found that the rules will produce astonishing spontaneous patterns in CA. These patterns directly correspond to extensive algorithms and logic systems. He thought that simple rules will result in a complexity similar to the real world.

ABM is a bottom-up modeling method. It is widely used to model complex systems containing spontaneous and interactive agents (Topping *et al.*, 2003; Qi, 2003; Qi and Zhang, 2002; Qi *et al.*, 2004; Li and Ma, 2006; Chen *et al.*, 2008). It is a powerful tool for analyzing global behaviors of complex systems. ABM models the dynamics of adaptive systems based on the adaptation mechanism of individuals. It is thoroughly different from differential equation modeling. ABM is considered as the third scientific methodology, in addition to induction and deduction. ABM is concerned with a lot of scientific fields, including complexity science, systems science, computational science, etc.

As systems become more and more complex, conventional simulation tools show obvious shortages. However, ABM will play important roles in such situations: (1) conventional methods have reached their limitations; (2) data are organized as databases, and they can be stored hierarchically in databases; and (3) computational capabilities have been largely enhanced and allow for extensive simulation.

ABM was originally used to propose hypothesis on behaviors and mechanisms of a system, to explain system phenomena, and now further propose management strategies. It has been successfully used in organizational simulation (risks, organizational design, molecular self-organizing), diffusion simulation (diffusion dynamics), flow simulation (traffic, flow management), market simulation (stock market), etc. (Bonabeau, 2002). For example, traditional risk analysis appears to be unrelated to ABM. However, the risk can be in some cases treated as a spontaneous property of a system. ABM is thus available for solving these problems.

ABM can be used in these cases (Bonabeau, 2002): (1) interactions between agents are complex, non-linear, and discrete; (2) interactions are complex and heterogeneous; (3) spatial factors are very important and the locations of agents are not fixed; (4) population is heterogeneous, and each individual is different from others; and (5) agents show complex and diverse behaviors, including learning and adaptation.

ABM can be jointly used with other modeling techniques like systematic dynamics, etc. Statistics, like PCA, is also useful in ABM.

3.2.1 *Agent and behaviors*

Agents can be defined in various ways. Any independent component, such as the software, a model, an individual, etc., can be considered as an agent (Bonabeau, 2002). The behavior of an independent component might be a simple and primitive response and decision, or even a complex adaptive intelligence. In general, the behaviors of an agent must be self-adaptive and the agent can learn from its changing environment (Mellouli *et al.*, 2003). Casti (1997) held that the behavioral rules of an agent must include two parts, basic rules and the high-leveled rules that govern basic rules (rule-changing rules). Basic rules define necessary responses to the environment, and rule-changing rules define adaptation. Jennings (2000) argued that the basic characteristic of an agent is that it can independently make decisions.

Different from differential equations, combining new agents and new agent types into ABM is easy. At lower levels, differential equations can be included to approximate system's behaviors. At higher levels, the interactions between agents are allowed to define aggregative behaviors. Aggregative behaviors are spontaneous behaviors generated by between-agent or agent-environment interactions.

ABM allows an agent and the entire system to consist of more simple agents at various levels. Unlike AI (artificial intelligence) agent, the environmental responses and influences of an agent in ABM are always not so complex.

Agents in ABM must satisfy these criteria (Macal and North, 2005):

(1) An agent is an independent and identifiable individual, which possesses a set of attributes and rules that govern its behaviors and decision ability. An agent is self-contained and independent. It has a boundary through which people can easily discern between outside and inside the agent or shared characteristic.
(2) Each agent locates in a certain spatial position and interacts with its adjacent agents. An agent has a set of protocols that govern its interactions with other agents, such as communication protocol, the capability to affect its environment, etc. An agent is able to identify and discern the characteristics of other agents.
(3) Agents are goal-directed. An agent behaves to realize some goals.
(4) Agents are independent, autonomous, and self-guided. At least within a finite range, an agent can independently operate in its environment.
(5) Agents are flexible. An agent is capable of accumulating experiences and learning from the environment and adjusting its behaviors. This requires some form of memory. An agent possesses some high-level rules to adjust its low-level behavioral rules. It has diverse attributes and behavioral rules.

To define agents, exactly specifying their behaviors and reasonably representing interactions between agents are fundamental to ABM. Once agents are defined, we should exactly define their behaviors. First, it is necessary to determine a theory on behaviors and existing behavioral theories can be used. Agents may use various behavioral models, including if-then rules and threshold rules.

Behaviors of agents are bases for an existing or supposed system. Knowledge engineering and participative simulation can be used in defining behaviors. Knowledge engineering includes a series of techniques collected for organizing experts' knowledge.

3.2.2 *Procedures of agent-based modeling*

The main procedures of ABM include:

(1) Determine various types of agents and define the behaviors of these agents.
(2) Identify relationships between agents, and construct interaction types between agents.
(3) Choose the platforms and environments for ABM, and set the strategies for ABM.
(4) Acquire necessary data for ABM.
(5) Validate the patterns of agents' behaviors and system's behaviors.
(6) Run ABM model, and analyze the output from the standpoint of linking the micro-scale behaviors of agents to the macro-scale behaviors of the system.

To design the ABM, the key is software design and model development. Development time spans several highly staggered phases (Macal and North, 2005). At the design phase, the structure and function of the model should be defined. At implementation phase, we should develop a model based on the designed plan. The model will be used during the practical operation. These phases are usually repeated several times to generate a more detailed model.

Grimma *et al.* (2006) has also proposed a standard protocol for describing ABM. The core of the protocol is to build the information about ABM in the same sequence. This sequence consists of seven elements, which can be grouped in three blocks: overview, design concepts, and details (Grimma *et al.*, 2006):

(1) The overview consists of three elements, including purpose, state variables and scales, process overview and scheduling. It provides

an overview of the overall purpose and structure of the model. It includes the declaration of all objects (classes) describing the models entities (different types of individuals or environments) and the scheduling of the model's processes.

(2) The design concepts describe the general concepts that underly the design of the model. The purpose of this element is to link model design to general concepts that have been identified in the field of complex systems. These concepts include questions about emergence, the type interactions among individuals (agents), whether individuals consider predictions about future conditions, or why and how stochasticity is considered.

(3) The details include three elements, i.e., initialization, input, and sub-models, which present the details that were omitted in the overview. The sub-models implementing the model's processes are particularly described in detail. All information required to completely re-implement the model and run the baseline simulations should be provided here.

The logic behind the protocol sequence is that the context and general information are provided first (overview), followed by more strategic considerations (design concepts), and finally supplied with more technical details (details).

3.2.3 *Tools and platforms for agent-based modeling*

Object-oriented paradigms are a useful basis of ABM, because agents can be treated as self-directed objects that have extraordinary action-selecting capacity. Consequently, almost all of the ABM tools are objected-oriented (Zhang, 2012).

A simple ABM might have dozens/hundreds of agents, which can be used to learn how to build an ABM model and test the model. ABM usually runs in interpretative environments and hence, compiling or DLLs are not required. One of the most used environments is Java.

A large scale ABM always uses thousands or millions of agents. These ABM are usually operated in computer simulated environments, which support such characteristics as time dispatcher, communication mechanism, flexible interactive topology, some selective structures, components used to store and display the states of agents, etc.

There are some standards and protocols for the development of ABM platforms. For example, the Foundation for Intelligent Physical Agents' (FIPA 2005) architecture specifications, the Object Management Group Agent Platform Special Interest Group, Agent UML (OMG 2005), and the Knowledge-able Agent-oriented System architecture (KAoS) (Bradshaw, 1997).

Up till now, a lot of ABM platforms are available for users, including Startlogo (Resnick, 1994), Netlogo, Topographica (Bednar *et al.*, 2004), Swarm, TRANSIMS, CompuCell (Izaguirre *et al.*, 2004), NEURON (Hines and Carnevale, 1997), GENESIS (Wilson *et al.*, 1988), ABNNSim (Bonabeau, 2002), Evoland (Guzy *et al.*, 2008), etc.

StarLogo allows users to model using draggable GUI (Graphic User Interface). NetLogo is a branch of StarLogo, which allows users to model under a new Java GUI and redesigned model language.

Swarm is the first general ABM tool based on Java. It is a software platform on the basis of agents, which provides some advantages for modeling, including a set of standardized object bases, task scheduling base, detection base, and some structural characteristics as inheritance, message delivery, packaging, and hierarchical structures.

Echo was developed as a general model of the dynamics of adaptive systems. Echo is intended to embody mechanisms common to all adaptive systems. Holland gives an overview of attributes and mechanisms common to a range of adaptive systems, and builds a model whose mechanisms embody these tenets. The resulting model has similarities to extant modeling approaches, and also provides some innovative mechanisms which make it of particular value in the study of many-species dynamics. Echo is intended as a class of models whose members represent increasing complexity. As with the lattice-based models, Echo is an agent-based, spatially explicit model. Unlike a lattice-based model with one agent per node, populations of many agents interact within a site, and may migrate to other sites in a world of user-defined geometry. In addition to spatial- and individual-based bookkeeping, Echo individuals have a genetic component. Agents have their own haploid genomes, which are subject to heritable variation and differential reproductive success. Thus, an analogy to evolution is possible, along the lines of genetic algorithm simulations of evolutionary processes. As they evolve from initial conditions, populations tend to be dominated by genotypes with greatest reproductive success. It is also possible to create novel genotypes, or to model formation of novel species. This is a stark contrast to extant many-species models, which either lists all species and determine their interactive attributes, or models evolution as a random branching process. Echo genotypes are subjected to an endogenous fitness function. Endogenous fitness is made possible by resource-limited reproduction and genetically-mediated behavior. Resources constitute the basic currency of Echo, and are used to construct agent genotypes. An agent can obtain resources from the environment or by interacting with other agents. An agent may self-reproduce only when it has gathered sufficient resources to copy its genome. Genetically-mediated behavior determines whether two agents can interact.

TRANSIMS and CompuCell allow users to analyze the effects of different initial conditions on system's behaviors.

NEURON and GENESIS can exactly simulate the electrical and chemical nature of individual neurons. However, the size of network that can be simulated is finite. GENESIS uses compartmental model to simulate biological neurons. Each section of neuron is simulated with path equation and the parameters in path equation include conductivity and capacity of input and output signals of cell. Similar to GENESIS, NEURON uses compartmental model to also simulate biological neurons. However, the focus of NEURON is modeling rather than compartment details. NEURON include higher GUI than GENESIS.

Topographica focuses on functions and large scale structures of neural networks.

Repast is a Java-based API (Application Programming Interface) which allows Java-based modeling. It is open-sourced and API components can be replaced or rewritten. Repast contains advanced

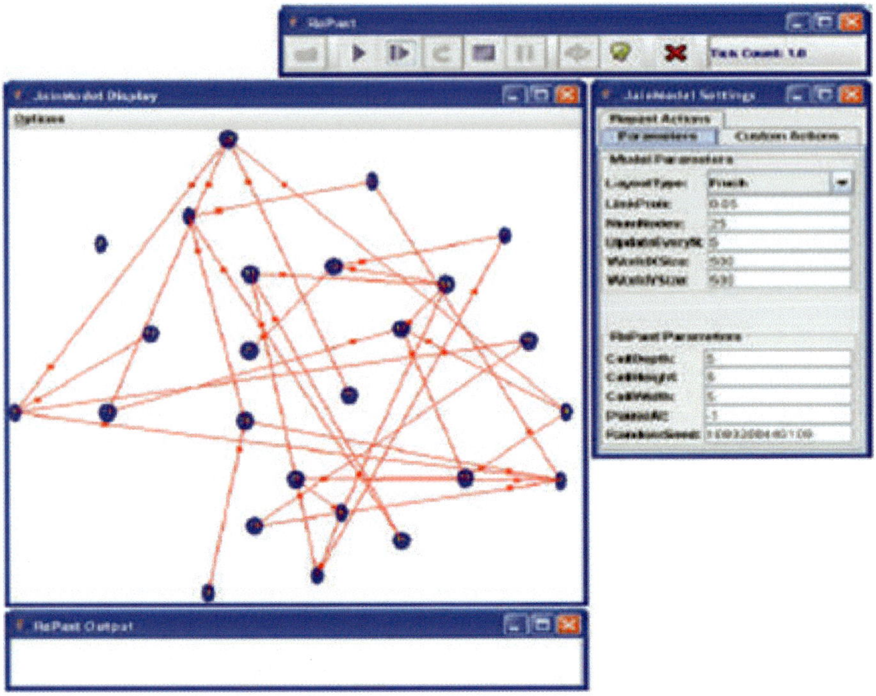

Fig. 1. Repast (Macal and North, 2005).

displaying and analyzing tools and allows fast simulation (Fig. 1). Its fine time scheduler makes the time interval of discrete events reach double precision of Java. Repast Py is a platform-independent and visual model development system and allows fast ABM. It enables users to develop models based on GUI, and write behaviors of agents with Python script language. Repast J is a pure Java modeling environment. It includes a variety of features such as a fully concurrent discrete integration with geographical information systems using both Lagrangian and Eulerian representations for modeling agents on real maps, and adaptive behavioral tools such as neural networks and genetic algorithms. Repast J supports large scale ABM.

ABNNSim (Schoenharl, 2005; Fig. 2) was designed on the basis of Reapst. It extends the electrical behaviors of neurons and includes the behavior for revising physical connections of a neuron to other neurons. Topological structure of network may thus change with time.

Evoland is used to simulate the changes of the environment under various policies in order to determine the regions that suffer from policy changes. The complete source codes, inputs guide, and other details of Evoland 3.5 can be found at: http://evoland.bioe.orst.edu/. In Evoland, an internal GIS manages various variables. GIS supports multiple languages, multiple selective desktop, and visualization pattern. Some attributes of Evoland are stored in the database. During the simulation, the attributes of the database change as the external environment changes. Agents in Evoland are defined by a series of parameters. These parameters are used to evaluate health and economic value of the ecosystem studied. The attributes and parameters of spatial database are governed by a specific mechanism which is also used in cell automata. Different from cell automata, however, Evoland uses multi-dimensional space rather than a set of cells. In Evoland, the view is the unit of analysis. Each view consists of various operations. For each view, periodic cycling controls individual cycling and other non-individual patterns. Some of the patterns are called automatic programs. They can run without the participation of individuals. These automatic programs include the population growth and vegetation succession resulted from the changes of LULC. Another set of patterns are landscape assessment models. These models calculate land's economic value and health condition of the ecosystem.

3.2.4 *NetLogo*

Hereafter some models/algorithms, concerning self-organization/ ABM, in Netlogo are described. NetLogo provides frameworks of computer codes of models/algorithms, and suggestions for extending these models/algorithms, which provides an alternative platform for self-organization modeling.

3.2.4.1 *Network algorithms*

(1) Giant component model (Janson *et al.*, 1993; Wilensky, 2005)
A component of the network refers to a group of nodes which are directly or indirectly connected to each other. A giant component

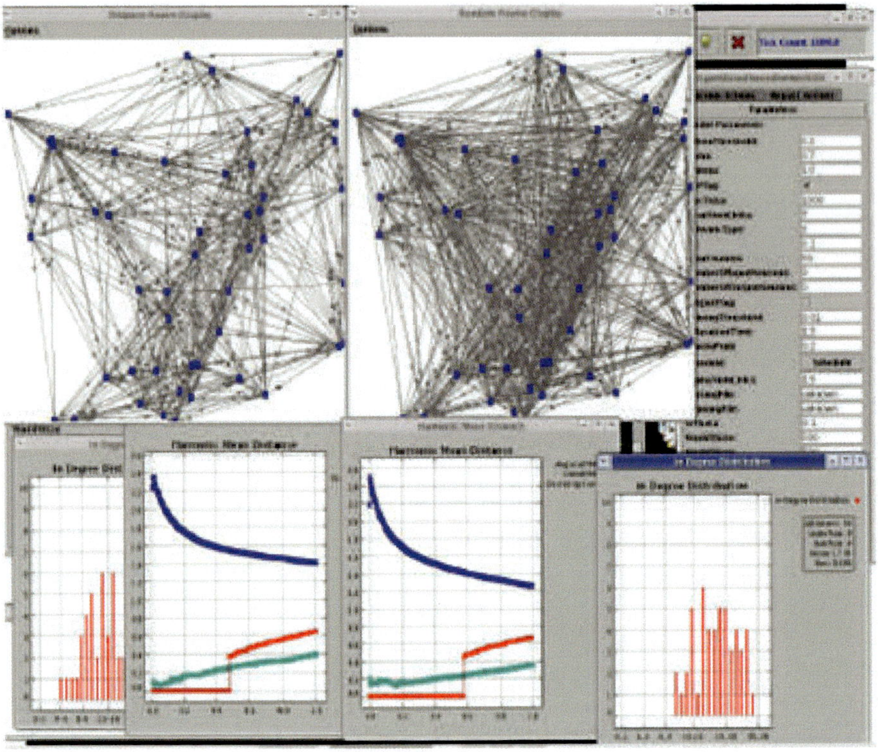

Fig. 2. ABNNSim (Schoenharl, 2005).

means that almost every node is reachable from almost every other node. This model demonstrates how quickly a giant component produces from a random network. In the model, the largest connected component produced by randomly connecting two nodes in the network each time step grows quickly after the average connections per node equals 1. In this model, initially we have nodes but no connections (edges) between them. At each step, two nodes are checked randomly and an edge between them is added if they were not directly connected. All possible connections have the same probability for producing. During the model running, small chain-like components, in which all nodes in each component are directly or indirectly connected, are produced. Two components merge into one if a connection is created between nodes in two different components. The

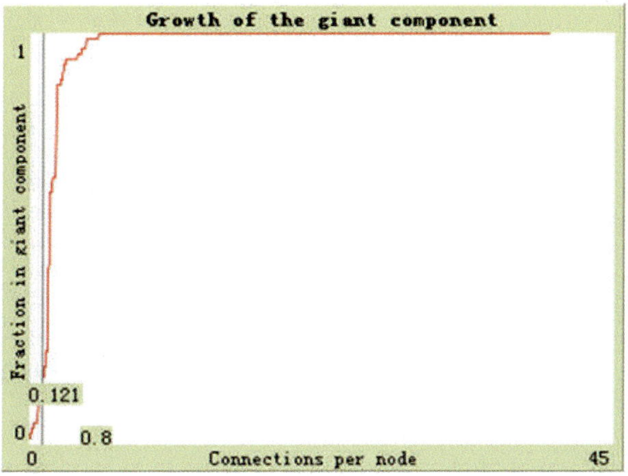

Fig. 3. Production of giant component in NetLogo.

component with the most nodes is defined as the giant component (Fig. 3).

(2) Diffusion in a directed network (Stonedahl and Wilensky, 2008a)
The model demonstrates how a quantity diffuses through a directed network. The quantity moves through nodes along established and directed edges between two nodes. Simple rules used in this model will result in various patterns regarding the topology, density, and stability of the network. In each step, each node shares some percentage of its value quantity with its neighboring nodes. A node will not share any of its value if it has no outgoing edges. It just accumulates any that its neighbors have provided through incoming edges.

(3) Preferential attachment (Barabasi and Albert, 1999; Barabasi, 2002)
In some of the networks, a few nodes have a lot of connections but other nodes have fewer. This model demonstrates how such networks produce. These networks are generated by a process called preferential attachment. In this process, new nodes tend to connect to existing nodes with more connections. It starts with two connected nodes. A new node is added at each step. It randomly, with different chances, picks an existing node to connect to. The chance is directly proportional to the number of connections of the node to be linked.

(4) Team assembly (Guimera *et al.*, 2005; Bakshy and Wilensky, 2007)
This model demonstrates how the behavior of individuals in assembling small teams for short-term projects can yield a variety of large-scale network structures over time (Bakshy and Wilensky, 2007). It is an adaptation of the team assembly model of Guimera *et al.* (2005). Team members are inexperienced newcomers or established incumbents. At each step, a new team is assembled. Each member is chosen sequentially. All members are linked to one another when a team is created. An agent and its links should be removed from the network if the agent does not participate in a new team for a prolonged period of time.

3.2.4.2 *Computer algorithms*

(1) Particle swarm optimization (Kennedy and Eberhart, 1995; Stonedahl and Wilensky, 2008b)
As described earlier, particle swarm optimization is a search method. It is usually used on multi-dimensional search spaces. This model demonstrates two-dimensional space. In the model, suppose there is an unknown function, fitness function $f(x, y)$ and we try to search x and y such that $f(x, y)$ is maximized. In particle swarm optimization, particles are placed in the search space and they move through the space based on some rules that take into account each particle's personal knowledge and swarm's knowledge. Through their movement, particles discover particularly high values for $f(x, y)$. This simple discrete model is based on the algorithm of Kennedy and Eberhart (1995). In this model, the particle swarm is trying to optimize a function that is determined by the values in the discrete cell grid. The procedure is described as follows (Stonedahl and Wilensky, 2008b): each particle has a position (x_{cor}, y_{cor}) in the search space and a velocity (v_x, v_y) at which it is moving through the space. Particles have a certain amount of inertia to keep them moving in the same direction they were moving previously. In addition, they have acceleration (change in velocity), which depends on two main things: (a) each particle is attracted towards the best position that it has personally found (personal best) previously in its history, and

(b) each particle is attracted towards the best position that any particle has ever found (global best) in the search space. The strength with which the particles are pulled in each of these directions is dependent upon the parameters ATTRACTION-TO-PERSONAL-BEST and ATTRACTION-TO-GLOBAL-BEST. As particles move farther away from these best positions, the attraction force increases its strength. There is also a random factor about how much the particle is pulled towards each of these positions (Stonedahl and Wilensky, 2008b). The model runs until some particle in the swarm has found the true optimum value (i.e., 1.00).

(2) Hexgonal cellular automata (Wilensky, 2007)

In NetLogo, the model of hexgonal cellular automata runs on a two-dimensional hexagonal grid of cells. Cells are alive or dead. Only the center cell is alive at the start. The switches determine which dead cells are to be revived. At each step, one candidate dead cell revives.

(3) Genetic algorithm (Holland, 1975; Stonedahl and Wilensky, 2008c)

This is a very simple genetic algorithm in NetLogo. It works by generating a random population of solutions to a problem, evaluating those solutions and using cloning, recombination, and mutation to create new solutions of the problem (Stonedahl and Wilensky, 2008c). In this model, Stonedahl and Wilensky (2008c) use the simple "ALL-ONES" problem to demonstrate the method. The idea of the "ALL-ONES" problem is to find a string of bits that contains all ones: "111111 ... 111". The genetic algorithm in this model includes the following procedures (Stonedahl and Wilensky, 2008c):

(i) Create a population of random solutions. In this step, each solution consists of a string of randomly mixed "1"s and "0"s.
(ii) Evaluate each solution based on how well it solves the problem. The goodness measure of the solution is called its fitness. As described, here the goal is simply to find a solution that consists of all "1"s.

(iii) Create a new generation of solutions from the old generation. Solutions with a higher fitness are more likely to be chosen as parent solutions than those with low fitness. In this step, the following cases are considered.

  (a) The "tournament selection" method is used in the model, with a tournament size of 3, i.e., three solutions are drawn randomly from the old generation, and the one with the highest fitness is chosen to become a parent.
  (b) Choose either one or two parents to create children. The child with one parent is a clone or copy of the parent. With two parents, the process is the digital analog of sexual recombination, i.e., the two children inherit part of their genetic material from one parent and part from the other.
  (c) Mutation will occasionally occur in the process. Some of the child's bits will be changed from "1"s to "0"s, and vice versa.

(iv) Repeat steps ii and iii, until a solution is found.

3.2.4.3 *Biological algorithms*

(1) Ants colony algorithm (Resnick, 1994; Wilensky, 1997a)

In this model, a colony of ants forages for food. Each ant follows a set of simple rules and the whole colony behaves in a more sophisticated way. When an ant finds a piece of food, it carries the food back to the nest and at the same time labels its path by releasing a chemical as it moves. When other ants detect the chemical, they will follow the chemical trail toward the food. As more ants carry food to the nest, they reinforce the chemical trail (Wilensky, 1997a).

(2) Termites model (Resnick, 1994; Wilensky, 1997b)

Termites gather wood chips into piles by following a set of simple rules. At first, each termite wanders randomly. If it bumps into a wood chip, it picks the chip up and continues to wander randomly. When it bumps into another wood chip, it finds a nearby empty space and puts its wood chip down. By doing so, the wood chips eventually end up in a single pile (Wilensky, 1997b; Fig. 4).

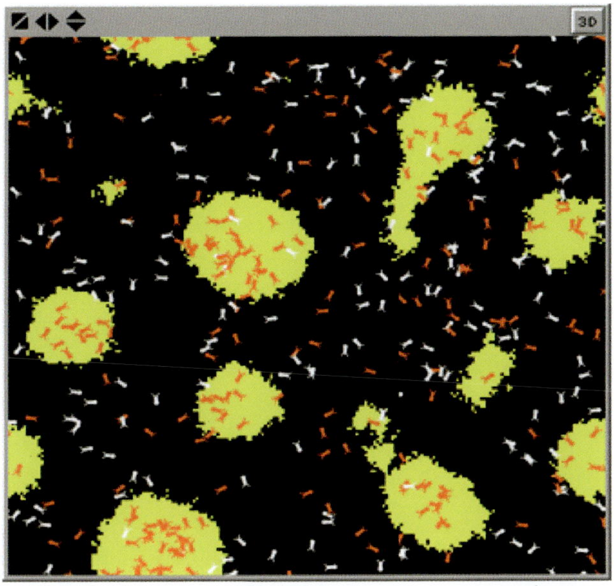

Fig. 4. Termite model in NetLogo.

(3) Tumor model (Wilensky, 1998)

This model aims to demonstrate how a tumor grows and how it resists chemical treatment. Generally a tumor contains two types of cells, stem cells and transitory cells. A stem cell can divide asymmetrically or symmetrically during mitosis. In asymmetric mitosis, one of the two daughter cells replaces its parent, remaining a stem cell. A stem cell never dies. Meanwhile, the other daughter cell changes into a transitory cell that moves outward. Young transitory cells may divide and breed other transitory cells. The transitory cells stop dividing at a certain age and change colors, eventually dying (Wilensky, 1998). Also, a stem cell can divide symmetrically into two stem cells. In this case, the original stem cell divides symmetrically only once. The first stem cell remains static but the second stem cell moves to the right (Fig. 5). This process is called metastasis. As the disease progresses, cells die younger and younger. The original Tumor model was contributed by Prof. Gershom Zajicek.

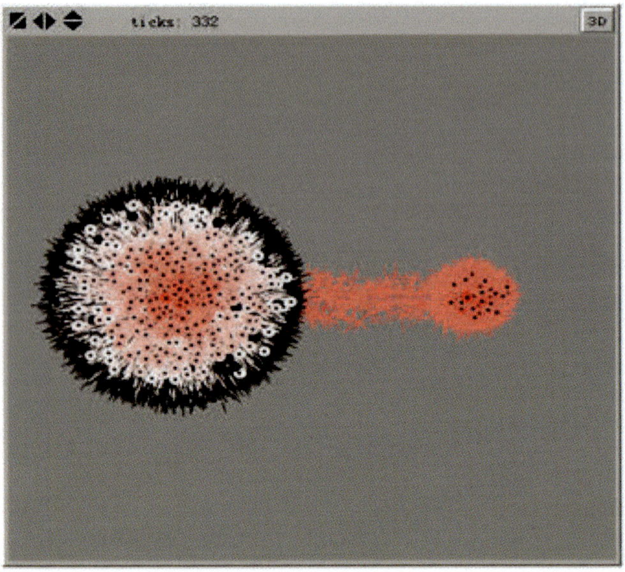

Fig. 5. Tumor model in NetLogo.

(4) Wolf-Sheep predation (Wilensky and Reisman, 1999)

This model is used to explore the stability of predator-prey ecosystems (Fig. 6). It has two main variations.

In the first variation, wolves and sheep wander randomly in the field and the wolves look for sheep to prey on. In each step, wolves must eat sheep in order to replenish their energy, or they die from running out of energy. To maintain the population, each wolf/sheep has a fixed probability of reproducing at each step (Wilensky and Reisman, 1999). This variation produces a population that is ultimately unstable. The second variation adds grass, in addition to wolves and sheep. The sheep must eat grass in order to maintain their energy, or they die from running out of energy. Once grass is eaten, it will only regrow after a fixed amount of time. This variation leads to a population that is generally stable.

3.2.4.4 *Social algorithms*

One of the social algorithms in NetLogo is cooperation model (EACH Unit, 1997; Wilensky, 1997c). This model is part of the EACH unit

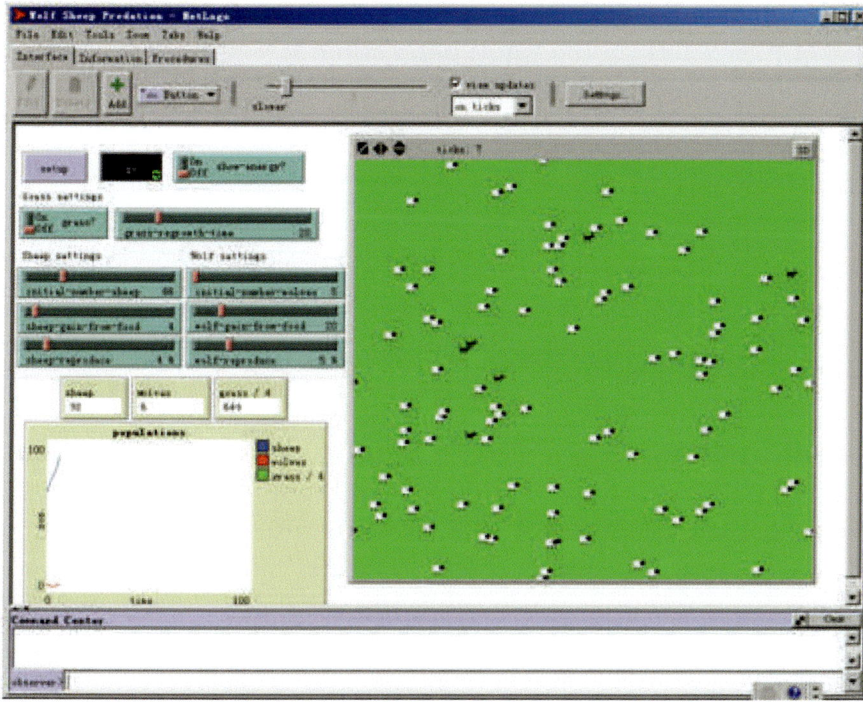

Fig. 6. Wolf-Sheep predation model model in NetLogo.

(Evolution of Altruistic and Cooperative Habits: Learning About Complexity in Evolution; http://ccl.northwestern.edu/cm/ EACH/). It is an evolutionary model. In this model, agents (cows) compete for natural resources (grass). Cows getting more grass reproduce more often and are thus more evolutionarily successful. This model includes two types of cows, i.e., greedy and cooperative cows. It demonstrates the behavior of the two strategies when they compete against each other within an evolving population (Wilensky, 1997c). In each step, each cow wanders around the current patch and eats a unit of grass. The greedy cows eat the grass regardless of the length of the grass. But the cooperative cows may not eat the grass below a certain height after which the grass will grow rapidly. The cooperative cows thus leave more food for the whole population at a cost of their individual well-being, and the greedy cows eat the grass regardless of their effect on the whole population.

3.2.4.5 *Algorithms in physics and chemistry*

(1) Crystallization basic (Wilensky, 2002)

A metal solidifies when it cools. The first atom solidifies at a random orientation. However the next atom will solidify at the oritention that the already solidified atom takes. Thus, they create a crystal grain. With more atoms solidified, the grains grow (Wilensky, 2002). All atoms in each grain have the same orientation; but different grains take different orientations. Deformations occur in the crystal structure when a metal is stressed. As the stress increases, deformations pass through the crystal structure and the metal tends to bend. However, grain boundaries will prevent deformations from passing through the metal. Thus the metal piece with fewer grain boundaries tends to be tough, while the metal piece with more grain boundaries tends to be brittle (Wilensky, 2002). This model demonstrates the formation process of grains when a metal crystallizes. In this model, a liquid metal is placed in a room temperature much lower than that of the metal. The metal begins to solidify as metal temperature drops. In this process, liquid atoms are free to rotate. If a liquid atom is next to a solid atom, it will orient itself with it, or else it will rotate randomly (Wilensky, 2002).

(2) Sand particles model (Wilensky, 1996)

This model demonstrates how sand particles interact with each other. In the sand environment, sand particles tend to move down if any of the following four rules apply (Wilensky, 1996): (a) If there is nothing directly beneath you, move down; (b) if there is something beneath you and to the lower right, move down and to the left; (c) if there is something beneath you and to your lower left, move down and to the right; and (d) if there is only something directly beneath you, move down and either left or right at random.

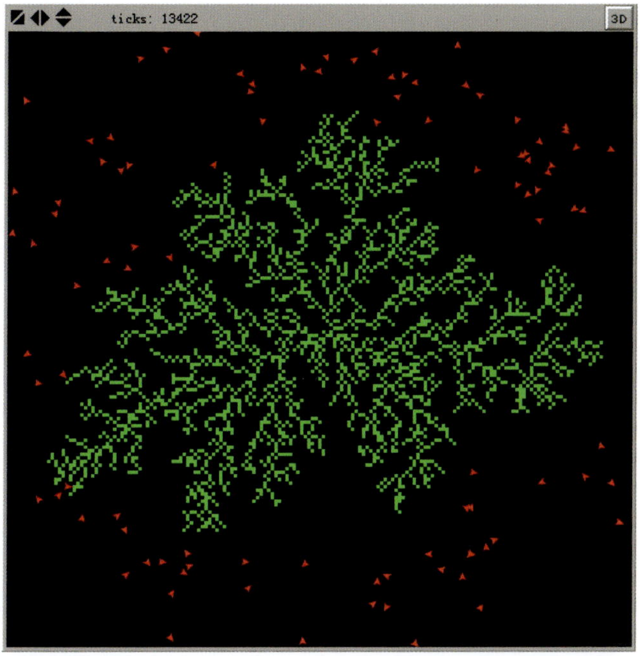

Fig. 7. Diffusion limited aggregation model in NetLogo.

(3) Diffusion limited aggregation (Wilensky, 1997)

This is a model to demonstrate diffusion-limited aggregation. In this model, randomly moving particles aggregate together to form tree-like fractal structures. The model begins with an initial seed in the center of the model view. Particles move around the view randomly and when a particle hits a square, it sticks and a new particle is then created to continue the process (Wilensky, 1997; Fig. 7).

Chapter 4

Intelligence Principles

Some of the research methods on artificial intelligence can be divided into two categories. The first category includes methods for symbol processing. They are based on the hypothesis on physical symbol system of Newell and Simon. Almost most of them are called "classic artificial intelligence." These methods use statements to express the knowledge of problem domain. These statements are based on, or substantially equivalent to the first-order logic statements. They use logical inference to derive results of the knowledge. These methods require sufficient knowledge of the problem domain, and thus are usually known as the knowledge-based approach. The second category comprises so-called "sub-symbols" methods. They usually adopt a bottom-up approach, i.e., address a problem from the bottom to the top. They also learn from the environment and stress the symbolic foundation. In the physical hypothesis, an agent does not use a centralized model, but uses different behavior modules to interact with the environment and conducts complex behaviors. An agent/component in the self-organizing systems usually possesses certain intelligence. The present chapter discusses some methods for artificial intelligence research.

4.1 State Space Search

Generalized problem solving is the focus of artificial intelligence. A problem can be expressed in a state space, in which each state is a set of some facts. We try to find a behavioral series and some production rules from the starting state to the destination state. The system to implement this process is usually called production system of the rule-operation series. The fundamental process is as follows (Hou and Deng, 1993):

(1) Data table (D) ← raw database;
(2) Until D satisfies the termination condition DO;
(3) Begin;
(4) Choose an appropriate rule R from the rules acting on data table D;
(5) Data table (D) ← results of D operated by R;
(6) End.

Two main types of search processes include blind search and heuristic search. Blind search refers to any information for problem inferencing that is not specified, for example, we need to only find a path to the destination. Among them are Breadth-first Search (BFS) and Depth-first Search (DFS). In the heuristic search, we specify the information to solve the problem.

4.1.1 *BFS* (*Breadth-first Search*)

A feature of BFS is that, once we have found the target node, we have also found a shortest path to the destination (Hou and Deng, 1993). BFS uses all operators to the start node to produce an explicit diagram of state space, and then uses all possible operators to all direct successors of the start node, and then to the successors of these successors, and so on. All possible operators at a step constitute a successor function. The search process will be extended outward from the start node. Applying the successor function to a node will produce a node set, which is called expanding. First, mark the start and the destination nodes. The expanding order of a node is represented by an ID number besides the node. Nodes with the same depth are expanded according to some fixed order. When expanding a node, apply the operators following the order of moving space to the left, up, right, and down. Although each step of movement is reversible, the arcs from the successor to its parents are deleted.

The procedures of BFS are as follows:

(1) Place the start node, s, in OPEN list, and let $\bar{g}(s) = 0$, where $\bar{g}(.)$ is the cost from the start node to node n.
(2) If the OPEN list is empty, the search fails and exits, or else continue.
(3) Get the node with the smallest g(.) in OPEN list and place it in CLOSED list, and name the node n.
(4) If node n is the destination node, track back by the pointer. A desired path is thus obtained, or else continue.
(5) Expand the node n. If it has no subnodes, return to step (2); or else, produce all of its subnodes, and for each subnode n_i, calculate $\bar{g}(n_i) = \bar{g}(n) + c(n, n_i)$, where $c(n, n_i)$ is the cost from the node n to its successor node n_i. Assign flag $\bar{g}(n_i)$ for n_i to return the pointer, and place them in the OPEN list.
(6) Return to the step (2).

Given a graph G = (V, E) and a start node s, using BFS will generate a tree that contains the root node s, and all reachable nodes of s. In the tree, the path from s to v is the shortest path containing the least edges. The following are pseudo-codes of BFS algorithm for graph search (Ku, 2011), in which all nodes are colored by white, gray, or black.

```
BFS(G, s)
for each node n ∈ V[G] − {s}
color[n] ← WHITE
d[n] ← ∞
π [n] ← NIL
color[s] ← GRAY
d[s] ← 0
π[s] ← NIL
Q ← φ
ENQUEUE(Q, s)
while Q ≠ φ
do n ← DEQUEUE(Q)
for each v ∈ Adj[n]
if color[v] = WHITE color[v] ← GRAY
```

```
d[v] ← d[n] + 1
π[v] ← n
ENQUEUE(Q, v)
color[n] ← BLACK
```

4.1.2 *DFS (Depth-first Search)*

DFS allows us to save only part of the search tree, which consists of the path currently being searched and the flags of nodes that have not been fully expanded on the path. For each step, DFS uses an operator to a node for producing a successor. Each node leaves a mark to indicate the additional operator for possible use. For a node, a decision is made to determine which operator should be firstly used. The successors of the current successor will be produced once the parent successor is produced. A depth bound is needed to constrain search depth, i.e., the successor is no longer produced if the depth exceeds the depth bound. Similar to BFS, when expanding a node, apply the operator following the order of moving space to the left, up, right, and down, and the arcs from the successor to its parents are omitted.

The procedures of DFS are as follows:

(1) Place the start node, s, in OPEN list, and let the depth of the node be 0.
(2) If the OPEN list is empty, the search fails and exits, otherwise continue.
(3) Get the first node in the OPEN list and place it in the CLOSED list, and name the node n.
(4) If the depth of node n reaches the depth bound, return to step (2), or else continue.
(5) Expand the node n. If it has no subnodes, return to step (2); or else, produce all of its subnodes, and for each subnode n_i, calculate the depth of n_i, and assign pointer for subnodes to return n, and place them in the OPEN list in arbitrary order.
(6) If a node in the subnodes is the destination node, track back by the pointer and a desired path is obtained, otherwise return to step (2).

The following are pseudo-codes of DFS algorithm for graph search (Ku, 2011), in which all nodes are colored by white, gray, or black.

```
DFS(G)
for each node n ∈ V [G]
color[n] ← WHITE
π[n] ← NIL
time ← 0
for each node n ∈ V [G]
if color[n] = WHITE DFS-VISIT(n)

DFS-VISIT(n)
color[n] ← GRAY
time ← time +1
d[n] time
for each v ∈ Adj[n]
do if color[v] = WHITE π[v] ← n
DFS-VISIT(v)
color[n] ← BLACK
time ← time +1
```

4.1.3 *Heuristic search*

Heuristic search is sometimes called best-first search. Different from the blind search, heuristic search chooses the node with the most possibility to reach the destination node for expanding. The possibility is represented by an evaluation function.

The procedures of a heuristic search are as follows:

(1) Place the start node, s, in OPEN list, and calculate $f(s)$.
(2) If the OPEN list is empty, the search fails and exits, or else continue.
(3) Get the node with the smallest $f(.)$ in the OPEN list and place it in the CLOSED list, and name the node n.
(4) If the node n is the destination node, track back by the pointer and a desired path is thus obtained, or else continue.

(5) Expand the node n. If it has no subnodes, return to step (2); or else, produce all of its subnodes, and for each subnode n_i, calculate $f(n_i)$.
(6) If a subnode n_i is not in both OPEN and CLOSED lists, assign n_i with the $f(n_i)$, calculated at step (5), and the pointer to n, and place it in the OPEN list.
(7) If a subnode n_i is already in the CLOSED list, then the node is not placed in the OPEN list. If n_i is in the OPEN list, compare previous $f(n_i)$ with the $f(n_i)$ calculated at step (5); if the previous is not greater than the later, no changes needed, or else change the previous $f(n_i)$ in the OPEN list to the $f(n_i)$ calculated in step (5), and the returning pointer is changed to n.
(8) Return to step (2).

A generalized heuristic search, GRAPHSEARCH (http://blog.csdn.net/v_july_v/article/details/6177380), allows users to customize state space search for heuristic-, breadth-, or depth-search. The procedures of its first version are as follows:

(1) Generate a search tree TREE which contains a start node s only. Place s in the ordered list, OPEN.
(2) Generate a list, CLOSED, which is initially empty.
(3) If OPEN is empty, then fail and exit.
(4) Get the first node (n) in the OPEN, remove it from OPEN, and place it in the CLOSED.
(5) If n is the target node, track back along the arcs of TREE, from n, to find a path back to s, and achieve the target arcs in step (6) and exit the search.
(6) Expand the node n and generate the successor nodes set S. Generate successors of n by finding the arcs from n to each node in S.
(7) Reorder OPEN list in any way or use heuristical method. For BFS, new nodes are placed in the end of OPEN; for DFS, new nodes are placed in the starting part of OPEN, and for heuristic search, reorder nodes in OPEN in a heuristic way.
(8) Return to step (3).

The revised algorithm of GRAPHSEARCH, A*, is as the following:

(1) Generate a search graph GRAPH which contains a start node s only. Place s in the list, OPEN.
(2) Generate a list, CLOSED, which is initially empty.
(3) If OPEN is empty, then fail and exit.
(4) Get the first node (n) in the OPEN, remove it from OPEN, and place it in the CLOSED.
(5) If n is the target node, track back along the arcs of GRAPH, from n, to find a path back to s, and achieve a tree in step (7) and exit the search.
(6) Expand the node n and generate the successor nodes set S. Generate successors of n by finding the arcs from n to each node in S. The ancestors of n should be excluded from S.
(7) Construct a pointer to n for each node in S but not in GRAPH, and add these nodes in OPEN list. For each node m of S, which is already in OPEN or CLOSED, if n is in the best path to m, then its pointer is made to point to n. For each node of S, which is already in CLOSED, reorient all its successors in S, in order to make them point to their ancestors along the currently best path.
(8) Reorder OPEN list in ascending order of $f(.)$.
(9) Return to step (3).

4.2 First-order Logic and Inference

First-order logic and inference will be useful in behavioral patterns of agents/components.

4.2.1 *First-order logic*

First-order logic can be used to express classic mathematical theory (Hou and Deng, 1993).

Atom is the fundamental unit in first-order logic. An atom is:

(1) A symbol for a constant;
(2) A symbol for a variable;

(3) A symbol for a function; or
(4) A symbol for a predicate.

A term is recursively defined as the follows:

(1) A constant is a term;
(2) A variable is a term;
(3) If f is the n-variable function, and $x_1, x_2, \ldots, x_n$, are terms, then $f(x_1, x_2, \ldots, x_n)$ is a term;
(4) Any term can be derived from limited utilization of the rules above.

If P is the n-predicate, and $x_1, x_2, \ldots, x_n$, are terms, then $P(x_1, x_2, \ldots, x_n)$ is an atom.

Formulas of first-order logic can be recursively defined as follows:

(1) An atom is a formula;
(2) If x and y are formulas, then $x \vee y, x \wedge y, \neg x, x \rightarrow y$, and $x \leftrightarrow y$ are also formulas;
(3) If F is a formula, and the variable x in F is free (i.e., x is not constrained), then $(\forall x)F$ and $(\exists x)F$ are also formulas;
(4) Any formula of first-order logic can be derived from limited utilization of the rules above.

For example, in the formula $F : (\forall x)(P(x) \rightarrow Q(f(x), c)), c$ is a constant symbol, f is a 1-variable function symbol, P is a 1-predicate symbol, and Q is a 2-predicate symbol.

A formula can be logically inferenced from several formulas. The clause that logically inferences a formula from several other formulas is called theorem, and the inference process is called a proof of the theorem.

An interpretation of the formula F comprises a non-empty domain D, and the following assignment to values of all constants, function symbols and predicate symbols in F:

(1) Assign an element of D to the constants and free variables in F;
(2) Assign a map from D^n to D for each n-variable function, where $D^n = \{(x_1, x_2, \ldots, x_n)|x_i \in D\}$;

(3) Assign a map from D^n to $\{F, T\}$ for each n predicate symbol, where F means false (□) and T means true (■) (Hou and Deng, 1993).

Two formulas G and F are equivalent, i.e., $G = F$, if and only if G and F have the same T-F value under every interpretation.

4.2.2 *Formalized expression of mathematical theorem*

A formula G is to be called compatible if and only if there is at least an interpretation I, such that G is true under I. If G is true under I, I is called a model of G, and I satisfies G. G is universally valid if and only if each interpretation of G satisfies G.

Formula G is the logic reasoning of formulas $F_1, F_2, \ldots$, and F_n, if and only if for every interpretation I, if $(F_1 \wedge F_2 \wedge \ldots \wedge F_n)$ is true under I, then G is true under I also. $F_1, F_2, \ldots$, and F_n are called the axiom of G. G is the logic reasoning of formulas $F_1, F_2, \ldots$, and F_n, if and only if $((F_1 \wedge F_2 \wedge \ldots \wedge F_n) \to G)$ is universally valid (Hou and Deng, 1993).

If G is the logic reasoning of formulas $F_1, F_2, \ldots$, and F_n, then $((F_1 \wedge F_2 \wedge \ldots \wedge F_n) \to G)$ is a theorem, and G is an inference of the theorem.

4.2.3 *Prenex normal form of first-order logic*

A formula F is called prenex normal form, if and only if F is as the following: $(Q_1x_1)\ldots(Q_nx_n)M$, where Q_ix_i is $\forall x_i$ or $\exists x_i$, and M is a formula without quantifiers. $(Q_1x_1), \ldots, (Q_nx_n)$, are called prenex, and M is the mother formula of F. For example, $(\forall x)(\exists y)(P(x,y) \to Q(x,y))$ is a prenex normal form.

Some of the transformation rules are as follows:

(1) $F \leftrightarrow G = (F \to G) \wedge (G \to F)$
(2) $F \to G = \neg F \vee G$
(3) $F \vee (G \wedge H) = (F \vee G) \wedge (F \vee H)$
(4) $F \wedge (G \vee H) = (F \wedge G) \vee (F \wedge H)$
(5) $\neg(F \vee G) = \neg F \wedge \neg G$
(6) $\neg(F \wedge G) = \neg F \vee \neg G$

(7) $(Qx)F(x) \vee G = (Qx)(F(x) \vee G)$
(8) $(Qx)F(x) \wedge G = (Qx)(F(x) \wedge G)$
(9) $\neg((\forall x)F(x)) = (\exists x)(\neg F(x))$
(10) $\neg((\exists x)F(x)) = (\forall x)(\neg F(x))$
(11) $(\forall x)F(x) \wedge (\forall x)G(x) = (\forall x)(F(x) \wedge G(x))$
(12) $(\exists x)F(x) \vee (\exists x)G(x) = (\exists x)(F(x) \vee G(x))$
(13) $(Qx)F(x) \wedge (Qx)G(x) = (Qx)(Qz)(F(x) \wedge G(z))$
(14) $(Qx)F(x) \vee (Qx)G(x) = (Qx)(Qz)(F(x) \vee G(z))$

Any formula of first-order logic can be transformed to its prenex normal form by using above rules.

4.2.4 *Skolem standard form and clause*

For the prenex normal form of a formula, e.g.,$(\exists x)(\forall y)(\forall z)(\exists u)(\forall v)$ $F(x, y, z, u, v, w)$, its skolem standard form is $(\forall y)(\forall z)(\exists v)F(a, y, z, f(y, z), v, g(y, z, v))$ (Hou and Deng, 1993). The Skolem standard form of a formula of first-order logic can be constructed following these steps:

(1) Transform a formula to the prenex normal form, of which the mother formula does not contain quantifiers and the prenex is a series of quantifiers.
(2) Transform the mother formula to the conjunctive normal form.
(3) Resolve existential quantifiers by using the Skolem functions. Here the constants and functions to substitute the variables of existential quantifiers are called Skolem functions.

F' is a Skolem standard form of the first-order logic formula F. F is universally contradictory if and only if F' is universally contradictory.

Atoms and negation of atoms are called literals, and the disjunctive of some literals is called a clause. A clause containing m literals is called m-literal clause. In a Skolem standard form, the mother formula is a conjunctive normal form, and the conjunctive normal form itself is the conjunctive of some clauses. The first-order logic formula F is universally contradictory if and only if its clauses set S is universally contradictory.

Terms, a set of terms, atoms, a set of atoms, literals, a set of literals, a clause, and a set of clauses, etc., are all called expressions.

4.2.5 *Resolution principle of first-order logic*

Resolution principle aims to examine whether the set of clauses S contains empty clause (i.e., $\square$). If S contains $\square$, S is unsatifiable; if S does not contain $\square$, examine if $\square$ can be derived from S. Resolution principle can be treated as a set of inference rules to construct new clauses in S (Hou and Deng, 1993).

A substitution is a limited set in the form of $\{t_1/v_1, t_2/v_2, \ldots, t_n/v_n\}$, where v_i are variables and all t_i are terms different from v_i. The substitution is called ground substitution if all t_i are ground terms. The substitution not containing any elements is called empty substitution, ε.

Suppose there is a substitution, $\theta = \{t_1/v_1, t_2/v_2, \ldots, t_n/v_n\}, E$ is an expression, $E\theta$ is an expression substituting all v_i from E with t_i. $E\theta$ is a ground instance of E.

Suppose there are two substitutions, $\theta = \{t_1/v_1, t_2/v_2, \ldots, t_n/v_n\}$, and $\delta = \{u_1/y_1, u_2/y_2, \ldots, u_m/y_m\}$. $\theta \cdot \delta$ is a substitution, and its elements are $\{t_1\delta/x_1, t_2\delta/x_2, \ldots, t_n\delta/x_n, u_1/y_1, u_2/y_2, \ldots, u_m/y_m\}$, by eliminating $t_i\delta/x_i$ that $t_i\delta = x_i$, and u_i/y_i that y_i is in the set $\{x_1, x_2, \ldots, x_n\}$.

A substitution θ is called the unifier of the expression set, $\{E_1, E_2, \ldots, E_m\}$, if and only if $E_1 \cdot \theta = E_2 \cdot \theta = \ldots = E_m \cdot \theta$. $\{E_1, E_2, \ldots, E_m\}$ is called unifiable if it has any unifiers. The unifier δ of expression set $\{E_1, E_2, \ldots, E_m\}$ is the most general unifier, if and only if for every unifier θ there exists a substitution λ, such that $\theta = \delta \cdot \lambda$.

If at least two literals of the clause C have the most general unifier $\delta, C\delta$ is called the factor of C. If $C\delta$ is a single clause, call C the unit factor clause.

Suppose C_1 and C_2 are clauses without sharing the same variables, L_1 and L_2 are literals in C_1 and C_2 respectively. If L_1 and L_2 have the most general unifier δ, the clause, $(C_1\delta - L_1\delta) \cup (C_2\delta - L_2\delta)$, is called resolvent, and L_1 and L_2 are literals on the resolvent. Resolution is a set of inference rules of producing resolvents from a set of clauses.

The complete resolvent of clauses C_1 and C_2 are one of the following resolvents:

(1) The resolvent of C_1 and C_2.
(2) The resolvent of C_1's factor and C_2.
(3) The resolvent of C_1 and C_2's factor.
(4) The resolvent of C_1's factor and C_2's factor.

The set of clauses S is unsatifiable, if and only if there is a deduction from S to empty clause.

A general method, using resolution principle, to prove a mathematical theorem is as follows (Hou and Deng, 1993):

(1) Calculate the resolvents of all clause pairs in the set of clauses S; add the new generated clauses to S, and thus form a new set of clauses S_1.
(2) Resolute new clauses in S_1 with each of the clauses in S; add new generated resolvents to S_1, and thus form a new set of clauses S_2.
(3) Repeat the above procedures, until the empty clause is obtained.

Chapter 5

Catastrophe Theory and Methods

Catastrophe theory is a science based on topology and stability theory (bifurcation theory and singularity theory) (Thom, 1969, 1975, 1977; Zeeman, 1976, 1977; Zhao and Yu, 1987). It studies the behavior of system at equilibrium points, uncovers the properties of sudden changes in state variables during the process of continuous and smooth change of control variables. In catastrophe theory, we study the uncontinuous properties of potential function around its critical points and develop the qualitative models.

The continuous change of control variables causes the continuous and smooth change of the potential function at non-critical points but catastrophic at critical points. Catastrophe phenomenon are popular in nature. Gene mutation, population outbreak, embryonic development, etc., are all catastrophe phenomenon. So far, catastrophe theory and methods have been widely used in physics, chemistry, biology, sociology and other sciences and engineering practices (Piyaratne *et al.*, 2014; Zhang, 2013a, 2014). Further details on catastrophe theory can be found in Thom (1969, 1975, 1977), Zeeman (1976, 1977), Zhao and Yu (1987). Details on topology can be found in Spanier (1966), Chan (1987), and Lin (1998).

The properties, e.g., jumps, bisteady states, lagging, divergency, inaccessibility, and multipath, etc., can be explored by catastrophe theory (Zhao and Yu, 1987): (1) multimodality. The system denoted by potential function may possess more than one minimum; (2) inaccessibility. There are state surfaces inaccessible by state variables driven by control variables (parameters); (3) jumps. A small change in control variables leads to a great change in state variables, and thus results in the jump of the system from a critical point to another critical point (catastrophe); (4) divergency. In the neighborhood of a degenerated critical point, a small change in control variables will result in a great change in state variables; (5) lagging. The jump point from critical point 1 to critical point 2 is different from the jump point from critical point 2 to critical point 1, which is called lagging; (6) multipath. A state in the equilibrium surface can be accessed by adjusting different trajectories of control variables.

5.1 Dynamic Systems and Catastrophe Phenomenon

As mentioned above, catastrophe theory tries to uncover the properties of sudden changes in state variables during the process of continuous and smooth change of control variables/factors. It uses the topological methods and other methods to analyze the singular properties in the junction area between continuous and uncontinuous areas. Thus, the discussion on differential equations that describe continuous dynamics is necessary in catastrophe theory.

A differential equation group can be expressed as (Zhao and Yu, 1987):

$$\begin{gathered} f_i(\varphi_j; C_h; t, d\varphi_j/dt, d^2\varphi_j/dt^2, d^3\varphi_j/dt^3, \ldots; \\ x_l; \partial\varphi_j/\partial x_l, \partial^2\varphi_j/\partial x_l \partial x_m, \ldots; \quad \int dx_l, \ldots) = 0 \\ 1 \le l, m \le N; \quad 1 \le h \le p; \quad j = 1, 2, \ldots \\ 1 \le i \le n \end{gathered}$$

where $x = (x_1, x_2, \ldots, x_N)$ is space position, t is time, and $\varphi_j = \varphi_j(t, x; C_h), j = 1, 2, \ldots$, are solution of the differential equation group. φ_j are state variables of the system, and C_h are control parameters.

A simple form of the differential equation group above is

$$\begin{gathered} f_i(\varphi_j; C_h; t, d\varphi_j/dt) = 0 \\ 1 \le i \le n; \quad 1 \le h \le p; \quad j = 1, 2, \ldots \end{gathered}$$

i.e., the dynamic system

$$\begin{gathered} d\varphi_i/dt = g_i(\varphi_j; C_h; t) \\ 1 \le i \le n; \quad 1 \le h \le p; \quad j = 1, 2, \ldots \end{gathered}$$

Assume it is an autonomus system, i.e., $g_i(\varphi_j; C_h; t) = g_i(\varphi_j; C_h)$, and

$$g_i = -\partial v(\varphi_j; C_h)/\partial \varphi_i$$

where $v(\varphi_j; C_h)$ is the potential function. Thus we have

$$d\varphi_i/dt + \partial v/\partial \varphi_i = 0$$

i.e., $d\varphi_i/dt = -\nabla_\varphi v$. If $\mathrm{d}\varphi_i/\mathrm{dt} = 0$, the system is in its steady state. The equation

$$\partial v(\varphi_j; C_h)/\partial \varphi_i = 0$$

is called the balance equation and its solution is the critical point of the system. The hidden function of the solution determines the balance hypersurface M.

In mathematical view, catastrophe theory tries to study: (1) the rules of how the equilibrium points $\varphi_j(C_h)$ of potential function $v(\varphi_j; C_h)$ change with the control parameters C_h, and (2) the relationship between $v(\varphi_j; C_h)$ and φ_j and C_h (Zhang and Yu, 1987).

As for how a potential function is obtained, here I take two examples. For the system

$$\partial u/\partial t = c^2(\partial^2 u/\partial x^2 + \partial^2 u/\partial y^2) + f(x, y)$$

the potential function is

$$v(u; c) = \int [-c^2(\partial^2 u/\partial x^2 + \partial^2 u/\partial y^2) - f(x, y)]du + a$$

For the system

$$dx/dt = f(x; c)$$

we have $f(x; c) = -\partial v(x; c)/\partial x = -dv(x; c)/dx$, and thus $dv(x; c) = -f(x; c)dx$, and the potential function is

$$v(x; c) = -\int f(x; c)dx + a$$

5.2 Local Properties of the Potential Function

5.2.1 *Representation of the potential function*

Suppose the potential function is $v = v(x; c), x \in R^n, c \in R^p$. The potential function can be represented by the following three forms (Zhao and Yu, 1987).

(1) Hidden function
Let $y_i = y_i(x_1, x_2, \ldots, x_n), i = 1, 2, \ldots, n$, and thus make $-\nabla v$ be non-zero along one direction only (e.g., along y_1), then the potential function can be represented by

$$v \approx y_1 + b$$

In this form, v is the smooth function with x, and for any point along y_1, the potential function can be expressed as a linear form.

(2) Morse form
If $\nabla v = 0$, the system is in its steady state. The system in steady state can be represented by a Hessen matrix (v_{ij})

$$v_{ij} = \partial^2 v / \partial x_i \partial x_j \quad i, j = 1, 2, \ldots, n$$

Thus, if $\nabla v = 0$ and $(v_{ij}) \neq 0$, the potential function can be represented by a quadratic form when state variables change smoothly and continuously

$$v \approx \sum_{i=1}^{n} \lambda_i y_i^2$$

where there are eigen values of Hessen matrix (v_{ij}). Let $v \approx M_i^n(\hat{y}) = \hat{y}_1^2 - \hat{y}_2^2 - \cdots - \hat{y}_i^2 + \hat{y}_{i+1}^2 + \cdots + \hat{y}_n^2$, by the transformation $\hat{y}_i = |\lambda_i|^{0.5} y_i$, $M_i^n(\hat{y})$ is the Morse form (i-order saddle function) of the potential function (Zhao and Yu, 1987). Only 0-order saddle function has the local stability because it has the local minimum when the system is in equilibrium state.

The properties of a critical point can be determined based on Hessen matrix: (a) $\nabla v \neq 0$, potential function is not in equilibrium state and no critical point; (b) $\nabla v = 0$ and $\det(v_{ij}) \neq 0$, potential function is in equilibrium state and there are some isolated critical points (Morse critical points); (c) $\nabla v = 0$ and $\det(v_{ij}) = 0$, potential function is in equilibrium state and there are some non-isolated critical points.

Taylor expansion of a function $v(x)$ in $x = 0$ is

$$v(x) = v(0) + x_i v_i(0) + x_i x_j v_j(0)/2! + \cdots$$

From the Taylor expansion of $v(x; c)$, we find that if $\nabla v \neq 0$, the properties of $v(x; c)$ is completely determined by the term of the first degree; if $\nabla v = 0$ and $\det(v_{ij}) \neq 0$, the properties of $v(x; c)$ is determined by the terms of the second degree, and further the terms of the second degree can be transformed to standard terms of the second degree, $\sum \lambda_i y_i^2$; if $\nabla v = 0$ and $\det(v_{ij}) = 0$, the properties of $v(x; c)$ is determined by the terms of the higher degree, e.g., the terms of k-th degree, and the terms of higher degree greater than k can be ignored, and this is called truncation of k-th degree.

The truncation of k-th degree of the Taylor expansion of a function $f(x)$ is

$$j^k f(x) = \sum_{r=0}^{k} \partial^r f(x)|_{x=0} x^r / r!$$

where $x = x(x_1, x_2, \ldots, x_n)$. If the type of a function is the same to that of arbitrary function with truncation of k-th degree, the function is called k-degree deterministic. Mathers' method for determining k is (Zhao and Yu, 1987): (a) suppose $f(x)$ is k-degree deterministic; (b) suppose $m(x)$ is the monomial series of x

$$m_j(x) = x_1, x_2, \ldots, x_l; \quad x_1^2, x_1 x_2, \ldots, x_l^2; \quad x_1^3, x_1 x_2 x_3, \ldots, x_l^3; \ldots$$

(c) compute the following polynomial

$$R_{ij}(x) = j^{k+1}\{(\partial f / \partial x_i) m_j(x)\}$$

(d) determine if the series $m_j(x)$ (choose the terms not greater than $k+1$ degree) can be expressed as the linear superposition of $R_{ij}(x)$ (it can contain ordinary coefficients). If $f(x)$ is k-degree deterministic, then each monomial of x can be expressed as the linear superposition of $R_{ij}(x)$, otherwise it is not true.

(3) Thom form

As mentioned above, the equilibrium points (critical points) that meet $\nabla v = 0$ and $\det(v_{ij}) \neq 0$ are non-isolated equilibrium points, if $\lambda_i(c) = 0$ $(i = 1, 2, \ldots, l)$ for some specific values of control parameters $c = (c_1, c_2, \ldots, c_p)$. The potential function can be divided into two parts (non-isolated equilibrium points, and isolated equilibrium points; Zhao and Yu, 1987)

$$v(x; c) \approx f_{NM}(y_1(x; c), y_2(x; c), \ldots, y_l(x; c)) + \sum_{i=l+1}^{n} \lambda_i(c)(y_i(x))^2$$

If the Hessen matrix of the second part (isolated equilibrium points) is not singular ($\det A_{n-l,n-l} \neq 0$), the Hessen matrix of the first part (non-isolated equilibrium points) $\partial^2 f_{NM}/\partial y_i \partial y_j$ $(i, j = 1, 2, \ldots, l)$ is singular. Choose the number of control parameters, the potential function can be expressed to be

$$v(x; c) \approx G(l) + \sum_{i=l+1}^{n} \lambda_i y_i^2$$

where $G(l)$ is the generated term of catastrophe. Thom (1975, 1977) proved that for a fixed $c = c^0$, if x^0 is the non-isolated critical point, then in an open neighborhood $(x^0; c^0)$, the potential function can be represented by

$$v(x; c) \approx \mathrm{Cat}(l, p) + \sum_{i=l+1}^{n} \lambda_i(c) y_i^2$$

where p is the number of control parameters, l is the spatial dimensionality of Hessen matrix (v_{ij}) at the non- isolated critical point, $\mathrm{Cat}(l, p)$ is the catastrophe function and $\mathrm{Cat}(l, p) = G(l) + \mathrm{Pert}(l, p)$. The catastrophe function $\mathrm{Cat}(l, p)$ can be treated as the Taylor expansion of $f_{NM}(y_1(x; c), y_2(x; c), \ldots, y_l(x; c))$ in the open neighborhood $(x^0; c^0)$. Pert (l, p) is the perturbation term.

The catastrophe functions corresponding to $p \leq 5$ and $l = 1$ or 2, are listed in Table 1.

Table 1. Catastrophe functions for various situations (Zhao and Yu, 1987).

Name	p	$G(l)$	Pert (l, p)
A_2	1	x^3	a_1x
$A_{\pm 3}$	2	$\pm x^4$	$a_1x + a_2x^2$
A_4	3	x^5	$a_1x + a_2x^2 + a_3x^3$
$A_{\pm 5}$	4	$\pm x^6$	$a_1x + a_2x^2 + a_3x^3 + a_4x^4$
A_6	5	x^7	$a_1x + a_2x^2 + a_3x^3 + a_4x^4 + a_5x^5$
D_{-4}	3	$x^2y - y^3$	$a_1x + a_2y + a_3y^2$
D_{+4}	3	$x^2y + y^3$	$a_1x + a_2y + a_3y^2$
D_5	4	$x^2y + y^4$	$a_1x + a_2y + a_3x^2 + a_4y^2$
D_{-6}	5	$x^2y - y^5$	$a_1x + a_2y + a_3x^2 + a_4y^2 + a_5y^3$
D_{+6}	5	$x^2y + y^5$	$a_1x + a_2y + a_3x^2 + a_4y^2 + a_5y^3$
$E_{\pm 6}$	5	$x^3 \pm y^4$	$a_1x + a_2y + a_3xy + a_4y^2 + a_5xy^2$

5.2.2 *Topological space*

In catastrophe theory, we focus on the properties of supersurface determined by $\nabla v(x, c) = 0$. To study the properties of supersurface in a space, we usually need to make some transformations and preserve the properties of the supersurface. Theories and methods on topological space, topological mapping, topological transformation, homeomorphism and topological invariants are thus necessary. Readers are encouraged to refer to Spanier (1966), Chan (1987), Lin (1998), and Zhang, (2012a) for more details.

5.3 Structural Stability

5.3.1 *Stability of potential function*

A potential function is structurally stable if arbitrary perturbations are not able to affect its quanlitative properties (Zhao and Yu, 1997). Suppose the potential function $v(x) = v(x_1, x_2, \ldots, x_n)$; the perturbation is $\varepsilon f(x) = \varepsilon f(x_1, x_2, \ldots, x_n)$, and $\nabla v = 0$ and $\det(v_{ij}) \neq 0$, i.e., the potential function has some equilibrium points. If the perturbed potential function, $v'(x') = v(x) + \varepsilon f(x)$, in a new coordinate system has the same structure with the old potential function, $v(x)$, in original coordinate system, i.e., $v'(x') \approx v(x)$, where

$x'_i = x'_i(x_1, x_2, \ldots, x_n), i = 1, 2, \ldots, n$, then the potential function $v(x)$ is called stable at the critical point $x^0 \in R^n$.

Actually a perturbation is equivalent to a topological mapping of the function in a neighborhood. In the m-dimensional supersurface, expanded by m control parameters, a point represents a specific function. Suppose f_p is the function for the point p. If there is a point q in the neighborhood of p, such that the function f_q of point q has the same form with f_p, the f_p is the generating function (structurally stable function) of the function family. In a word, if Q is the neighborhood of p and f_Q is the neighborhood of f_p, then f_p is structurally stable and f_Q is the surjection (epimorphism) of f_p (Zhao and Yu, 1987). The set of p points corresponding to generating functions is called the subset of generating points and its complimentary set is called the set of bifurcation points. The set of bifurcation points is the mapping $R^m \to R^2$, and it is a topological invariant.

In general, if a function is in both original function family and perturbed function family, and the topological structure of its set of bifurcation points is preserved, the function is structurally stable. It is an isomohism in topology.

An unstable function may become stable by adding some terms (perturbations), i.e., extension method (Zhao and Yu, 1987). Suppose the following is true

$$f(x) = F(x; a)|_{a=0}$$

where $x = (x_1, x_2, \ldots, x_n), a = (a_1, a_2, \ldots, a_r)$. $F(x; a)$ is called r-dimensional extension of $f(x)$. If the other r-dimensional extensions of $f(x)$ can be derived from $F(x; a)$ by the following transformation

$$x'_i = x'_i(x; a); a'\alpha = a'\alpha(a), \alpha = 1, 2, \ldots, r'$$

$F(x; a)$ is called the complete extension of $f(x)$. Furthermore, $F(x; a)$ is the universal extension of $f(x)$ if it is the extension with least dimensionality.

Mathers' method for determining the extension of $f(x)$ is (Zhao and Yu, 1987): (a) obtain the k-degree determinant, $j^k f(x)$, of $f(x)$; (b) suppose $n(x)$ is the monomial series of x

$$n_j(x) = 1, x_1, x_2, \ldots, x_l; \quad x_1^2, x_1x_2, \ldots, x_l^2; \quad x_1^3, x_1x_2x_3, \ldots, x_l^3; \ldots$$

(c) suppose $F(x;a)$ is the r-dimensional extension of $j^k f(x)$, and compute the following

$$T_j(x) = \partial j^{k+1} F(x;a)/\partial x_j|_{a=0}$$

(d) compute the following polynomials

$$S_{ij}(x) = j^k\{(\partial j^k f(x)/\partial x_j)n_j(x)\}$$

(e) check if each monomial of $n_j(x)$ not greater than k-degree can be expressed as the following: $n_j(x) \leq k = \sum E_{ij}S_{ij}(x) + \sum T_j(x)$; if so, $F(x;a)$ is the complete extension of $j^k f(x)$.

(f) if $T_j(x)$ is the least combination, $F(x;a)$ is the universal extension of $j^k f(x)$.

If $F(x;a)$ is the universal extension of $j^k f(x), F(x;a)$ can be expressed as the linear combination of $j^k f(x)$ and $T_j(x)$, i.e., $F(x;a) = j^k f(x) + \sum_{j=1}^r a_j T_j(x)$.

5.3.2 *Dissection lemma*

Consider the Taylor's expansion of function $f(x)$ (generally with the single variable x, or two variables x, y). A function is called n-degenerated, if all derivatives less than n-degree are zeros and thus the properties of the function (e.g., properties of critical points) must be determined by checking its n-degree derivatives. In this case, n extention parameters should be added to preserve the stability of the function.

The dissection lemma is stated as the following. Suppose $f(x) = f(x_1, x_2, \ldots, x_n)$ is a n-variable function with critical point at the origin $(0, 0, \ldots, 0)$. All partial devariatives of the first degree of $f(x)$ are zeros at the origin. The Hessen matrix of the second degree devaritives is

$$v_{ij}(x) = \partial^2 f/\partial x_i \partial x_j$$
$$i, j = 1, 2, \ldots, n$$

If $\det(v_{ij}) \neq 0$, we have

$$f(x) = \sum_{i=1}^{n} \lambda_i x_i^2 + \sum_{i,j,k} f_{ijk} x_i x_j x_k + \cdots$$

where $\text{sign}(\lambda_i) = \pm 1$, and the number of positive *vs.* negative signs represents the type of critical point, and $f(x)$ is structurally stable. If $\text{rank}(v_{ij}) = n - l$, and $l > 0$, there is a transformation such that

$$f(x_1, x_2, \ldots, x_n) \approx f_{NM}(x_1, x_2, \ldots, x_l) + \sum_{i=l+1}^{n} \lambda_i x_i^2$$

Structural instability is only limited to $x_1, x_2, \ldots, x_l$, and thus the remaining variables, $x_{l+1}, x_{l+2}, \ldots, x_n$, can be ignored.

The function $f(x)$ is thus dissected into two parts, the Morse part with substanial variables related to structural stability, and the non-Morse part with non-substanial variables to structural stability, by the lemma. Thus the number of possible catastrophe types is only determined by the number of substanial variables, l, i.e., the corank of the Hessen matrix or $f(x)$. l is the number of parameters to be added for stabilizing the function (Zhao and Yu, 1987).

If $\det(v_{ij}) = 0$, $f(x_1, x_2, \ldots, x_n)$ is unstable, i.e., it is degenerative along the n directions.

Codimension of a geometrical object is an invariant in dimension-reducing method. It is the difference between the dimensionality of the object and its space. Codimension is the number of equations to describe the object. Codimension has two properties: (a) dissection-ability, e.g., codimension = 1 means the dimensionality of the object is 1 less than its space and it can dissect the space into two parts; (2) invariability, i.e., ignoring non-substantial variables will reduce the same dimensionality of the object and its space, and thus the codimension is invariable. The invariability of codimension means that preservation of the properties we are interested in, and further means that the number of state variables is not important or even unknown in catastrophe theory. If corank is n, the codimension should be $n(n+1)/2$.

5.3.3 *Singularity and equilibrium surface*

To obtain the singularity (singular points), the topological transformation, in particular differentiable homeomorphism (diffeomorphism), is needed. The diffeomorphism presents when the topological transformation between two geometrical objects is homeomorphic and the momeomorphism is continuously differentiable. For example, a sphere and an ellipsoid is diffeomorphic, but the sphere and a cube is not diffeomorphic (Zhao and Yu, 1987). Diffeomorphism is an elementary transformation. It ensures the smooth property of a function. A momeomorphism can be expressed, for example, as

$$x_i \to \varphi_i(x_1, x_2, \ldots, x_n)$$

where $\varphi_i(1 \leq i \leq n)$ are continuously differentiable functions.

In a catastrophe function, if the state variable x and control variable c are diffeomorphic, their equivalent equilibrium surfaces and bifurcation sets will also be diffeomorphic.

5.4 Basic Types of Catastrophe

Given a potential function v, its equilibrium surface M is $\nabla_x v = 0$, and its singularity set S can be derived by the equation: $\Delta H =$ det $H = 0$, where H is the Hessen matrix of v. Combine the two equations and eliminate the state variable $x = (x_1, x_2, \ldots, x_n)$, we obtain an equation of control variables. In other words, singularity set S is mapped to the space of control variables, and the trajectory is the bifurcation set B. The control variables in B will result in uncontinuity and lead to the catastrophe of potential function v (Zhao and Yu, 1987).

5.4.1 *Folding catastrophe*

Folding is the simplest catastrophe (A_2 in Table 1). Its state space is (x, a) and the potential function is

$$V(x) = x^3 + ax$$

Let $V'(x) = 0$, the equilibrium surface is thus $3x^2 + a = 0$. The equation of singularity set S, derived by $\Delta_x v = 0$, is $6x = 0$. If $a < 0$, there are two critical points, a maximum point and a minimum point. If $a > 0$, the potential function v is unstable.

5.4.2 *Cusp catastrophe*

Cusp catastrophe is the most used catastrophic type (A_{+3} in Table 1). Its state space is (x, a, b) and the potential function is

$$V(x) = x^4 + ax^2 + bx$$

The equilibrium surface is $4x^3 + 2ax + b = 0$. The equation of singularity set S is $12x^2 + 2a = 0$. Combine the two equations, the equation of bifurcation set is $8a^3 + 27b^2 = 0$.

5.4.3 *Swallowtail catastrophe*

Swallowtail catastrophe is type A_4 in Table 1. Its state space is (x, a, b, c) and the potential function is

$$V(x) = x^5/5 + ax^3/3 + bx^2/2 + cx$$

The equilibrium surface is $x^4 + ax^2 + bx + c = 0$. The equation of singularity set S is $4x^3 + 2ax + b = 0$.

5.4.4 *Unbilical point catastrophe of elliptic type*

Unbilical point catastrophe of elliptic type is type D_{-4} in Table 1. Its state space is (x, y, a, b, c) and the potential function is

$$V(x) = x^2y - y^3/3 + a(x^2 + y^2) + bx + cy$$

The equilibrium surface is $2xy + 2ax + b = 0; x^2 - y^2 + 2ay + c = 0$. The equation of bifurcation set B is $x^2 + y^2 = a^2$.

5.4.5 *Unbilical point catastrophe of hyperbolic type*

Unbilical point catastrophe of hyperbolic type is type D_{+4} in Table 1. Its state space is (x, y, a, b, c) and the potential function is

$$V(x) = x^2y + y^3/3 + a(y^2 - x^2) + bx + cy$$

The equilibrium surface is $2xy - 2ax + b = 0; x^2 + y^2 + 2ay + c = 0$. The equation of singularity set S is $4(y^2 - x^2 - a^2) = 0$.

5.4.6 *Butterfly catastrophe*

Butterfly catastrophe is type A_{+5} in Table 1. Its state space is (x, a, b, c, d) and the potential function is

$$V(x) = x^6 + ax^4 + bx^3 + cx^2 + dx$$

The equilibrium surface is $6x^5 + 4ax^3 + 3bx^2 + 2cx + d = 0$. The equation of singularity set S is $15x^4 + 6ax^2 + 3bx + c = 0$.

5.4.7 *Unbilical point catastrophe of parabolic type*

Unbilical point catastrophe of parabolic type is type D_5 in Table 1. Its state space is (x, y, a, b, c, d) and the potential function is

$$V(x) = y^4 + x^2y + ax^2 + by^2 + cx + dy$$

The equilibrium surface is $2xy + 2ax + c = 0; x^2 + 4y^3 + 2by + d = 0$. The equation of singularity set S is $2(y + a)(6y + b) - 2x^2 = 0$.

5.5 An Example

According to Piyaratne *et al.* (2014), catastrophe phenomena are frequent in insect ecology, especially in aphid populations. They developed a swallowtail model to describe the catastrophe. First, a modified logistic model was given as

$$dN/dt = rN(1 - eN^2/K) - PK(N - N_m)/((N - N_m) + d) \quad N > N_m$$

where N and P: the pest and predator population densities respectively, r and k: the rate of population increase and the rate of predation respectively, K: the carrying capacity which is affected by crop condition. e: the weather factor; N_m and d: the minimum pest population where the predation is possible and the half saturated prey consumption by predator respectively.

The swallowtail model was constructed from the population density of aphids and the weather factor, crop condition (carrying capacity) and predator effect are used as coefficients of the potential function. The potential function with three unfolding parameters were given by

$$v(x) = x^5 + ux^3 + vx^2 + wx$$

In this equation, x: the behavior or state variable, u, v and w: the control variables. The equilibrium surface M was given by

$$v'(x) = 5x^4 + 3ux^2 + 2vx + w$$

the singularity set S which is all singular points was given by

$$v''(x) = 20x^3 + 6ux + 2v$$

and the bifurcation set was given by

$$u(81u^3 + 540v^2)w - 360u^2w^2 + 400w^3 = v^2(27u^3 + 135v^2)$$

Chapter 6

Self-adaptation and Control Systems

6.1 Self-adaptation

We usually meet some interesting biological phenomena, e.g., some brightly colored mushrooms are considered toxic; a chameleon can change its body color and plant leaves will fall to the ground, etc. All of these result from biological adaptation. As early as in the 19th century, biological adaptation has been mentioned in the Lamarckian theory. Biological adaptation was further improved in the theory of Darwinian evolution. Thus, biological adaptation was generally based on the Darwinian theory. In the Darwinian era, biological adaptation was interpreted at the phenotypic level, e.g., the structural, physiological, and behavioral aspects of biological phenotype. Since the beginning of the 20th century, as molecular biology and molecular genetics advance, people began to focus on biological adaptation at the molecular level, i.e., interpreting biological adaptation to the environment through gene regulation. Biological adaptation has significant implications and has been widely used in industry, aerospace, transportation, machinery, electronics, control technology, signal processing, etc., to design various adaptive systems that are mainly controlled by computer and mathematical procedures. Adaptive control and adaptive filtering are two of the most important applications which have been used in automatic control and chaos elimination. So far there are a few special literatures on biological adaptation. Here we review the development of the theory of biological adaptation, adaptation at both phenotypic and molecular levels, and adaptation in engineering science. Details of self-adaptation in this section can also be found in Luo and Zhang (2014).

6.1.1 *Evolution and biological adaptation*

6.1.1.1 *Lamarckian theory*

As early as in the 19th century, the famous French naturalist Lamarck proposed the terminology "evolution" in his book, *Philosophy of Zoology*. He was the first scientist who first proposed biological evolution (Li, 2009). Lamarckian theory argues that all the world's major organisms have originated from evolution, rather than from the creation of God. It holds that organisms evolve from simple to complex. Changes in environmental conditions will result in the relative adaptations of organisms. The diversity of the environment facilitates biological diversity. Moreover, Lamarck proposed two famous rules in evolutionary biology, "Use and disuse" and "Inheritance of acquired characteristics." He thought the two rules are not only causes of biological variation but also the processes of adaptation formation. "Use and disuse" means that frequently used organs evolve, and not oftenly used organs will gradually degenerate. "Inheritance of acquired characteristics" means that the acquired biological traits can be inherited by subsequent generations. For example, giraffes with a longer neck will in general give birth to offsprings with a longer neck (Kronfeldner, 2007).

Lamarck believed that adaptation is the main process of biological evolution. He is also the first to propose the viewpoint that organisms adapt to the environment. Later, in Darwin's theory of biological evolution, biological adaptation to the environment was discussed in more detail.

6.1.1.2 *Darwinism*

During the mid-19th century, after collecting a lot of basic evidence, Darwin published the book, *The Origin of Species*, in which the theory of biological evolution, with natural selection as the core, was created. It gave a comprehensive interpretation on the occurrence and evolution of the entire biosphere. Darwin proposed the central rules of his theory, natural selection and survival of the fittest (Lewens, 2010). He believed that all organisms compete with each other, and the organisms best suited to the environment will survive and others will be eliminated. Following these rules, organisms evolve from simple to complex based on natural selection (Bradon, 1990; Kutschera, 2009).

The evolution of a birch moth in Britain can best interpret the concept of Darwinism. The birch moth was of gray type before the year 1850. However, its black mutant was found in Manchester in 1850. After the late 19th century, with the development of industrialization, H_2S in the exhaust gas killed gray lichen on the tree bark, and coal smoke blackened trunk. As a result, the gray moths, originally resting on gray lichen for protection, were easily predated by birds when they rested on black trunk, while the black type of moths survived and evolved. A rapid increase in the frequency of black type, and decline in gray type was thus recorded. By the end of 19th century, the former increased from less than 1% to more than 90%, and the latter decreased from over 90% to less than 5% (Gu, 2007). This is a classic example of Darwinian biological adaptation. We can find that the environment is inseparable from biological adaptation.

Modern theory of biological evolution is also called modern Darwinism. It is a theory that appeared in the 20th century. It is based on the Darwin's theory of natural selection, except that it argues that biological evolution resulted from changes in the genetic frequencies of population. It further clarifies the essence of heredity and variations, and the mechanism of natural selection (Zou and He, 2001). Because adaptation is the result of natural selection, this theory interprets biological adaptation as directional changes of genetic frequencies in population.

6.1.1.3 *Adaptationism and anti-adaptationism*

Adaptationism has emerged since Darwin's theory of evolution was published and the thought that organisms evolve from adaptation to the environment and natural selection is widely accepted. Sponsored by Wallace and Weismann, it started to become popular in the early

19th century to the 20th century (Dong, 1999). Adaptationism holds that adaptation is the sole result of natural selection, and the formation of almost all biological traits is the result of adaptation (Resnik, 1997; Yu and Li, 2012).

Anti-adaptationism argues that adaptation/natural selection is not the only reason for biological evolution (Yu and Li, 2012). It holds that the behaviors, structure, and functionalities of organisms could not be described by only adaptation. Among these, neutralism is the major theory.

6.1.1.4 *Theory of molecular biology*

Before the 20th century, people always interpreted biological adaptation by natural selection. Since the beginning of the 20th century, as the development of molecular biology and molecular genetics, molecular biologists have described biological adaptation with the genetic variations of organisms (Radwan and Babik, 2012), of which the most famous is the neutral theory of molecular evolution (Jukes, 2000).

6.1.2 *Biological adaptation*

6.1.2.1 *Definition*

Biological adaptation is a driving force of evolution. The occurrence of biological adaptation is to counter the changing environment. From Darwinian theory of evolution to the theory of molecular evolution, biological adaptation has been discussed at phenotypic level and molecular level (Chen and Deng, 2001), which indicates that the mechanism of biological adaptation is very complex. Nevertheless, these theories have a common basis, i.e., organisms change themselves — at morphological, physiological, behavioral, and molecular level — in order to better survive in a changing environment (Smith, 1979; Zhang and Liu, 2006; Fenichela *et al.*, 2011). The process of adapting to the environment is called biological adaptation. Biological adaptation acts at different levels, from molecular to phenotypic levels, and its objective is to coordinate with the environment.

6.1.2.2 *Adaptation of organisms to the environment*

There are various unpredictable and uncontrollable environmental factors. Organisms must adapt to the environment for survival, otherwise they will become extinct (Zheng, 2008). All organisms have their own adaptation traits in behaviors, morphological structures, or physiological functionalities, etc.

On the other hand, biological adaptation is not absolute. The adaptation of an organism to the previous environment may become non-adaptive to the changed environment. The stability of genetic molecules is a reason for the relativity of adaptation. Genetic traits are steadily inherited by the next generation, so these traits and genetic molecules are unable to simultaneously follow the rapid changes in the environment. For example, domesticated rabbits evolved from wild rabbits. Burrowing habit of domesticated rabbits, inherited from wild rabbits due to the stability of genetic molecules, is not conducive to escape predators, as functioned by wild rabbits.

6.1.2.3 *Interaction between organisms and the environment*

Organisms adapt to the environment in order to survive. On the other hand, the environment may be changed by organisms. For example, the earthworm excretes nitrogen, phosphorus, and potassium, which can increase the fertility of the soil and improve the soil's nutrients.

6.1.3 *Phenotypic adaptation*

6.1.3.1 *Phenotypic adaptation of plants*

Plants adapt to various environments, e.g., cold, salty, warm, and other harsh environments, through morphological and structural changes in roots, stems, leaves, flowers, fruits, etc. To adapt to the environment, they must store water and oxygen, and resist other harsh conditions. For example, in order to store more water in its storing cells, xeric plants, such as cacti, have particularly well-developed palisade tissue (Shields, 1951). Formation of mangrove's breathing roots is to better reserve atmosphere. Leaves of tropical plants have relatively larger area for better transpiration, and so on.

6.1.3.2 *Phenotypic adaptation of animals*

Morphological adaptation of animals includes camouflage, mimicry, warning coloration, convergent adaptation, divergent adaptation: (1) camouflage. It means that animals share the same or similar body color as the environment. For example, chameleons can change their body color to that of the environment to protect themselves from predators. (2) Mimicry refers to the appearance of an animal that is similar to other biotic/abiotic phenomenon, so that they can confuse predators and protect themselves. Flies and bees, for example, can imitate birds and escape predation. (3) Warning coloration refers to animals yielding the body color (very bright or terrified color) that is in contrast to the environment to attract other animals' attention and thus warn them. For example, the ladybug's bright appearance reminds predators on the production of a disgusting smell. (4) Convergent adaptation means that different species of animals, in the same living environment, have the same or similar morphological structure (Ma *et al.*, 2006). For example, the xeric plants living in the same arid desert environment, share similar morphological appearance, internal structure, life history characteristics, and physical characteristics, under the same or similar selection pressures. (5) Divergent adaptation refers to the biological adaptation of animals that the same species living in different environments evolves to form distinct morphological appearance and internal structure and so on.

Physiological functions of animals are very important to their survival. Water, light, air, soil, temperature, and other environmental factors affect the physiological functions. For example, camels live in arid environments. Their kidneys have strong ability of re-absorption and a special protein in the blood can reserve water under the condition of little water supply. Some animals' hibernation is an adaptation to extreme temperatures, which helps to reduce metabolism in order to survive in winter.

The formation of some behaviors of animals is attributed to biological adaptation. For example, the gecko always breaks its tail to escape. Behavioral adaptation usually involves biological stress. Stress refers to the biological response to external stimulation, e.g., the paramecium adapts to different temperatures by distinct behaviors (Duncan, 2011).

6.1.4 *Biological adaptation at molecular level*

6.1.4.1 *Neutral theory of molecular evolution*

Neutral theory of molecular evolution was founded by the Japanese geneticist Kimura in 1968. The neutral theory of molecular evolution applies only for evolution at the molecular level. The theory holds that most evolutionary changes and most of the variation within/between species is not caused by natural selection, but by random drift of neutral mutant alleles (Kimura, 1968). A neutral mutation does not affect an organism's ability to survive and reproduce. The theory allows for the possibility that most mutations are deleterious, but holds that because these are rapidly purged by natural selection, they are not significant to molecular variation within/between species. Non-deleterious mutations are assumed to be mostly neutral (Masatoshi *et al.*, 2010). The theory also assumes that the fate of neutral mutations is determined by the sampling processes described by specific models of random genetic drift (Kimura, 1983).

6.1.4.2 *Adaptation mechanisms at molecular level*

Gene expression is regulated in the process that synthesizes protein from DNA, which includes transcription and translation. In this process, proteins are synthesized to maintain the desired mechanism for life activities. Genes express according to different needs, which makes organisms adaptable to changing environments. For example, when a living body has a sufficient glucose supply, bacteria can utilize glucose as an energy and carbon source; they do not need to synthesize enzymes that use other sugars. Nevertheless, without enough glucose supply, bacteria will activate more genes to use other hydrocarbons present in the environment, such as lactose, galactose, arabinose, etc., in order to meet the needs of the body, or else they will die.

Both gene mutations and duplications are popular in molecular evolution. Gene mutation refers to the additions, deletions, or changes in the order of arrangement of base pairs in genomic DNA, which causes changes in genetic structure. Gene duplication refers to the duplication of a DNA fragment containing the gene, also known as duplicated gene. Changes in genetic structure from mutations, production of new genetic functions and increase in expression of genes, contribute to the adaptation of organisms (Kondrashov, 2012).

In terms of gene mutations, researchers have found two new gene mutations (EGLN1 and PPARA) that help the mountainous Tibetan to contain less hemoglobin in the blood, in order to use oxygen more efficiently than people living in low-altitude areas (Simonson *et al.*, 2012). As for gene duplications, white arm langur monkeys mainly eat leaves. It is found that the pancreatic ribonuclease gene of the leaf monkey, which is responsible for normal digestion, has duplicated. One of two genes is responsible for the production of pancreatic ribonuclease, and another duplicated gene produces other ribonuclease to digest other foods (Li *et al.*, 2010).

6.1.4.3 *Fitness*

Fitness refers to the degree that organisms fit environmental conditions. Fitness is mainly measured by comparing the biological genotypes. Genotypes with great fitness will more likely be passed on to the offspring (Ma, 1995).

6.1.5 *Adaptation in engineering science*

Biological adaptation has been recognized since Darwin's era. Currently, people are using the principles of biological adaptation to develop various self-adjusting techniques. These adaptive techniques try to adjust the system's structure and behaviors by themselves, in response to the changes of environmental conditions. In the early 1990s, German Aerospace Research Institute started to study adaptive technology in aerospace science (Li *et al.*, 2010). Since then, adaptive technology has attracted much attention from various areas. At first, adaptive technology is mainly used in transportation and mechanics. Up till now it has been used in solid state physics, materials science, mechatronics, medical technology, aerospace, optical communications, machinery and equipment manufacturing, robotics, transportation machinery, electronic equipment, control lines, signal processing, and other industries.

6.1.5.1 *Adaptive control*

Adaptive control means that in view of the uncertainty of the system, an appropriate control technique to achieve the best performance of the system will be designed. Uncertainty of the system refers to structural parameters that may be influenced by the environment, and other random factors. Dynamic models, usually (stochastic) differential/difference equations that include various feedback mechanisms, are developed to achieve adaptive control. By using adaptive control, the system automatically adjusts its parameters or structure to maintain a good performance.

Adaptive control was proposed in the early 1950s. In adaptive control, the system operates by comparing the deviation between expected and observed outputs, i.e., feedback control. By the 1970s, with the advancement of computer technology and the development of adaptation theory, adaptive control has been greatly developed, not only in the machinery industry, but also in other areas, such as chemical industry, medical science, and physics, etc. (Liu, 2007). The main types of adaptive control systems include self-tuning regulator, model reference adaptive control systems, self-stabilization system, self-optimizing systems, self-organizing systems, and learning control systems (Zhang, 2013a, b). Of which, the model reference adaptive systems and self-tuning adaptive control systems are the most classic adaptive control systems.

A model reference adaptive system is mainly composed of reference model and controller. The output of the reference model is set as the expected output. By comparing the deviation between reference model and observed outputs, the controller fixes the state or parameters of the system, so that the output of the controlled system can maximally approximate the expected. Model reference adaptive control system is initially based on local optimization of parameters. Now it is used in the autopilot of aircraft and ship, toptical tracking servo-system of telescope, speed control system of SCR, and robotic control systems, etc. (Liu *et al.*, 2004).

The self-tuning system is designed based on the minimum variance theory. It is mainly composed of regulator, reference model, subtractor, and self-tuning regulator. Subtractor is used to estimate the deviation between the expected (reference model) and observed outputs, and self-tuning regulator corrects regulator's parameters according to the deviation calculated by the substractor (Alexandridis and Sarimveis, 2005). Currently this technology has been successfully used in the paper industry, chemical industry, metallurgy, and the autopilot system of aircraft, etc.

There are still some problems in the design of adaptive control: (1) stability is a major topic in the operation of adaptive control, which has not been well solved; (2) convergence of system algorithm still needs to be improved; (3) adaptive control is used in only a few systems; (4) robustness of the system is easily impaired by the changes of external factors (Liu *et al.*, 2004; Sanei and French, 2004; Alexandridis and Sarimveis, 2005; Liu, 2007).

6.1.5.2 *Adaptive filtering*

The adaptive filtering system adjusts the parameters and signals of filters by using the adaptive algorithm. The adaptive algorithm calculates the time-varying coefficients for processing signals. Currently, the most widely used adaptive algorithms include Least Mean Square (LMS) and Recursive Least Square (RLS) adaptive algorithms (Khalili and Tinati, 2010).

LMS adaptive algorithm takes least mean square error as the best criterion. It is a gradient steepest descent method. It estimates gradient vector from output signal, and thus minimizes the mean square error between the desired signal and the observed signal (Sheu *et al.*, 2012). LMS is simple and easy to operate, but the convergence of the algorithm is low. RLS is complex but it has a better performance in convergence. Currently they are mainly used in noise control, signal prediction, system identification, echo cancellation, and beam-forming of antenna array.

Adaptive filtering was first proposed in the early 1940s. Wiener developed a linear adaptive filter, i.e., Wiener filter, to eliminate noise. By the early 1960s, people began to develop various nonlinear filters, and Kalman filtering theory was thus developed (Lippuner and Moschytz, 2004). Both filters depend upon known signals and they have fixed adaptive coefficients. By the mid-1970s, the adaptive filtering theory was basically developed. The theory of optimal filtering design was then proposed.

Biological adaptation has always been the focus of biological evolution, both are inextricably linked. Darwinian evolution focuses on the biological adaptation at visible level, including morphology, physiology, and behaviors. The theory of molecular evolution stresses the role of gene regulation and gene mutations in biological adaptation. Both natural selection and molecular interpretation are indispensible when describing biological adaptation. Biological adaptation is a complex phenomenon. The mechanism of adaptation, for example, the correlation between the adaptation at phenotypic and molecular levels, will still be a topic for biologist for a long time. Biological adaptation-inspired applications, such as adaptive control and adaptive filtering, have made great achievements in the past few years and are attracting more attention from around the world.

6.2 Control Systems

For a non-self-organized system, we can use some theories/methods in control theory to make it a self-organizing system.

6.2.1 *Linear feedback control*

A linear ordinary system is certainly not stable. It will become stable by adding a negative feedback in the system. The linear feedback control system is (Qian and Song, 1983):

$$\begin{aligned} dx/dt &= Ax + Bu, \quad x(t_0) = x_0 \\ u &= -Dx \\ y &= Cx \end{aligned}$$

where $x = (x_1, x_2, \ldots, x_n), u = (u_1, u_2, \ldots, u_m), y = (y_1, y_2, \ldots, y_p)$; A: system matrix, B: control matrix, C: output matrix, D: feedback matrix, and x_0: initial state. The closed system for the linear feedback control system is

$$dx/dt = (A - BD)x$$

The solution for closed system is

$$x(t) = e^{(A-BD)t}x(t_0)$$

As a general case, if A, B and C are time dependent, then the solution is

$$x(t) = \Phi(t, t_0)x(t_0) + \int_{t0}^{t} \Phi(t, \tau)B(\tau)u(\tau)d\tau$$

and

$$y(t) = C(t)\Phi(t, t_0)x(t_0) + \int_{t0}^{t} c(t)\Phi(t, \tau)B(\tau)u(\tau)d\tau$$

Let $h = t(k+1) - t(k)$, then the output is

$$\begin{aligned} x(t_{k+1}) &= e^{(A-BD)h}x(t_k) \\ y(t_{k+1}) &= Cx(t_{k+1}) \\ k &= 0, 1, 2, \ldots \end{aligned}$$

and the control dynamics is

$$\begin{aligned} u(t_{k+1}) &= -Dx(t_{k+1}) \\ k &= 0, 1, 2, \ldots \end{aligned}$$

where

$$e^{(A-BD)h} = E + (A-BD)h + ((A-BD)h)^2/2! + ((A-BD)h)^3/3! + \ldots$$

here $h = 1$.

For the closed system, if all the eigenvalues for the equation

$$|A - BD - \lambda^* E| = 0$$

have negative real parts, the system will be asymptotic steady; if all of them are not great than 0 and the eigenvalues equivalent to 0 are first-order, then the system is steady; otherwise, the system is not steady.

The following are some of the Java codes for the linear feedback control system:

```
h=1;
t0=tt-h;
tt=t0;
for(i=1;i<=n;i++) x1[i]=x[i];
for(i=1;i<=n;i++) for(j=1;j<=n;j++) for(k=1;k<=m;k++)
a[i][j]-=b[i][k]*d[k][j];
px=n;
substeady();
n=px;
for(i=1;i<=n;i++) for(j=1;j<=n;j++) a[i][j]*=h;
subexp();
while (tt<tm) {
tt=tt+h;
for(i=1;i<=n;i++) {x[i]=0;for(j=1;j<=n;j++) x[i]+=g[i][j]*x1[j];}
for(i=1;i<=pp;i++) {y[i]=0;for(j=1;j<=n;j++) y[i]+=c[i][j]*x[j];}
for(i=1;i<=m;i++) {u[i]=0;for(j=1;j<=n;j++) u[i]-=d[i][j]*x[j];}
for(i=1;i<=n;i++) x1[i]=x[i];
editt1.appendText(``t=''+String.valueOf((int)tt)+''
for(i=1;i<=n;i++) {
editt1.appendText(``x''+String.valueOf(i)+``=''+String.valueOf((int)
(x[i]*10000)/10000.00)+`` '');
xg[i][tt-t0]=x[i]; }
editt1.appendText(''
for(i=1;i<=pp;i++) {
editt1.appendText(``y''+String.valueOf(i)+``=''+String.valueOf((int)
(y[i]*10000)/10000.00)+`` '');
yg[i][tt-t0]=y[i]; }
editt1.appendText(''
for(i=1;i<=m;i++) {
editt1.appendText(``u''+String.valueOf(i)+``=''+String.valueOf((int)
(u[i]*10000)/10000.00)+`` '');
ug[i][tt-t0]=u[i]; }
editt1.appendText(''
label5.setText(``Running Time:
```

```
''+String.valueOf((System.currentTimeMillis()-start)/1000.0)+'' seconds'');
editt1.appendText(''
}
for(i=1;i<=n+m+pp;i++) {
xmax[kp]=tm;
umax[kp]=-1e+50;
umin[kp]=1e+50;
for(j=1;j<=tm-t0;j++) {
if (i<=n) {
if (xg[i][j]<umin[kp]) umin[kp]=xg[i][j];
if (xg[i][j]>umax[kp]) umax[kp]=xg[i][j]; }
else if ((i>n) & (i<=n+pp)) {
if (yg[i-n][j]<umin[kp]) umin[kp]=yg[i-n][j];
if (yg[i-n][j]>umax[kp]) umax[kp]=yg[i-n][j]; }
else {
if (ug[i-n-m][j]<umin[kp]) umin[kp]=ug[i-n-pp][j];
if (ug[i-n-m][j]>umax[kp]) umax[kp]=ug[i-n-pp][j]; }
xx[kp][j]=j+t0;
if (i<=n) yy[kp][j]=xg[i][j];
else if ((i>n) & (i<=n+pp)) yy[kp][j]=yg[i-n][j];
else yy[kp][j]=ug[i-n-pp][j]; }
if (umin[kp]<0) umax[kp]-=umin[kp];
else umin[kp]=0;
if (i<=n) pk[kp]=i;
else if ((i>n) & (i<=n+pp)) pk[kp]=i-n;
else pk[kp]=i-n-pp;
kp++; }

public double[][] subexp() {
int n1=1;
double ff=1;
for(i=1;i<=n;i++)
for(j=1;j<=n;j++)
{if (i==j) g[i][j]=1;
else g[i][j]=0;
l[i][j]=0;
```

```
e[i][j]=a[i][j]; }
while (n1<51) {
for(i=1;i<=n;i++)
for(j=1;j<=n;j++)
for(k=1;k<=n;k++) e[i][j]+=l[i][k]*a[k][j];
for(i=1;i<=n;i++)
for(j=1;j<=n;j++) l[i][j]=e[i][j];
for(i=1;i<=n;i++)
for(j=1;j<=n;j++) {
g[i][j]+=e[i][j]/ff;
e[i][j]=0; }
n1+=1;
ff=n1*ff; }
return g; }

public void substeady()
double bb=0,ep,yy,ee,ff,p1,q1,q,r,ss,vv=0,ww,xx,cc1,ll1;
int n2=n*2,n3=n2+1,n4=n*n2,ll,aa,cc,it,dd,gg;
for(i=1;i<=n;i++)
for(j=1;j<=n;j++)
{s[i+(j-1)*n2]=a[i][j];
p[i+(j-1)*n2]=a[i][j];}
w[n+1]=1;
ll=1;
while (ll<=n) {
bb=0;
for(i=1;i<=n4;i+=n3) bb-=p[i];
bb=bb/ll;
w[n+1-ll]=bb;
for(i=1;i<=n4;i+=n3) p[i]+=bb;
if (ll==n) break;
for(i=1;i<=n;i++) {
aa=1;
for(j=i;j<=n4;j+=n2) {
bb=0;cc=j-i;
for(k=i;k<=n4;k+=n2) {
```

```
cc+=1;
bb+=p[k]*s[cc]; }
t[aa]=bb;
aa+=1; }
aa=1;
for(j=i;j<=n4;j+=n2) {
p[j]=t[aa];aa+=1; }}
ll+=1; }
ep=1e-05;
aa=50;
p1=0;
q1=0;
for(i=1;i<=n+1;i++) t[i]=w[n-i+2];
for(i=1;i<=n+1;i++) {
w[i]=t[i];t[i]=0;}
it=0;
yy=w[1];
for(i=1;i<=n;i++) w[i]=w[i+1]/yy;
nm5: while (n>=0) {
if (n<=1) {
it++;
dd=2*it;
f[dd]=0;
f[dd-1]=-w[1];
n=it;
eigen();
break nm5; }
if (n<=2) {
ee=w[1];
ff=w[2];
yy=-0.5*ee;
it+=2;
ww=ff-Math.pow(yy,2);
xx=Math.sqrt(Math.abs(ww));
dd=2*it;
if (ww>0) {
```

```
f[dd]=-xx;f[dd-1]=yy;
f[dd-2]=xx;
f[dd-3]=yy;
n-=2;
if (n>0) {
for(i=1;i<=n;i++)
w[i]=t[i];
continue nm5; }
if (n<=0) {
n=it;
eigen();
break nm5; }}
if (ww<=0) {
f[dd]=0;f[dd-1]=yy-xx;
f[dd-2]=0;f[dd-3]=yy+xx;
n-=2;
if (n>0) {
for(i=1;i<=n;i++)
w[i]=t[i];
continue nm5; }
if (n<=0) {
n=it;
eigen();
break nm5; }}
}
if (n>2) {
ee=p1;
ff=q1;
gg=1;
t[1]=w[1]-ee;
t[2]=w[2]-ee*t[1]-ff;
dd=n-1;
v[1]=t[1]-ee;
v[2]=t[2]-ee*v[1]-ff;
if (dd==2) dd=3;
for(j=3;j<=dd;j++)
```

```
{t[j]=w[j]-ee*t[j-1]-ff*t[j-2];v[j]=t[j]-ee*v[j-1]-ff*v[j-2];}
dd=n-1;
ll1=v[dd]-t[dd];
cc1=-ll1;
if (n!=3)
cc1*=v[n-3];
cc1=cc1+Math.pow(v[n-2],2);
if (cc1<=1e-08) {
n=it;
break nm5; }
nm4: while (cc1!=0) {
t[n]=w[n]-ee*t[n-1]-ff*t[n-2];
q=-t[n];
if (n!=3) q*=v[n-3];
q=(q+t[n-1]*v[n-2])/cc1;
r=(t[n]*v[n-2]-ll1*t[n-1])/cc1;
ee+=q;
ff+=r;
ss=Math.abs(q)+Math.abs(r);
if (gg==1) vv=ss;
if ((gg==5) & (ss>vv)) {
n=it;
break nm5; }
if (ss>ep) {
if (gg<aa) {gg++;
t[1]=w[1]-ee;
t[2]=w[2]-ee*t[1]-ff;
dd=n-1;
v[1]=t[1]-ee;
v[2]=t[2]-ee*v[1]-ff;
if (dd==2) dd=3;
for(j=3;j<=dd;j++) {
t[j]=w[j]-ee*t[j-1]-ff*t[j-2];
v[j]=t[j]-ee*v[j-1]-ff*v[j-2];}
dd=n-1;
ll1=v[dd]-t[dd];
```

```
cc1=-ll1;
if (n!=3) cc1*=v[n-3];
cc1=cc1+Math.pow(v[n-2],2);
if (cc1<=1e-08) {
n=it;break nm5; }
continue nm4; }
if (gg>=aa) {
n=it;
break nm5; }}
yy=-0.5*ee;
it+=2;
ww=ff-Math.pow(yy,2);
xx=Math.sqrt(Math.abs(ww));
dd=2*it;
if (ww>0) {
f[dd]=-xx;f[dd-1]=yy;
f[dd-2]=xx;
f[dd-3]=yy;
n-=2;
if (n>0) {
for(i=1;i<=n;i++)
w[i]=t[i];
continue nm5; }
if (n<=0) {
n=it;eigen();
break nm5; }}
if (ww<=0) {
f[dd]=0;
f[dd-1]=yy-xx;
f[dd-2]=0;
f[dd-3]=yy+xx;n-=2;
if (n>0) {
for(i=1;i<=n;i++)
w[i]=t[i];
continue nm5; }
if (n<=0) {
```

```
n=it;
eigen();
break nm5; }}
}}}
}

public void eigen() {
int sd,zw=0;
editt1.appendText(''Eigenvalues of Matrix\n'');
editt1.appendText(''Real part''+''  ''+''Image part\n'');
i=1;
while (i<=2*n) {
editt1.appendText(String.valueOf((int)(f[i]*10000)/10000.00)+''
''+String.valueOf((int)(f[i+1]*10000)/10000.00)+''\n'');
i+=2;}
sd=0;
for(i=1;i<=2*n;i+=2) {
if (f[i]<0) sd+=1;
else if (f[i]>0) {
editt1.appendText(''System is unsteady\n'');
zw=1;
break; }}
if ((sd<n) & (zw==0)) editt1.appendText(''System is steady\n'');
else if ((sd>=n) & (zw==0)) editt1.appendText(''System is asymptotic
      steady\n''); }
```

6.2.2 *Linear optimization control*

The linear optimal control system is (Qian and Song, 1983)

$$dx/dt = Ax + Bu$$

where $x = (x_1, x_2, \ldots, x_n), u = (u_1, u_2, \ldots, u_m)$; A: system matrix, B: control matrix. The objective general function is

$$\min J = \int_0^\infty ((x(t)Qx(t) + u(t)Ru(t))dt)$$
$$u(t) = -Dx(t)$$

where D: feedback matrix, Q, R: general function matrices for state x and control u respectively.

The closed loop system for the system is

$$dx/dt = (A - BD)x$$

First, to solve Riccati equation

$$Az + zA + Q - zFz = 0$$

where $F = B^* R^{-1*} B, z = (z_1, z_2, \ldots, z_n)$, the iterative method is

$$z(0) = \left(\int_0^h (e^{-At} F e^{-At} dt) \right)^{-1}$$

where h is a constant. Then,

$$(A - Fz(n-1))z(n) + z(n)(A - Fz(n-1)) + Q + z(n-1)Fz(n-1) = 0$$

To solve $z(n), n = 1, 2, \ldots$, until

$$\max|(z(n-1)_{ij} - z(n)_{ij}| <= E$$

where $1 \leq i, j \leq n$, then the Riccati solution is $z = z(n)$, and optimal control matrix is $D = R^{-1} * B * z$, and the optimal closed loop matrix is $A - Fz$, where

$$e^{At} = E + At + (At)^2/2! + (At)^3/3! + \ldots$$

If all the eigenvalues for

$$|A - BD - \lambda^* E| = 0 \text{ (closed system)}$$

or

$$|A - \lambda^* E| = 0 \text{ (open system)}$$

have negative real parts,the system will be asymptotic steady; if all of them are not great than 0 and the eigenvalues equavalent to 0 are first-order, then the system is steady; otherwise, the system is not steady.

The following are some of the Java codes for the linear optimization control system:

```
pp=m;
for(i=1;i<=m;i++)
for(j=1;j<=m;j++) c[i][j]=r[i][j];
subpro();
for(i=1;i<=n;i++)
for(j=1;j<=m;j++) v[j][i]=b[i][j];
for(i=1;i<=m;i++)
for(j=1;j<=n;j++) b1[i][j]=v[i][j];
for(i=1;i<=n;i++)
for(j=1;j<=m;j++)
{b2[i][j]=0;
for(kk=1;kk<=m;kk++) b2[i][j]+=b[i][kk]*c[kk][j];}
for(i=1;i<=n;i++)
for(j=1;j<=n;j++)
{k[i][j]=0;
for(kk=1;kk<=m;kk++) k[i][j]+=b2[i][kk]*b1[kk][j];}
pp=n;
for(i=1;i<=n;i++)
for(j=1;j<=n;j++) c[j][i]=a[i][j];
hh=1;
int bn=10*(int)Math.round(hh+1);
double bh=(double)hh/(double)bn;
for(i=1;i<=n;i++)
for(j=1;j<=n;j++) {s[i][j]=a[i][j];h[i][j]=0; }
for(int ll=1;ll<=bn;ll++) {
for(i=1;i<=n;i++)
for(j=1;j<=n;j++) a[i][j]=-s[i][j]*ll*bh;
subexp();
label5.setText(''Running Time:
''+String.valueOf((System.currentTimeMillis()-start)/1000.0)+'' seconds'');
for(i=1;i<=n;i++)
for(j=1;j<=n;j++) {
e1[i][j]=g[i][j];
a[i][j]=-c[i][j]*ll*bh; }
```

```
subexp();
for(i=1;i<=n;i++)
for(j=1;j<=n;j++) {
l[i][j]=0;
for(kk=1;kk<=n;kk++)
l[i][j]+=e1[i][kk]*k[kk][j]; }
for(i=1;i<=n;i++)
for(j=1;j<=n;j++) {
e[i][j]=0;
for(kk=1;kk<=n;kk++)
e[i][j]+=l[i][kk]*g[kk][j];}
for(i=1;i<=n;i++)
for(j=1;j<=n;j++) h[i][j]+=e[i][j]; }
for(i=1;i<=n;i++)
for(j=1;j<=n;j++) {
a[i][j]=s[i][j];
c[i][j]=(k[i][j]+2*h[i][j]-e[i][j])*bh/2; }
pp=n;
subpro();
for(i=1;i<=n;i++)
for(j=1;j<=n;j++) x[i][j]=c[i][j];
subgra();
label5.setText(``Running Time:
``+String.valueOf((System.currentTimeMillis()-start)/1000.0)+'' seconds'');
pp=m;
for(i=1;i<=m;i++)
for(j=1;j<=m;j++) c[i][j]=r[i][j];
subpro();
for(i=1;i<=m;i++)
for(j=1;j<=n;j++) {
v[i][j]=0;
for(kk=1;kk<=m;kk++)
v[i][j]+=c[i][kk]*b1[kk][j]; }
for(i=1;i<=m;i++)
for(j=1;j<=n;j++) {
d[i][j]=0;
```

```
for(kk=1;kk<=n;kk++) d[i][j]+=v[i][kk]*x[kk][j]; }
editt1.appendText(''Optimal control matrix D\n'');
for(i=1;i<=m;i++) {
for(j=1;j<=n;j++)
editt1.appendText(String.valueOf((int)(d[i][j]*10000)/10000.00)+'' '');
editt1.appendText(''\n''); }
editt1.appendText(''Optimal closed loop matrix A-BD\n'');
for(i=1;i<=n;i++) { for(j=1;j<=n;j++)
editt1.appendText(String.valueOf((int)(h[i][j]*10000)/10000.00)+'' '');
editt1.appendText(''\n''); }
an=n;
editt1.appendText(''Stablity of open loop system\n'');
substeady();
editt1.appendText(''\n'');
for(i=1;i<=n;i++)
for(j=1;j<=n;j++) a[i][j]=h[i][j];
n=an;
editt1.appendText(''Stablity of closed loop system\n'');
substeady();

public double[][] subexp() {
int n1=1;
double ff=1;
for(i=1;i<=n;i++)
for(j=1;j<=n;j++)
{if (i==j) g[i][j]=1;
else g[i][j]=0;
l[i][j]=0;
e[i][j]=a[i][j]; }
while (n1<51) {
for(i=1;i<=n;i++)
for(j=1;j<=n;j++)
for(kk=1;kk<=n;kk++) e[i][j]+=l[i][kk]*a[kk][j];
for(i=1;i*=n;i++)
for(j=1;j<=n;j++) l[i][j]=e[i][j];
for(i=1;i<=n;i++)
```

```
for(j=1;j<=n;j++) {
g[i][j]+=e[i][j]/ff;
e[i][j]=0; }
n1+=1;
ff=n1*ff;}
return g; }

public void substeady() {
double bb=0,ep,yy,ee,ff,p1,q1,qq,rr,ss,vv=0,ww,xx,cc1,ll1;
int n2=n*2,n3=n2+1,n4=n*n2,ll,aa,cc,it,dd,gg;
for(i=1;i<=n;i++)
for(j=1;j<=n;j++)
{y[i+(j-1)*n2]=a[i][j];
p[i+(j-1)*n2]=a[i][j]; }
w[n+1]=1;
ll=1;
while (ll<=n) {
bb=0;
for(i=1;i<=n4;i+=n3) bb-=p[i];
bb=bb/ll;
w[n+1-ll]=bb;
for(i=1;i<=n4;i+=n3) p[i]+=bb;
if (ll==n) break;
for(i=1;i<=n;i++) {
aa=1;
for(j=i;j<=n4;j+=n2) {
bb=0;
cc=j-i;
for(kk=i;kk<=n4;kk+=n2) {
cc+=1;
bb+=p[kk]*y[cc]; }
t[aa]=bb;
aa+=1;}
aa=1; for(j=i;j<=n4;j+=n2) {
p[j]=t[aa];
aa+=1;}}
```

```
||+=1; }
ep=1e-05;
aa=50;
p1=0;
q1=0;
for(i=1;i<=n+1;i++) t[i]=w[n-i+2];
for(i=1;i<=n+1;i++) {
w[i]=t[i];t[i]=0; }
it=0;
yy=w[1];
for(i=1;i<=n;i++) w[i]=w[i+1]/yy;
nm5: while (n>=0) {
if (n<=1) {
it++;
dd=2*it;
f[dd]=0;f[dd-1]=-w[1];n=it;
eigen();
break nm5; }
if (n<=2) {
ee=w[1];
ff=w[2];
yy=-0.5*ee;
it+=2;
ww=ff-Math.pow(yy,2);
xx=Math.sqrt(Math.abs(ww));
dd=2*it;
if (ww>0) {
f[dd]=-xx;
f[dd-1]=yy;
f[dd-2]=xx;
f[dd-3]=yy;
n-=2;
if (n>0) {
for(i=1;i<=n;i++) w[i]=t[i];
continue nm5; }
if (n<=0) {
```

```
n=it;
eigen();
break nm5; }}
if (ww<=0) {
f[dd]=0;
f[dd-1]=yy-xx;
f[dd-2]=0;
f[dd-3]=yy+xx;
n-=2;
if (n>0) {
for(i=1;i<=n;i++)
w[i]=t[i];
continue nm5; }
if (n<=0) {
n=it;
eigen();
break nm5; }}
}
if (n>2) {
ee=p1;
ff=q1;
gg=1;
t[1]=w[1]-ee;
t[2]=w[2]-ee*t[1]-ff;
dd=n-1;
u[1]=t[1]-ee;
u[2]=t[2]-ee*u[1]-ff;
if (dd==2) dd=3;
for(j=3;j<=dd;j++) {
t[j]=w[j]-ee*t[j-1]-ff*t[j-2];
u[j]=t[j]-ee*u[j-1]-ff*u[j-2]; }
dd=n-1;
ll1=u[dd]-t[dd];
cc1=-ll1;
if (n!=3) cc1*=u[n-3];
cc1=cc1+Math.pow(u[n-2],2);
```

```
if (cc1<=1e-08) {
n=it;
break nm5; }
nm4: while (cc1!=0) {
t[n]=w[n]-ee*t[n-1]-ff*t[n-2];
qq=-t[n];
if (n!=3) qq*=u[n-3];
qq=(qq+t[n-1]*u[n-2])/cc1;
rr=(t[n]*u[n-2]-ll1*t[n-1])/cc1;
ee+=qq;
ff+=rr;
ss=Math.abs(qq)+Math.abs(rr);
if (gg==1) vv=ss;
if ((gg==5) & (ss>vv)) {
n=it;
break nm5; }
if (ss>ep) {
if (gg<aa) {
gg++;
t[1]=w[1]-ee;
t[2]=w[2]-ee*t[1]-ff;
dd=n-1;
u[1]=t[1]-ee;
u[2]=t[2]-ee*u[1]-ff;
if (dd==2) dd=3;
for(j=3;j<=dd;j++) {
t[j]=w[j]-ee*t[j-1]-ff*t[j-2];
u[j]=t[j]-ee*u[j-1]-ff*u[j-2]; }
dd=n-1;
||1=u[dd]-t[dd];
cc1=-||1;
if (n!=3) cc1*=u[n-3];
cc1=cc1+Math.pow(u[n-2],2);
if (cc1<=1e-08) {
n=it;
break nm5; }
```

```
continue nm4; }
if (gg>=aa) {
n=it; break nm5; }}
yy=-0.5*ee;
it+=2;
ww=ff-Math.pow(yy,2);
xx=Math.sqrt(Math.abs(ww));
dd=2*it;
if (ww>0) {
f[dd]=-xx;
f[dd-1]=yy;
f[dd-2]=xx;
f[dd-3]=yy;
n-=2;
if (n<0) {
for(i=1;i<=n;i++)
w[i]=t[i];
continue nm5;}
if (n<=0) {
n=it;
eigen();
break nm5; }}
if (ww<=0) {
f[dd]=0;
f[dd-1]=yy-xx;
f[dd-2]=0;
f[dd-3]=yy+xx;
n-=2;
if (n>0) {for(i=1;i<=n;i++)
w[i]=t[i];
continue nm5;}
if (n<=0) {
n=it;
eigen();
break nm5; }}
}}}
```

```
}

public void eigen() {
int sd,zw=0;
editt1.appendText(``Eigenvalues of Matrix\n'');
editt1.appendText(``Real part``+''  ``+''Image part\n'');
i=1;
while
(i<=2*n){editt1.appendText(String.valueOf((int)(f[i]*10000)/10000.00)+``
''+String.valueOf((int)(f[i+1]*10000)/10000.00)+``\n'');
i+=2;}
sd=0;
for(i=1;i<=2*n;i+=2) {
if (f[i]<0) sd+=1;
else if (f[i]>0) {
zw=1;
editt1.appendText(``System is unsteady\n'');
break; }}
if ((sd<n) & (zw==0)) editt1.appendText(``System is steady\n'');
else if ((sd>=n) & (zw==0))
editt1.appendText(``System is asymptotic steady\n''); }
public double[][] subpro() {
for(i=1;i<=pp;i++)
for(j=1;j<=pp;j++)
if (i==j) c[i][j+pp]=1;
else c[i][j+pp]=0;
for(i=1;i<=pp-1;i++)
for(kk=i+1;kk<=pp;kk++)
for(j=i+1;j<=2*pp;j++)
c[kk][j]-=c[kk][i]*c[i][j]/c[i][i];
for(i=1;i<=pp;i++)
for(j=1;j<=2*pp;j++) g[i][j]=c[i][j]/c[i][i];
for(i=pp;i>=2;i--)
for(kk=i-1;kk>=1;kk--)
for(j=2*pp;j>=1;j--)g[kk][j]-=g[kk][i]*g[i][j]/g[i][i];
for(i=1;i<=pp;i++)
```

```
for(j=pp+1;j<=2*pp;j++) c[i][j-pp]=g[i][j];
return c; }
public double[][] subgra() {
int w1,w2,d1,d2;
double k1=1;
while(k1>1e-06) {
for(i=1;i<=n;i++)
for(j=1;j<=n;j++) {h[i][j]=0;
for(kk=1;kk<=n;kk++) h[i][j]+=a[i][j]-k[i][kk]*x[kk][j];}
for(i=1;i<=n;i++)
for(j=1;j<=n;j++) {e1[i][j]=0;
for(kk=1;kk<=n;kk++) e1[i][j]+=x[i][kk]*k[kk][j]; }
for(i=1;i<=n;i++)
for(j=1;j<=n;j++) {s[i][j]=0;
for(kk=1;kk<=n;kk++)
s[i][j]+=q[i][j]+e1[i][kk]*x[kk][j];
x1[i][j]=x[i][j]; }
for(i=1;i<=n;i++)
for(j=1;j<=n;j++) e[j][i]=h[i][j];
pp=n*n;
for(i=1;i<=pp;i++)
point1: for(j=i;j<=pp;j++)
if (i==j) {
w1=(int)Math.round((i-1)/n)+1;
w2=j-(int)Math.round((j-1)/n)*n;
c[i][j]=e[w1][w2]+h[w1][w2];continue point1; }
else {if (j<=((int)Math.round((i-1)/n)+1)*n)
{c[i][j]=e[(int)Math.round((i-1)/n)+1][j-(int)Math.round((j-1)/n)*n];
continue point1; }
if (j==(int)Math.round((j-1)/n)*n+(int)Math.round((i-1)/n)+1)
{c[i][j]=h[(int)Math.round((i-1)/n)+1][j-(int)Math.round((j-1)/n)*n];
continue point1; }
c[i][j]=0; }
for(i=1;i<=pp;i++)
for(j=i;j<=pp;j++) c[j][i]=c[i][j];
subpro();
```

```
for(j=1;j<=pp;j++) {
d1=(int)Math.round((j-1)/n)+1;
d2=j-(int)Math.round((j-1)/n)*n;
x[d1][d2]=0;
for(kk=1;kk<=pp;kk++)
x[d1][d2]-=c[j][kk]*s[(int)Math.round((kk-1)/n)+1]
      [kk-(int)Math.round((kk-1)/n)*n];}
k1=0;
for(i=1;i<=n;i++)
for(j=1;j<=n;j++)
if (Math.abs(x1[i][j]-x[i][j])>k1) k1=Math.abs(x1[i][j]-x[i][j]);
}
editt1.appendText(''Solution z of Riccati Equation\n'');
for(i=1;i<=n;i++) {
for(j=1;j<=n;j++)
editt1.appendText(String.valueOf((int)(x[i][j]*10000)/10000.00)+''  '');
editt1.appendText(''\n''); }
return x; }
```

6.2.3 *Number of controllable states of linear ordinary systems*

It is necessary to determine the number of controllable states of a system in order to effectively design the control system. Consider a linear ordinary system

$$dx/dt = Ax + Bu$$

where $x = (x_1, x_2, \ldots, x_n), u = (u_1, u_2, \ldots, u_m); A$: system matrix, B: control matrix. The number of controllable states is the rank of the matrix

$$(B\ AB \ldots A^{n-1}B)$$

If the rank is n, then the system is completely controllable. The following are some of the Java codes for detecting the number of controllable states:

//The first row in array A,B,C,D should be 0 for each element(array index)

```
import java.io.*;
class controlnumber {
public static int n,m,i,j,k,r,c,nr;
public static double a[][];
public static double l[][];
public static double e[][];
public static double w[][];
public static double t[][];
public static double b[][];
public static void main(String args[]) throws IOException {
double p,ff;
n=3;m=2;
a=new double[n+2][n+2];
l=new double[n+2][n+2];
e=new double[n+2][n+2];
w=new double[n*m+2][n*m+2];
t=new double[n*m+2][n*m+2];
b=new double[n+2][m+2];
System.out.println(``Enter state coefficient array A(n,n):'');
for(i=1;i<=n;i++)
for(j=1;j<=n;j++) {
a[i][j]=value();System.out.println(``a(``+i+'',``+j+'')=''+a[i][j]);}
System.out.println(``Enter output coefficient array B(n,m):'');
for(i=1;i<=n;i++)
for(j=1;j<=m;j++) {
b[i][j]=value();System.out.println(``b(``+i+'',``+j+'')=''+b[i][j]);}
nr=m;
for(i=1;i<=n;i++) {
for(j=1;j<=n;j++) {e[i][j]=a[i][j];l[i][j]=0; }
for(j=1;j<=m;j++) t[i][j]=b[i][j];}
while (nr<(n-1)*m) {
for(i=1;i<=n;i++)
for(j=1;j<=n;j++)
for(k=1;k<=n;k++) e[i][j]+=l[i][k]*a[k][j];
```

```
for(i=1;i<=n;i++)
for(j=1;j<=m;j++) {t[i][j+nr]=0;
for(k=1;k<=n;k++) t[i][j+nr]+=e[i][k]*b[k][j]; }
for(i=1;i<=n;i++)
for(j=1;j<=n;j++) {l[i][j]=e[i][j];e[i][j]=0; }
nr+=m;}
nr+=m;
r=0;
for(i=1;i<=nr;i++)
for(j=1;j<=n;j++) w[j][i]=t[i][j];
c=n;
n=nr;
nr=c;
for(i=1;i<=nr;i++) for(j=1;j<=n;j++) t[i][j]=w[i][j];
point2: for(i=1;i<=nr;i++) {
if (Math.abs(t[i][i])>1e-08) {
subr();r++;continue point2;}
else {for(int l=i+1;l<=nr;l++) {for(j=1;j<=n;j++) {
ff=t[l][j];t[l][j]=t[i][j];
t[i][j]=ff;}
if (Math.abs(t[i][i])>1e-08) {
subr();
r++;continue point2; }}
if (Math.abs(t[i][i])<1e-08) {
subp();
System.exit(0); }
else {subr();r++;continue point2; }}}
subp(); }

public static double[][] subr() {
double p;
for(j=i+1;j<=n;j++) t[i][j]/=t[i][i];
t[i][i]=1;
point1: for(k=i+1;k<=nr;k++) {
if (t[k][i]==0) continue point1;
else for(j=i+1;j<=n;j++) {
```

```
p=t[k][j];t[k][j]=p-t[k][i]*t[i][j]; }}
return t; }

public static void subp() {
System.out.println(''Rank=''+r);
if (r==c) System.out.println(''System is controllable'');
else System.out.println(''System is uncontrollable'');
}}
```

6.2.4 *Number of observable states of linear ordinary systems*

Suppose the linear ordinary system is

$$dx/dt = Ax$$
$$y = Cx$$

where $x = (x_1, x_2, \ldots, x_n), y = (y_1, y_2, \ldots, y_m)$; A: system matrix, C: output matrix. The number of observable states is the rank of the matrix

$$(C\ CA \ldots CA^{n-1})$$

If the rank is n, then the system is completely observable. The following are some of the Java codes for detecting the number of observable states:

//The first row in array A,B,C,D should be 0 for each element(array index)

```
import java.io.*;
class observenumber {
public static int n,m,i,j,k,r,c1,nr;
public static double a[][];
public static double l[][];
public static double e[][];
public static double w[][];
public static double t[][];
public static double c[][];
public static void main(String args[]) throws IOException {
double p,ff;
```

```
n=3;m=2;
a=new double[n+2][n+2];
l=new double[n+2][n+2];
e=new double[n+2][n+2];
w=new double[n*m+2][n*m+2];
t=new double[n*m+2][n*m+2];
c=new double[m+2][n+2];
System.out.println(''Enter state coefficient array A(n,n):'');
for(i=1;i<=n;i++)
for(j=1;j<=n;j++) {
a[i][j]=value();System.out.println(''a(''+i+'',''+j+'')=''+a[i][j]); }
System.out.println(''Enter output coefficient array C(m,n):'');
for(i=1;i<=m;i++)
for(j=1;j<=n;j++) {
c[i][j]=value();System.out.println(''c(''+i+'',''+j+'')=''+c[i][j]); }
nr=m;
for(j=1;j<=n;j++) {for(i=1;i<=n;i++) {
e[i][j]=a[i][j];l[i][j]=0;}
for(i=1;i<=m;i++) t[i][j]=c[i][j];}
while (nr<(n-1)*m) {
for(i=1;i<=n;i++)
for(j=1;j<=n;j++)
for(k=1;k<=n;k++)
e[i][j]=l[i][k]*a[k][j];
for(i=1;i<=m;i++)
for(j=1;j<=n;j++) {
t[i+nr][j]=0;
for(k=1;k<=n;k++) t[i+nr][j]+=c[i][k]*e[k][j]; }
for(i=1;i<=n;i++)
for(j=1;j<=n;j++) {
l[i][j]=e[i][j];
e[i][j]=0; }
nr+=m;}
nr+=m;
r=0;
c1=n;
```

```
if (n>nr) {
for(i=1;i<=nr;i++) for(j=1;j<=n;j++) w[j][i]=t[i][j];
c1=n;
n=nr;
nr=c1;
for(i=1;i<=nr;i++)
for(j=1;j<=n;j++) t[i][j]=w[i][j];}
point2: for(i=1;i<=nr;i++) {
if (Math.abs(t[i][i])>1e-08) {
subr();
r++;
continue point2; }
else {for(int l=i+1;l<=nr;l++) {
for(j=1;j<=n;j++) {
ff=t[l][j];t[l][j]=t[i][j];
t[i][j]=ff; }
if (Math.abs(t[i][i])>1e-08) {
subr();
r++;
continue point2; }}
if (Math.abs(t[i][i])<1e-08) {
subp();
System.exit(0); }
else {subr();r++;continue point2;}}}
subp(); }

public static double[][] subr() {
double p;
for(j=i+1;j<=n;j++) t[i][j]/=t[i][i];
t[i][i]=1;
point1: for(k=i+1;k<=nr;k++) {
if (t[k][i]==0) continue point1;
else for(j=i+1;j<=n;j++) {
p=t[k][j];
t[k][j]=p-t[k][i]*t[i][j]; }}
return t; }
```

```
public static void subp() {
System.out.println(''Rank=''+r);
if (r==c1) System.out.println(''System is observable'');
else System.out.println(''System is unobservable'');
}}
```

6.2.5 *Discrete control systems*

Consider a control system

$$\begin{aligned} dx/dt &= f(x,u,t) \\ y(t) &= g(x,t) \end{aligned}$$

where
$x = (x_1, x_2, \ldots, x_n), u = (u_1, u_2, \ldots, u_m), y = (y_1, y_2, \ldots, y_p), f = (f_1, f_2, \ldots, f_n)$. Let

$$dx/dt \approx (x(k+1) - x(k))/T$$

the system becomes (Wang, 1987)

$$\begin{aligned} (x(k+1) - x(k))/T &= f(x(k), u(k), k) \\ k &= 0, 1, 2, \ldots \end{aligned}$$

and the discrete control system is

$$\begin{aligned} x(k+1) &= x(k) + Tf(x(k), u(k), k) \\ y(k) &= g(x(k), k) \\ k &= 0, 1, 2, \ldots \end{aligned}$$

For the linear ordinary system above, the discrete control system is

$$\begin{aligned} x(k+1) &= x(k) + T(Ax(k) + Bu(k)) \\ y(k) &= Cx(k) \\ k &= 0, 1, 2, \ldots \end{aligned}$$

The general form of discrete linear system is

$$x(k+1) = A(k)x(k) + B(k)u(k))$$

$$y(k) = C(k)x(k)$$
$$k = 0, 1, 2, \ldots$$

The solution is

$$x(k) = \Phi(k,0)x(0) + \sum_{j=0}^{k-1} \Phi(k,j+1)B(j)u(j)$$

and the output is

$$y(k) = C(k)\Phi(k,0)x(0) + \sum_{j=0}^{k-1} C(k)\Phi(k,j+1)B(j)u(j)$$

For the linear ordinar system, the solution of discrete control system is

$$x(k) = A^k x(0) + \sum_{j=0}^{k-1} A^{k-(j+1)} Bu(j)$$

and the output is

$$y(k) = CA^k x(0) + \sum_{j=0}^{k-1} A^j Bu(k-(j+1))$$

Given the discrete system

$$x(k+1) = f(x(k), u(k), k)$$
$$x(k_0) = x_0$$

The optimization problem is to solve the control series

$$u(k_0), u(k_0+1), u(k_0+2), \ldots, u(k_f-1)$$

such that the objective function

$$J = \sum_{j=k0}^{kf-1} L(x(i), u(i), i) + \theta(x(k_f), k_f)$$

be minimized or maximized.

6.2.6 *Distributed parameter control systems*

We have developed a spatial and temporal model for population diffusion (Zhang and Gu, 2011)

$$\begin{aligned}\partial z/\partial t &= (p(t)+q(t)z)\partial^2 z/\partial x^2 + (p(t)+q(t)z)\partial^2 z/\partial y^2 \\ &\quad -B(t)z^2 + D(t)z \\ (x,y) &\in G, 0 < t < T \\ z(x,y,0) &= W(x,y), (x,y) \in G \\ z(x,y,t) &= V(x,y,t), (x,y) \in \Gamma, 0 \leq t < T\end{aligned}$$

A control system can be

$$\begin{aligned}\partial z/\partial t &= (p(t)+q(t)z)\partial^2 z/\partial x^2 + (p(t)+q(t)z)\partial^2 z/\partial y^2 \\ &\quad -B(t)z^2 + D(t)z + E(t)u \\ (x,y) &\in G, 0 < t < T \\ z(x,y,0) &= W(x,y), (x,y) \in G \\ z(x,y,t) &= V(x,y,t), (x,y) \in \Gamma, 0 \leq t < T\end{aligned}$$

and the control is

$$u(x,y,t) = -a(t)\int_0^t \partial^2 z(x,y,\tau)/(\partial x \partial y) d\tau$$

or

$$u(x,y,t) = -a(t)z(x,y,t)$$

Chapter 7

Cellular Automata and Spatial Diffusion Models

7.1 Cellular Automata

Cellular automata are a kind of dynamic models with discrete time, discrete space, and discrete states. They are finite state machines, i.e., each cell in the model has a finite number of states. Cellular automata were proposed by von Neumann in 1950s, which were used to simulate the self-reproduction of biological cells. Since then, many scientists have studied cellular automata. Wolfram (2002) analyzed the effect of 256 rules on the evolution outcomes of cellular automata, and derived the classifications of cellular automata based on their evolution behaviors.

7.1.1 *Classification of cellular automata*

7.1.1.1 *Evolution behavior based classification*

Cellular automata are divided into four types according to their behavioral patterns, homogeneous, periodic, chaos, and edge of chaos (Wolfram, 2002; Wang *et al.*, 2007).

(1) Homogeneous. It refers to almost all patterns evolving to a steady and homogeneous state. The steady state does not change with time, and the initial random pattern disappears gradually. In the steady state, some cells are alive, or all cells die.
(2) Periodic. Almost all patterns evolve to a relatively stable oscillation structure. The states of some cells change periodically with time. Some stochasticity in initial states can be filtered and some

of them remain. A local change of initial pattern may only produce a local effect.

(3) Chaos. Almost all initial patterns evolve to chaotic states. Any temporarily stable structure occurred in evolution will be terminated by the environmental chaos. The changes in the initial pattern may be unlimitedly propagated to entire cellular automaton. Chaotic cellular automata are completely stochastic and unordered, without producing valuable structures.
(4) Edge of chaos. Almost all initial patterns evolve to complex but not chaotic structures.

In general, the final outcome of evolution of these cellular automata is similar to that of chaos. However, the steady states or periodic states may occur in some local evolution. Such cellular automata are the interim model from periodic to chaotic ones.

To better describe the four categories of cellular automata, Wolfram defined a parameter, $\lambda(0 < \lambda < 1)$, i.e., the proportion of a set of rules that transformed to non-zero states, to classify cellular automata. In his definition, $\lambda = (m^{2r+1} - n_q)/m^{2r+1}$, where, m: the number of states in set of states $\mathrm{S} = \{s_1, \ldots, s_m\}$, r: radius of neighbors, n_q: the number of outputs with 0. $\lambda = 0 \sim 0.1$, homogeneous cellular automata; $\lambda \approx 0.2$, periodic cellular automata; $\lambda = 0.3 \sim 0.6$, the cellular automata at edge of chaos; $\lambda \geq 0.6$, cellular automata of chaos (Wang *et al.*, 2007).

7.1.1.2 *Space dimensionality based classification*

Cellular automata are divided into three types according to their spatial dimensionality.

(1) One-dimensional. In this model, cells distribute along a line, which is similar to the Turing machine.
(2) Two-dimensional. It is a mostly used cellular automata. According to the different definitions of states of adjacent cells, such cellular automata are included, i.e., von Neumann type, Moore type, and Margolus type, etc.
(3) Three or higher dimensional cellular automata are seldom used for their complexity (Tomoyoshi and Tsutomu, 2002).

In addition, cellular automata can be divided by their boundary conditions, into periodic, random, etc.

7.1.2 *Algorithms of cellular automata*

Following some common rules, all cells in a cellular automaton update their states synchronously during evolution, given the initial states of cells.

A cellular automaton contains some rules constructed by a series of models. It can be expressed as a conceptual structure as the following:

$$A = (L, d, S, N, F)$$

where L: space of cells, d: spatial dimensionality, S: the set of finite and discrete states of cells, N: the set of cells in the adjacent area, and f: local rules (Amoroso and Patt, 1972).

For example, in the elementary cellular automaton (one-dimensional) of Wolfram, $S = \{0, 1\}$, the radius is that there is only one adjacent cell, and the local mapping rules are

$$S_i^{t+1} = f(S_{i-1}^t, S_i^t, S_{i+1}^t)$$

where i: the present cell, $i + 1$ and $i - 1$: the adjacent cells, t and $t + 1$: the present time and the next time. This means that the state of present cell in the next time depends on its current state and the current states of its adjacent cells.

In the two-dimensional cellular automata, different neighborhood patterns may result in different outcomes of evolution. Local rules of a two-dimensional cellular automaton can be

$$S_i^{t+1} = f(S_i^t, S_N^t)$$

where S_N^t is the combination of states of neighborhood cells.

7.1.3 *Applications of cellular automata*

The nature of parallel computation, locality, and conformity of cellular automata makes them an effective dynamic model to understand the global behavior and complex phenomena of various systems. They have been widely used in a variety of areas.

Cellular automata are used to simulate traffic breakdown, highway capacity, and synchronized flow (Kerner *et al.*, 2011), real-world urban processes (Sante *et al.*, 2010), and evacuation process with obstacles (Alizadeh, 2011). They are used to simulate reaction process of atom and molecular. For example, in the Belousov-Zhabotinsky reaction, the feedbacks and interactions between various moleculars may yield an oscillation in which two colors occur alternatively. The universe is considered as a multi-dimensional cellular automaton (Ilachinski, 2001). High-quality random numbers can be generated by using two-dimensional cellular automata (Tomassini *et al.*, 2000).

Biology is the science for the origination of concept of cellular automata, and is the most widely used area for cellular automata. Natural cellular automata are sometimes found in organisms and biology. For example, the shell pattern of a sea shell, *Conus textile*, is similar to a one-dimensional Wolfram cellular automaton (Coombes, 2009). Stomata on the plant leaves show complex and collective dynamics and perform emergent and distributed computation through communications and interactions between them (Peak, 2004). They are similar to a two-dimensional cellular automaton. Jiao and Torquato (2011) used the emergent behaviors from a cellular automaton model to simulate invasive tumor growth in heterogeneous microenvironments. Also, cellular automaton has been used to simulate dynamics of the HIV infection under antiretroviral therapy (González *et al.*, 2013). Based on the protein sequence only, various properties of proteins can be predicted using cellular automata (Xiao *et al.*, 2011).

7.2 Cellular Automata for Population Diffusion

7.2.1 *Diffusion in homogeneous environment*

Cellular automata have been used to describe population diffusion in a homogeneous rectangular area (Qi, 2003; Qi *et al.*, 2004; Zhang *et al.*, 2015a).

Suppose the area has $c \times f$ cells. Given a cell, individuals in the cell diffuse to adjacent four cells; or, individuals in the adjacent four cells diffuse to this cell. Population diffuses until all cells have the same number of individuals. At this time, the number of individuals for immigration and migration are the same in each cell. The time step for diffusion is ∂t and the diffusion coefficient is a. The diffusion rules are thus,

$$\begin{aligned}
\partial u(1,1,t)/\partial t &= a^*(u(1,2,t) + u(2,1,t) - 2^*u(1,1,t)) \\
\partial u(1,c,t)/\partial t &= a^*(u(1,c-1,t) + u(2,c,t) - 2^*u(1,c,t)) \\
\partial u(1,j,t)/\partial t &= a^*(u(1,j-1,t) + u(2,j,t) + u(1,j+1,t) \\
&\quad - 3^*u(1,j,t)) \\
\partial u(f,1,t)/\partial t &= a^*(u(f-1,1,t) + u(f,2,t) - 2^*u(f,1,t)) \\
\partial u(f,c,t)/\partial t &= a^*(u(f-1,c,t) + u(f,c-1,t) - 2^*u(f,c,t)) \\
\partial u(f,j,t)/\partial t &= a^*(u(f,j-1,t) + u(f-1,j,t) + u(f,j+1,t) \\
&\quad - 3^*u(f,j,t)) \\
\partial u(i,1,t)/\partial t &= a^*(u(i-1,1,t) + u(i,2,t) + u(i+1,1,t) \\
&\quad - 3^*u(i,1,t)) \\
\partial u(i,c,t)/\partial t &= a^*(u(i-1,c,t) + u(i,c-1,t) + u(i+1,c,t) \\
&\quad - 3^*u(i,c,t)) \\
\partial u(i,j,t)/\partial t &= a^*(u(i-1,j,t) + u(i,j-1,t) + u(i+1,j,t) \\
&\quad + u(i,j+1,t) - 4^*u(i,j,t)) \\
&\qquad i = 1,2,\ldots,f; j = 1,2,\ldots,c
\end{aligned}$$

if $u(i,j,t) < 0$, then let $u(i,j,t) = 0$.

Regardless of diffusion process, the natural population dynamics in a cell can be described by Malthus model, Logistic model, or oscillation model:

$$\begin{aligned}
\partial u(i,j,t)/\partial t &= ru \\
\partial u(i,j,t)/\partial t &= ru(K-u)/K \\
\partial u(i,j,t)/\partial t &= -2\pi/T^* A^* \sin(2\pi t/T + \varphi)
\end{aligned}$$

where, $u = u(x,y,t)$: population size of the cell (x,y) at time t; r: intrinsic rate, $-\infty < r < \infty$; K: environmental capacity of a cell; φ: phase, $0 \le \varphi \le 2\pi$; $A = (u-b)/\cos(\varphi)$, and b is the mean number of individuals per cell; T: oscillation cycle, $0 < T < \infty$.

Suppose diffusion coefficient $a = 0.5$, and environmental capacity per cell $K = 30$. Initial distribution of population in the area (10×8 cells) is

3	18	21	3	1	0	0	0
1	1	0	2	1	0	0	0
0	1	2	6	1	1	0	0
1	0	0	1	0	1	0	1
0	0	1	1	0	12	24	0
2	1	2	2	0	1	0	0
0	0	0	0	1	0	1	0
0	0	1	0	0	1	0	0
0	0	0	2	0	1	0	0
0	0	0	0	0	0	0	0

Assume the population grows in Logistic pattern. Let intrinsic rate $r = 0.1, 0.2, 0.3, 0.4$ and 0.5, respectively. Simulated dynamics of total population is shown in Fig. 1.

Overall the population grows in the Logistic pattern. A greater intrinsic rate will result in the larger population size. On the other hand, population size of each cell will not exceed the environmental capacity, $K = 30$. Due to the effect of diffusion, the population dynamics in a given cell is not certainly a Logistic growth, as indicated in Fig. 2.

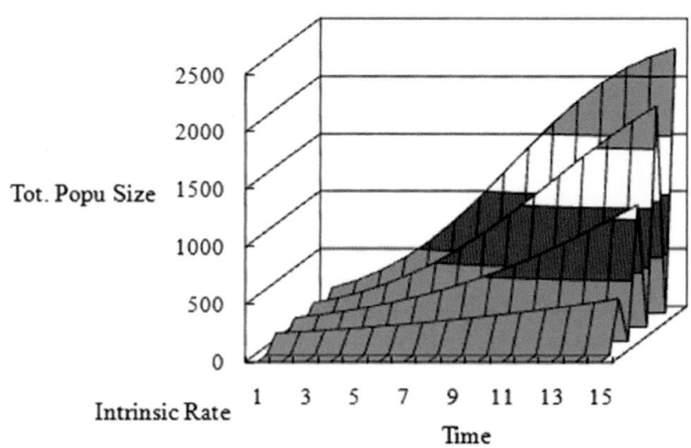

Fig. 1. Comparison of total population using different intrinsic rates in the model.

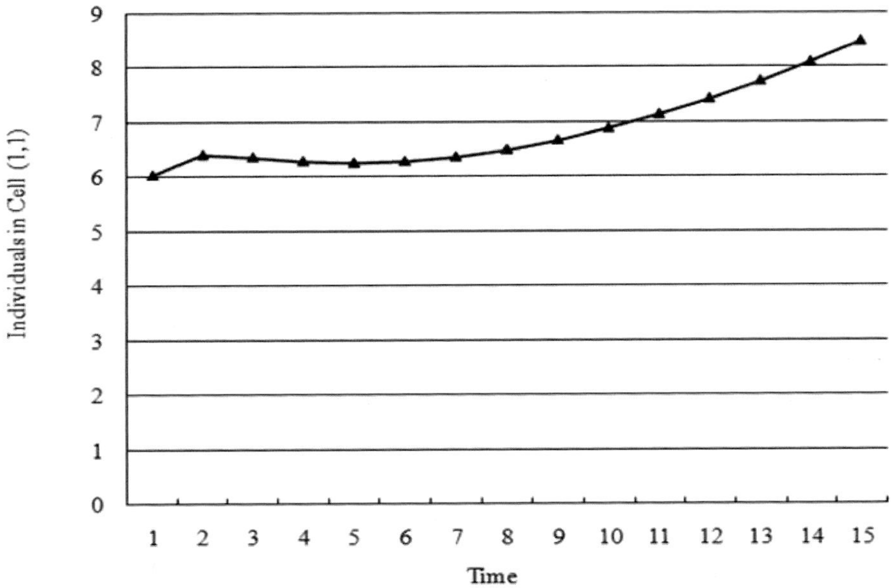

Fig. 2. Population dynamics in the cell (1, 1), with intrinsic rate $r = 0.1$.

Figure 3 shows the comparison of distribution of individuals. Although the initial distribution is highly heterogeneous, it will tend to be uniform after a period of diffusion due to the limitation of given uniform environmental capacity and the little intrinsic rate compared to diffusion coefficient (in this case, the latter is more significant).

Population will reach a steady state, with the same number of individuals (30 in the situation of Logistic growth with environmental capacity, $K = 30$) in each cell and, population will change if a cell is disturbed, or the environmental capacity is changed, until the steady state is reached. It is a self-organizing process.

For a population dynamics with oscillation pattern (generally found in the situation of year by year changes of biological population), let oscillation cycle $T = 5$, phase $\varphi = 0$, mean number of individuals per cell $b = 10$. The initial distribution of population is the same as the mentioned above. For a given cell, the oscillation amplitude decreases with the increase of diffusion coefficient (Fig. 4; Table 1).

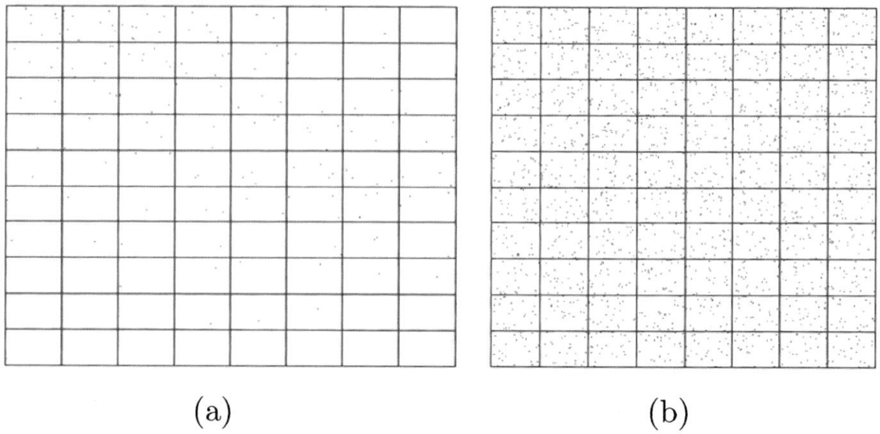

Fig. 3. Distribution of individuals during Logistic growth, intrinsic rate $r = 0.2$. (a) $t = 2$; (b) $t = 25$.

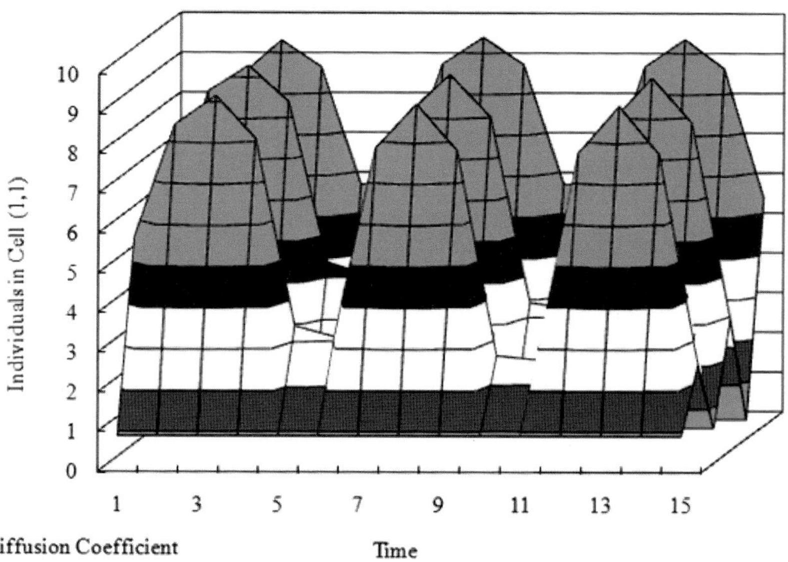

Fig. 4. Population dynamics in the cell (1, 1), for oscillative dynamics.

The following are some of the Java codes for above cell automaton.

Table 1. Population dynamics in the cell (1, 1), for oscillative dynamics with different diffusion coefficients (a).

a	u	a	u	a	u
0.9	5.6113	0.5	5.616	0.1	4.0637
0.9	8.438	0.5	8.6589	0.1	8.424
0.9	9.1551	0.5	9.2958	0.1	9.3119
0.9	8.0858	0.5	8.3945	0.1	8.6432
0.9	3.3623	0.5	4.3504	0.1	5.6938
0.9	3.0944	0.5	4.0322	0.1	5.7591
0.9	7.8625	0.5	8.1275	0.1	8.7288
0.9	8.9579	0.5	9.0763	0.1	9.3905
0.9	7.7845	0.5	8.0165	0.1	8.7122
0.9	2.6219	0.5	3.3382	0.1	5.7077
0.9	2.5293	0.5	3.205	0.1	5.6292
0.9	7.7313	0.5	7.9233	0.1	8.6609
0.9	8.9093	0.5	8.996	0.1	9.3492
0.9	7.7056	0.5	7.8773	0.1	8.6144
0.9	2.4221	0.5	2.9579	0.1	5.3648

```
kn=1;
for(j=1;j<=f;j++)
for(i=1;i<=c;i++) {
if (!((j==f) & (i==c)))
{u[j][i]=(Double.valueOf(sp.substring(0,sp.indexOf(' ')))).doubleValue();
u2[j][i]=u[j][i];}
else {u[j][i]=(Double.valueOf(sp)).doubleValue(); break;}
sp=(sp.substring(sp.indexOf(' '))).trim();}
delt=1;
det=100;
for(nt=1;nt<=tm;nt++){
for(kk=1;kk<=det;kk++) {
for(j=1;j<=c;j++)
for(i=1;i<=f;i++) {
if (i==1) {
if (j==1) u1[1][1]=u[1][1]+delt*dif/det*(u[1][2]+u[2][1]-2*u[1][1]);
else if (j==c) u1[1][c]=u[1][c]+delt*dif/det*(u[1][c-1]+u[2][c]-2*u[1][c]);
else u1[1][j]=u[1][j]+delt*dif/det*(u[1][j-1]+u[2][j]+u[1][j+1]-3*u[1][j]);
```

```
  continue; }
if (i==f) {
if (j==1) u1[f][1]=u[f][1]+delt*dif/det*(u[f-1][1]+u[f][2]-2*u[f][1]);
else if (j==c)
u1[f][c]=u[f][c]+delt*dif/det*(u[f-1][c]+u[f][c-1]-2*u[f][c]);
else
u1[f][j]=u[f][j]+delt*dif/det*(u[f][j-1]+u[f-1][j]+u[f][j+1]-3*u[f][j]);
continue; }
if (j==1) {
u1[i][1]=u[i][1]+delt*dif/det*(u[i-1][1]+u[i][2]+u[i+1][1]-3*u[i][1]);
continue; }
if (j==c) {
u1[i][c]=u[i][c]+delt*dif/det*(u[i-1][c]+u[i][c-1]+u[i+1][c]-3*u[i][c]);
continue; }
u1[i][j]=u[i][j]+delt*dif/det*(u[i-1][j]+u[i][j-1]+u[i+1][j]
+u[i][j+1]-4*u[i][j]);
if (u1[i][j]<0) u1[i][j]=0; }
for(j=1;j<=c;j++)
for(i=1;i<=f;i++) {
switch (choose) {
       case 1:u[i][j]=u1[i][j]**(1+(double)delt/det*r);break;
       case 2:u[i][j]=u1[i][j]*(1+(double)delt/det*r*((ke-u1[i][j])/ke));
         break;
       case
3:u[i][j]=u1[i][j]-(double)delt/det*2*3.1415926/pe*(u1[i][j]-barr)/Math.
cos(d)*Math.sin(2*3.1415926/pe*(nt+(kk-1)/det-1)+d); }
if (u[i][j]<0) u[i][j]=0; }}
for(i=1;i<=c;i++)
for(j=1;j<=f;j++) gr[kn][j][i]=u[j][i];
tt[kn]=nt;
kn++; }
```

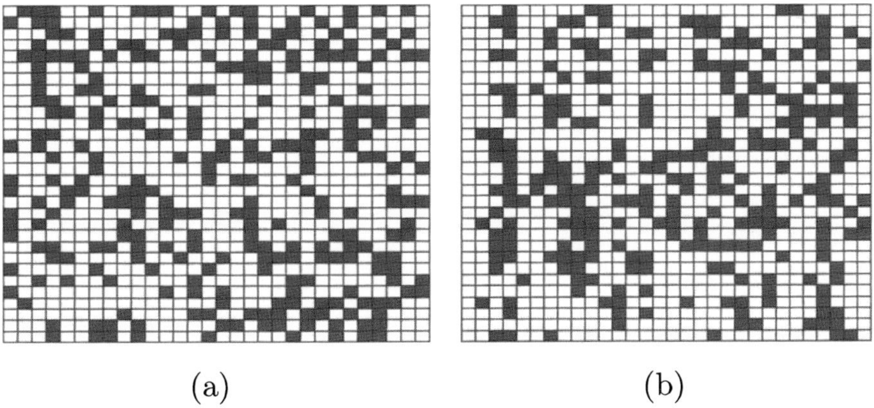

Fig. 5. Population diffusion in a heterogeneous environment. (a) $t = 2$; (b) $t = 29$.

7.2.2 *Diffusion in heterogeneous environment*

The cellular automata, described by Qi and Zhang (2002), Qi (2003), and Zhang (2012), can be used in the simulation of population diffusion in heterogeneous environment (Fig. 5).

Suppose the area is divided into $c \times f$ cells with the same area, $a_{ij}, i = 1, 2, \ldots, c; j = 1, 2, \ldots, f$. Population diffusion occurs in this region (Qi, 2003; Qi and Zhang, 2002; Zhang, 2012a).

Given the sensitivity level of each cell against population diffusion, $q_k(q_k \in \boldsymbol{R}, k = 1, 2, \ldots, m)$, the corresponding probability of being diffused $p_k(p_k \in [0, 1], k = 1, 2, \ldots, m)$, the initial distribution of population diffusion (i.e., whether or not the population occurs in a cell before the simulation, $t = 0$). The initial distribution is

$$\begin{aligned} v_{ij} &= 1, \text{ if cell } a_{ij} \text{ has been diffused} \\ v_{ij} &= 0, \text{ if cell } a_{ij} \text{ is not yet diffused} \\ i &= 1, 2, \ldots, c; j = 1, 2, \ldots, f \end{aligned}$$

In addition,

$$\begin{aligned} v_{0i} = v_{f+1i} &= 0, \qquad i = 1, 2, \ldots, c+1; \\ v_{i0} = v_{ic+1} &= 0, \qquad i = 1, 2, \ldots, f+1 \end{aligned}$$

Sensitivity distribution is also known.

$$u_{ij} = q_k, \text{ if the sensitivity of cell } a_{ij} \text{ is } q_k (k = 1, 2, \ldots, m)$$
$$i = 1, 2, \ldots, c; j = 1, 2, \ldots, f$$

Probability distribution of population is thus,

$$p_{ij} = p_k, \quad \text{if } u_{ij} = q_k, \quad k = 1, 2, \ldots, m$$
$$i = 1, 2, \ldots, c; j = 1, 2, \ldots, f$$

Assume each diffusion process occurs at unit time, and population diffuses to those cells with which the present cell has adjacent edges. For an undiffused cell x, the population will diffuse to the cell x with diffusion probability p_i from the adjacent diffused cells. For sensitivity k, calculate the number of cells available for the diffusion, N_k

$$N_k = \sum_i \sum_j b_{ij}$$

where

$$b_{ij} = 1, \text{ if } (v_{ij} = 0) \wedge (p_{ij} = p_k) \wedge ((v_{i-1j} = 1) \vee (v_{i+1j} = 1)$$
$$\vee (v_{ij-1} = 1) \vee (v_{ij+1} = 1)) \text{ is true}$$
$$b_{ij} = 0, \text{ if } (v_{ij} = 0) \wedge (p_{ij} = p_k) \wedge ((v_{i-1j} = 1) \vee (v_{i+1j} = 1)$$
$$\vee (v_{ij-1} = 1) \vee (v_{ij+1} = 1)) \text{ is false}$$

The number of actually diffused cells with sensitivity k is

$$n_k = \mathrm{INT}((100 p_k N_k + 0.5)/100), k = 1, 2, \ldots, m$$

A diffused cell will restore to an undiffused one at the restoration rate r, i.e., the proportion of cells that can restore from diffused state during unit time

$$v_{ij}, \quad \text{if } 1 - r + R \geq 1$$
$$v_{ij} = 0, \quad \text{if } 1 - r + R < 1$$
$$i = 1, 2, \ldots, c; \quad j = 1, 2, \ldots, f$$

where R is a random value, $0 \leq R \leq 1$.

Population dynamics shows different patterns with the change of restoration rate, diffusion probability, sensitivity levels, etc., and as partially indicated in Fig. 6. Various equilibrium states may be reached with different combinations of parameters.

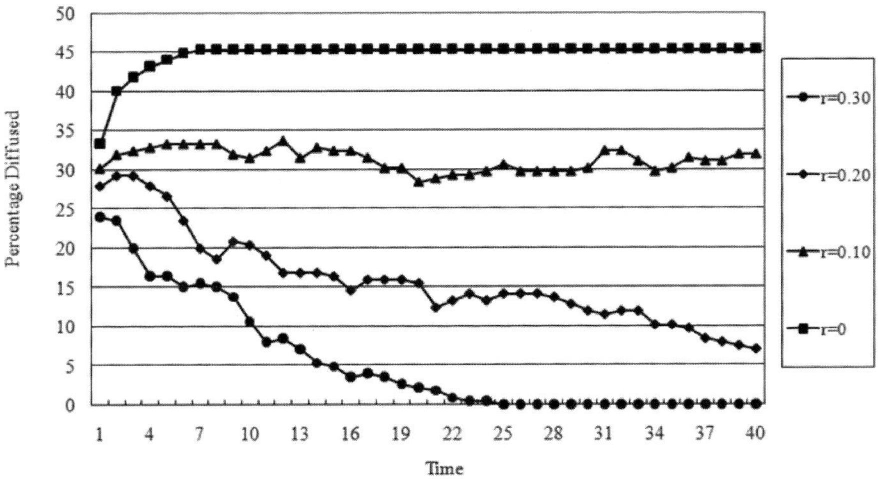

Fig. 6. Population dynamics with different restoration rate, r.

7.3 Spatial Diffusion Models

Rules derived from some spatial diffusuion models are expected in self-organizing research. Here I describe some diffusion models.

7.3.1 *One-dimensional diffusion equation*

Population diffuses along a direction x, i.e., a transect. The diffusion equation is (Qi, 2013; Qi *et al.*, 2014)

$$\partial u/\partial t = a^2 \partial^2 u/\partial x^2 \qquad x > 0, t > 0$$
$$u(x,0) = 0, u(0,t) = f(t)$$

where, $u = u(x,t)$ is the population size in position x at time t, and a is the diffusion coefficient. The analytical solution is

$$u(x,t) = x/(2^* a^* \pi^{1/2})^* \int_0^t f(\tau)/(t-\tau)^{1.5} \times e^{-x^* x/(4^* a^* a^* (t-\tau))} d\tau$$

Using the Chebyshev method, let

$$s = \int_0^t f(\tau) d\tau$$

then

$$s = t/2^* \int_{-1}^{1} f(t^*\tau/2 + t/2)d\tau$$

and

$$2/n^* \sum_{k=0}^{n-1} f(t^*T_k/2 + t/2)$$

where, $n = 5$, $T_0 = -0.8324975$, $T_1 = -0.3745414$, $T_2 = 0$, $T_3 = 0.3745414$, $T_4 = 0.8324975$.

Suppose the population diffuses within the interval $(0, \infty)$; the environment along the direction x is homogeneous, and growth and mortality of the population are ignored. Given the boundary condition $u(0, t) = f(t)$, i.e., the population sizes at some time points in the position x, $f(t)$, is known. Once starting position, $x_{min} > 0$, terminating position, $x_{max} < \infty$, and step length $x_{step} > 0$, are given, the simulation can be conducted across the transect with cell length x_{step}.

Suppose $x_{min} = 1, x_{max} = 10$, and $x_{step} = 1$. There are 6 sets of boundary data $(t, f(t))$: (2, 2), (3, 5), (4, 7), (5, 4), (6, 2), and (7, 3) (Qi, 2013; Qi *et al.*, 2014). The diffusion coefficient $a = 0.9$. The simulated diffusion dynamics is indicated in Fig. 7.

7.3.2 *Two-dimensional diffusion equation*

A dynamic model can be used to describe time-space dynamics of population over a plane (Zhang and Gu, 2001). Suppose $u(x, y, t)$ is the population size on position (x, y) at time t, and $J(u)$ is population dynamics, we have

$$\partial u/\partial t = J(u)$$

In general, let $J(u) = a(x, y, t)u + b(x, y, t)$, then

$$\partial u/\partial t = a(x, y, t)u + b(x, y, t)$$

where $b(x, y, t)$ is the diffusion rate, and

$$a(x, y, t) = e(x, y, t) - d(x, y, t) + c(x, y, t)$$

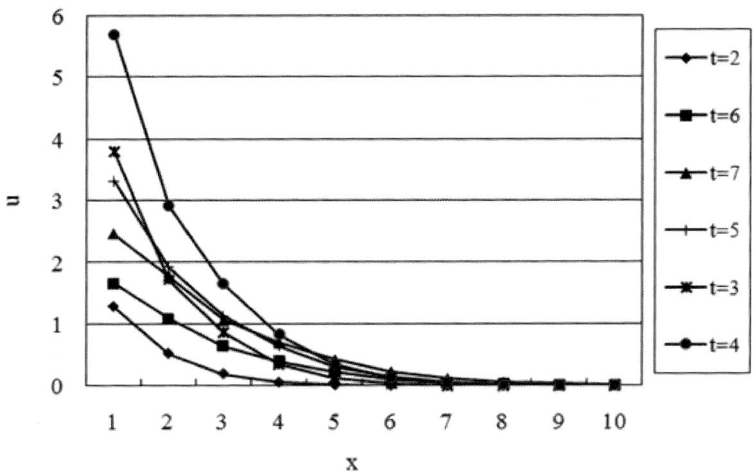

Fig. 7. Simulated diffusion dynamics.

where $e(x, y, t), c(x, y, t)$, and $d(x, y, t)$ are borning velocity, migrating velocity and dying velocity, respectively. Finally we obtain

$$\partial u/\partial t = b(x, y, t) + c(x, y, t)u + e(x, y, t)u - d(x, y, t)u$$

Migrating velocity and diffusion rate are dependent upon the density-space relationship and motion property. Therefore, let

$$b(x, y, t) = p(x, y, t)(\partial^2 u/\partial x^2 + \partial^2 u/\partial y^2)$$
$$c(x, y, t) = q(x, y, t)(\partial^2 u/\partial x^2 + \partial^2 u/\partial y^2)$$

Moreover, the borning velocity and dying velocity are density-dependent. Therefore we have

$$d(x, y, t) = f(x, y, t) + g(x, y, t)u$$
$$e(x, y, t) = h(x, y, t) - k(x, y, t)u$$

The partial differential equation is thus (Zhang and Gu, 2001)

$$\begin{aligned}\partial u/\partial t = {} & (p(x, y, t) + q(x, y, t)u)\partial^2 u/\partial x^2 \\ & + (p(x, y, t) + q(x, y, t)u)\partial^2 u/\partial y^2 \\ & - (k(x, y, t) + g(x, y, t))u^2 \\ & + (h(x, y, t) - f(x, y, t))u\end{aligned}$$

where, $p(x, y, t)$: diffusion coefficient relating to motion property; $q(x, y, t)$: migration coefficient that denote response property of individuals to within-population crowding, which is related to motion property; $f(x, y, t), h(x, y, t)$: density-indepedent dying velocity and borning velocity, which are related to species and the environment; $g(x, y, t), k(x, y, t)$: crowding coefficients which reflect the effect of density dependency of population fluctuation.

Assume the environment is uniform; let $p(x, y, t) = p(t)$, $q(x, y, t) = q(t)$, $k(x, y, t) = k(t)$, $g(x, y, t) = g(t)$, $k(t) + g(t) = B(t)$, $h(x, y, t) = h(t), f(x, y, t) = f(t), h(t) - f(t) = D(t)$, then a simple model is obtained:

$$\partial u/\partial t = (p(t)+q(t)u)\partial^2 u/\partial x^2+(p(t)+q(t)u)\partial^2 u/\partial y^2-B(t)u^2+D(t)u$$

where, $p(t)$: diffusion coefficient; $q(t)$: migration coefficient; $B(t)$: crowding coefficient; $D(t)$: coefficient for species and the environment.

Suppose the population diffuses across a rectangular area,

$$G = \{(x, y)|0 < x < F + 1, 0 < y < C + 1\}$$

$W(x, y)$ is the initial distribution, $V(x, y, t)$ is the boundary condition, T is the termination time for diffusion, and Γ is the boundary for diffusion. The problem is thus,

$$\begin{aligned} \partial u/\partial t &= (p(t) + q(t)u)\partial^2 u/\partial x^2 + (p(t) + q(t)u)\partial^2 u/\partial y^2 \\ &\quad - B(t)u^2 + D(t)u \\ (x, y) &\in G, 0 < t < T \\ u(x, y, 0) &= W(x, y), (x, y) \in G \\ u(x, y, t) &= V(x, y, t), (x, y) \in \Gamma, 0 \leq t < T \end{aligned}$$

Let h and l are step length along x and $y; \tau$ is step length of time t, and $U_{jk}^n = U(jh, kl, n\tau)$. According to the Peaceman-Rachford method, the difference solution of above problem, i.e., the rules that are followed by population to diffuse across the area with $h \times l$ cells, is

$$\begin{aligned} (U_{jk}^{n+0.5} - U_{jk}^n)/(0.5\tau) &= (p(n) + q(n)U_{jk}^n)(U_{j-1k}^{n+0.5} - 2U_{jk}^{n+0.5} \\ &\quad + U_{j+1k}^{n+0.5})/h^2 + (p(n) + q(n)U_{jk}^n)(U_{jk-1}^n \\ &\quad - 2U_{jk}^n + U_{jk+1}^n)/l^2 - B(n)(U_{jk}^n)^2 \\ &\quad + D(n)U_{jk}^n \end{aligned}$$

$$\begin{aligned}(U_{jk}^{n+1} - U_{jk}^{n+0.5})/(0.5\tau) = {} & (p(n+0.5) + q(n+0.5)U_{jk}^{n+0.5})(U_{j-1k}^{n+0.5} \\ & - 2U_{jk}^{n+0.5} + U_{j+1k}^{n+0.5})/h^2 + (p(n+0.5) \\ & + q(n+0.5)U_{jk}^{n+0.5})(U_{jk-1}^{n+1} - 2U_{jk}^{n+1} \\ & + U_{jk+1}^{n+1})/l^2 - B(n+0.5)(U_{jk}^{n+0.5})^2 \\ & + D(n+0.5)U_{jk}^{n+0.5}\end{aligned}$$

7.4 Epidemic System

We have developed the mathematical model for insect-nucleopolyhedroviruscrop system dynamics (Zhang *et al.*, 2011). The processes for temperature time course, crop growth, polyhedron inactivation, virus contamination by cadavers, virus contamination due to secretions by infected larvae, infection by horizontal transmission and polyhedron ingestion, infection by vertical transmission, insect population dynamics, and crop injury were included in this model. Immigration and emigration of adults were also incorporated in the model. The model includes a group of ordinary differential equations to describe insect population dynamics, and some models to describe temperature dynamics, crop growth, and virus dynamics, etc. Given the crop, virus, insect related parameters, and initial conditions, the model will produce the uninfected and infected insect populations, crop growth, virus accumulation dynamics, etc.

The population dynamics of insect are simulated with a group of ordinary differential equations which are developed according to the principles of compartment modeling of Zhang *et al.* (1997).

The main state variables in the model are the densities of the healthy and infected stages of the insect, the densities of viral entities (polyhedron inclusion bodies; PIBs) on the leaves, and the leaf area of the crop.

Densities of m healthy larval stages, L1–Lm, are denoted as $x_1, \ldots, x_m$. The number of larval instars is flexible and can be input as a parameter (m). The densities of pupae, adults and eggs are denoted as, respectively, x_{m+1}, x_{m+2}, and x_{m+3}. The model takes only female individuals into account. The sex ratio is thus applied to reduce egg laying from total eggs to female eggs. Male adults are assumed to be sufficient in number to fertilize all female adults. The densities of the $m+3$ stages of infected individuals are written down as $y_1, \ldots, y_{m+3}$. Virus-killed individuals are indicated by the symbol z, indexed by stage. A portion of cadavers dying from infection is lost from leaves by falling to the ground, and does not contribute to the horizontal transmission.

Some equations for epidemic system are as follows.

Temperature over time is simulated with trigonometric functions

$$\begin{aligned}
T &= (T_0 - T_1)/2 \times \cos(\pi(1 + t - t_{max})/ \\
&\quad (1 - t_{max} + t_{min})) + (T_0 + T_1)/2, && 0 \le t \le t_{min} \\
T &= (T_2 - T_1)/2 \times \sin(\pi(t - (t_{max} \\
&\quad + t_{min})/2)/(t_{max} - t_{min})) + (T_2 + T_1)/2, && t_{min} \le t \le t_{max} \\
T &= (T_2 - T_3)/2 \times \cos(\pi(t - t_{max})/ \\
&\quad (1 - t_{max} + t_{min})) + (T_2 + T_3)/2, && t_{max} \le t \le 1
\end{aligned}$$

where t is time (d) in a day, T is temperature (°C) at time t, t_{max} is the time (d) with the highest temperature during a day, t_{min} is the time (d) with the lowest temperature during a day, T_0 is yesterday's highest temperature, while T_2 and T_1 are today's highest temperature and lowest temperature, respectively. Finally, T_3 is tomorrow's lowest temperature.

In a situation without insect injury, the growth of leaf area is in logistic form:

$$dL/dt = rL(1 - L/L_{min})$$

where t is cumulative effective day degrees, L is leaf area (m^2 leaf per m^2 ground) at physiological time t, L_{max} is the maximum leaf area, r is the relative growth rate of leaf area, $r > 0$.

Inactivation of sprayed polyhedra can be described by assuming a constant relative inactivation rate

$$V = V_0 e^{-rt}$$

where t is time (d), V is the density per m^2 ground of infective polyhedra on leaves (PIBs per m^2 ground) at time t, V_0 is the initial polyhedra on leaves, and r is the relative inactivation rate of polyhedra.

The dynamics of virus-killed larvae is described as

$$dz_i/dt = b_i'' \times y_i, \qquad i = 1, 2, \ldots, m$$

where z_i is the cumulative number of dead virus-killed larvae, b_i'' is the dying rate of the ith instar infected larvae due to infection (1/d), while y_i is the number of infected larva per m^2 ground, $i = 1, 2, \ldots, m$.

Dynamics of contaminated leaf area

$$dP/dt = \sum_{i=1}^{m} dz_i/dt \times (1 - F_i) \times G_i \times (1 - P/L_p)$$

where $\mathrm{d}P/\mathrm{d}t$ is the rate at which contaminated leaf area (P) increases, L_p is the present leaf area per m^2 ground, P is the contaminated leaf area at time t, F_i is the fraction loss of bodies of virus-killed larva due to falling from leaves and bodies that did not break down on leaves, and G_i is the leaf area (m^2) contaminated per dying larva of instar i, when it is naturally broken down, $i = 1, 2, \ldots, m$.

The rate of increase of the amount of polyhedra released by cadavers on contaminated leaves:

$$dV/dt = \sum_{i=1}^{m} dz_i/dt \times (1 - F_i) \times E_i$$

where E_i is the polyhedron production of ith instar per virus-killed larva (PIBs per virus-killed larvae), $i = 1, 2, \ldots, m$. V is the cumulative number of polyhedra on contaminated leaves by time t.

The cumulative leaf injury (per m^2 ground) over time up till time t:

$$L_{loss} = \sum_{i=1}^{m} \int_{t0}^{t+dt} (C_i \times x_i(\tau) + C_i' \times y_i(\tau)) d\tau$$

Crop leaf area (per m^2 ground) at time $t + dt$ is calculated as the leaf area without insect injury, L, minus the cumulative leaf area loss, L_{loss}, i.e., $L - L_{\mathrm{loss}}$.

Healthy L1s

$$dx_1/dt = e_{m+3}x_{m+3} - (b_1 + e_1)x_1 - d_1x_1$$

The first term in the differential equation for healthy L1s denotes egg hatch, the second term attrition plus development into L2, and the third term infection.

Healthy L2s–Lms

$$dx_i/dt = e_{i-1}x_{i-1} - (b_i + e_i)x_i - d_ix_i \quad i = 2, \ldots, m$$

The meaning of terms in the equations for L2–Lm is analogous to those for L1.

Healthy pupae

$$dx_{m+1}/dt = e_mx_m - (b_{m+1} + e_{m+1})x_{m+1}$$

The meaning of terms in the equation for pupae is analogous to those for the larvae. The third term is lacking because pupae do not feed and therefore do not contract infection.

Healthy adults

$$dx_{m+2}/dt = e_{m+1}x_{m+1} - (b_{m+2} + W_1)x_{m+2} + R_1$$

The first term in the equation for healthy adults indicate emergence of pupae. The second term includes attrition and emigration. The third term is immigration.

Healthy eggs

$$\begin{aligned} dx_{m+3}/dt = {} & \beta_1e_{m+2}x_{m+2} - (b_{m+3} + e_{m+3})x_{m+3} \\ & + \beta_1'(1-h)e_{m+2}'y_{m+2} \end{aligned}$$

The first term in the equation for eggs is the production of female eggs. The second term includes attrition and development into L1. The third term includes the production of uninfected eggs by infected female adults.

Infected L1s

$$dy_1/dt = e_{m+3}'y_{m+3} - (b_1' + e_1')y_1 - b_1''y_1 + d_1x_1$$

The first term in the differential equation for infected L1s indicates the hatching of infected eggs into infected L1s. The second term includes attrition and development into infected L2. The third term is the death rate of infected L1s due to virus infection. The final term is the recruitment of new infected L1s, due to horizontal or spray-induced infection.

Infected L2s–Lms

$$dy_i/dt = e'_{i-1}y_{i-1} - (b'_i + e'_i)y_i - b''_i y_i + d_i x_i; \quad i = 2, \ldots, m$$

The terms in the equation for infected L2–Lm are analogous to those in the equation for infected L1.

Infected pupae

$$dy_{m+1}/dt = e'_m y_m - (b'_{m+1} + e'_{m+1})y_{m+1} - b''_{m+1}y_{m+1}$$

The equation for infected pupae is analogous to those for larvae, without the last term, which quantifies horizontal and spray-induced infection in larvae.

Infected adults

$$dy_{m+2}/dt = e'_{m+1}y_{m+1} - (b'_{m+2} + W'_1)y_{m+2} - b''_{m+2}y_{m+2} + R'_1$$

The equation for infected adults has four terms. The first indicates emergence of the infected pupae. The second term includes attrition and emigration of infected adults. The third term is virus-induced death. The fourth term is immigration.

Infected eggs

$$dy_{m+3}/dt = h\beta'_1 e'_{m+2}y_{m+2} - (b'_{m+3} + e'_{m+3})y_{m+3}$$

The equation for infected eggs has two terms. The first term indicates laying of infected eggs (vertical transmission) by infected females, taking into account the probability of passing the virus on to the egg (h) and the sex ratio (β'_1). The second term covers attrition and development into L1.

The software package was developed by using JDK (Java), Javascript, and HTML. It comprises of computation class, graphics class, result showing class, parameter inputting class, warning showing class, homepage, etc.

Simulation results showed that epidemic system behaved as a periodic dynamic system in some combinations of parameters. In general, disease incidence of population tends to be steady after a longer period. Therefore, it is a conditionally self-organizing system.

Chapter 8

Artificial Neural Networks

Artificial neural networks learn from the environment in an unsupervised (for example, selforganizing neural networks) or supervised way (training samples; for example, feedback neural networks). Principles, rules and algorithms of artificial neural networks, in particular the self-organizing and feedback neural networks, are of importance to self-organization research. This chapter describes self-organizing and feedback neural networks, learning rules, architecture determination, and mechanism interpretability of neural networks. More details can be found in Zhang (2010), Yang and Zhang (2000), and many others.

8.1 Self-organizing Neural Networks

Self-organizing neural networks can be used to explore mechanisms and relationships hidden in redundant data, and adjust themselves to adapt to the environment (Fecit, 2003). The self-organizing neural network may be a feedforward multilayer network in which self-organizing process is conducted layer by layer (Yang and Zhang, 2000; Zhang, 2010). A self-organizing neural network composes of two layers, input layer and output layer. In the output layer, between-layer forward connections and lateral connections can be found (Zhang and Li, 2007). The changes of connection weights only depend upon the states of neighborhood neurons (Yang and Zhang, 2000). They are called local interactions. Stochastic local interactions may result in the global order. Local interactions in a self-organizing process follows three principles (Von der Malsburg, 1990): (1) connection strength tends to be powered by itself (i.e., Hebb rule); (2) all neurons compete with each other, and the strongest neuron is activated and other neurons are inhibited (i.e., winner take-all); and (3) coordination exists among neurons because the single neuron could not function by itself.

Self-organizing neural networks have different architectures (Kohonen, 1982, 1988, 1990, 1995). Here I describe several kinds of self-organizing neural networks (Zhang, 2010): self-organizing feature map network, and self-organizing competitive learning network, etc.

8.1.1 *Self-organizing feature map neural network*

Self-organizing feature map (SOM) neural network is a map from higher dimensional space to two-dimension (a two-dimension curvature in n-dimension space, i.e., principal curvature) or one-dimension space (a curve in n-dimension space, i.e., principal curve) (Zhang, 2010). SOM is somewhat the generalization of PCA and thus functions as the feature extraction. However the map is not unique — it is dependent upon the squence of samples provided and initial weights (Bian and Zhang, 2000). In a self-organizing feature map neural network, the neurons of a layer learn from the environment to represent different regions in the sample space and the neighborhood neurons learn to respond to similar samples (Kohonen, 1982, 1995). SOM can learn both the topology of sample space and the distribution of samples (Song *et al.*, 2007; Zhang, 2007a, b; Zhang and Li, 2007). Moreover, it can be used to recognize additional samples.

The principle of SOM is as the following (Kohonen, 1982, 1988, 1990, 1995; Zhang, 2010). Know the input $x = (x_1, x_2, \ldots, x_n)^T$, and the weight vector of the neuron $j, w_j = (w_{j1}, x_{j2}, \ldots, x_{jn})^T$, $j = 1, 2, \ldots, N$. Firstly, input a sample x, and find the neuron i (i.e., winner neuron that matchs x optimally by finding the maximum $w_j^T x$). It is equivalent to solve

$$i(x) = \arg\min_j ||x - w_j||, \quad j = 1, 2, \ldots, N$$

if all $w_j, j = 1, 2, \ldots, N$, are normalized to fixed Euclidean norms. Secondly, determine the neighborhood $\delta_i(k)$ of the winner neuron i, which may change as the increase of iterations. Finally, determine the rule of changes of connection weights. The algorithm of SOM is summarized as the following (Yan and Zhang, 2000; Zhang, 2010):

(1) Assign small random values to initialize weights $w_j(0), j = 1, 2, \ldots, N$.
(2) Choose a sample x randomly from the sample set, and use x as the input.
(3) In step k, find the neuron i that matches x optimally, i.e.,

$$i(x) = \arg\max_j w_j^T x, \quad j = 1, 2, \ldots, N$$

or

$$i(x) = \arg\min_j ||x - w_j||, \quad j = 1, 2, \ldots, N$$

(4) Find the neighborhood $\delta_i(k)$ of the winner neuron i.
(5) Modify connection weights following the rule

$$w_j(k+1) = w_j(k) + \eta(k)[x(k) - w_j(k)], \quad j \in \delta_i(k)$$
$$w_j(k+1) = w_j(k), \quad j \notin \delta_i(k)$$

(6) Stop training when a desirable network is achieved, otherwise $k = k + 1$, and go to (2).

Finding the optimal neighborhood $\delta_i(k)$ and learning rate $\eta(k)$ are the key to train a desirable network (Kohonen, 1988, 1990). Generally, in the first 1000 iterations $\eta(k)$ should be around 1.0 and decrease (but not less than 0.1) thereafter. For two-dimension SOM, the neighborhood $\delta_i(k)$ may be choosn as a square, or a hexagon, etc. In earlier iterations, $\delta_i(k)$ is large enough even to contain all neurons and then shrinks step by step, until it contains 1 or 2 adjacent neurons. In the convergence stage, $\delta_i(k)$ contains the nearest neuron or even itself only (Zhang, 2010).

SOM algorithm has been revised and thus produced a lot of variants. For example, Amari (1983) described a theory of self-organizing neural fields and addressed the problem of continuous SOM. In addition, SOM was treated as a nonparametric regression, or constrained topological map (Cherkassky and Lari-Najafi, 1991).

In Matlab, the function for SOM is newsom (Mathworks, 2002).

8.1.2 *Self-organizing competitive learning neural network*

In a self-organizing competitive learning neural network, the neurons in a competitive layer learn from samples to represent different regions in the sample space. The network can thus learn the distribution of samples (Zhang, 2010). The competitive network may be used to classify samples into natural classes (Mathworks, 2002). In the network, the competitive layer is a classifier and each neuron in this layer represents a different category. Moreover, additional samples can be recognized using the trained network (Zhang, 2007a, 2007b, 2010; Watts and Worner, 2009).

The Matlab function of the self-organizing competitive learning neural network is newc (Mathworks, 2002).

8.1.2.1 *Hamming network*

Hamming network is a simple competitive neural network. It can be used to find the class with the smallest Hamming distance to existing vectors. A Hamming network composes of two layers, the feedforward layer and the recursive layer (output layer). The feedforward layer links input vector with prototype vector, and the recursive layer uses a competitive algorithm, through lateral inhibition among neurons in the layer, to find the prototype vector that is the nearest to input vector (Hagan *et al.*, 1996; Zhang, 2010).

The algorithm for Hamming network is as the following (Yan and Zhang, 2000; Zhang, 2010). Given m vectors, s^j, $j = 1, 2, \ldots, m$, where $s^j = (s_1^j, s_2^j, \ldots, s_n^j)^T$, and $w_j^1 = (w_{j1}^1, w_{j2}^1, \ldots, w_{jn}^1)^T$, $j = 1, 2, \ldots, m$; Hamming distance between input $x, x = (x_1, x_2, \ldots, x_n)^T$, and s^j is

$$d_h(x, s^j) = (n - x^T s^j)/2$$

Let $w^1 = (s^1, s^2, \ldots, s^m)^T/2$, the threshold be $n/2$, the activation function of neurons is

$$f(u_j) = u_j/n, \quad j = 1, 2, \ldots, m$$

where $u_j = n - d_h(x, s^j)$. For weight matrix $w^2 = (w^2_{ij}), w^2_{ij} = -\varepsilon$, if $i \neq j$, and $w^2_{ij} = 1$, if $i = j$. The output is

$$y(k+1) = \Lambda(w^2 y(k))$$

where Λ is a diagonal operator and its element is

$$f(u) = u, \quad \text{if } u \geq 0$$
$$f(u) = 0, \quad \text{if } u < 0$$

8.1.2.2 *WTA network*

Winner take-all (WTA) network is a simple competitive neural network (Yang and Zhang, 2000; Zhang, 2010). In this algorithm, given the weight vectors $w_j = (w_{j1}, w_{j2}, \ldots, w_{jn})^T, j = 1, 2, \ldots, m$, then

(1) Normalize connection weights

$$\hat{w}_j = w_j/||w_j||, \quad j = 1, 2, \ldots, m$$

(2) Find a w_l, such that

$$\max_j \ \hat{w}_j^T x$$

(3) Modify connection weights

$$\hat{w}_l(k+1) = \hat{w}_l(k) + \eta(k)[x - \hat{w}_l(k)]$$
$$\hat{w}_j(k+1) = \hat{w}_j(k), \quad j \neq l$$

(4) Stop training if the desirable performance is reached, otherwise repeat the training process.

8.1.2.3 *LVQ neural network*

Learning vector quantization (LVQ) network is a competitive neural network. It is a mixed network that produces classification by both supervised and unsupervised learning (Kohonen, 1990; Zhang, 2010).

LVQ network composes of two layers, the competitive layer and the linear layer. Several neurons are assigned to the same class in the competitive layer, and each class is assigned to a neuron in the linear layer. The competitive layer learns classification, i.e., subclass, to input vectors and the linear layer transforms the classification produced in the competitive layer to the categories (i.e., targeted class) defined by the user (Zhang, 2010). The number of neurons in the competitive layer is usually larger than that in the linear layer.

The algorithm of LVQ network is (Yan and Zhang, 2000; Zhang, 2010):

(1) Find a neuron i with the largest output in output set when a vector x is delivered to network.
(2) Suppose that x belongs to class p, and i belongs to q in the last learning, then conduct the rule

$$
\begin{aligned}
&w_i(k+1) = w_i(k) + \eta(k)[x(k) - w_i(k)], \quad p = q \\
&w_i(k+1) = w_i(k) - \eta(k)[x(k) - w_i(k)], \quad p \neq q \\
&w_j(k+1) = w_j(k), j \neq i
\end{aligned}
$$

Some revised algorithms of LVQ are developed, for example, the GLVQ network (Pal *et al.*, 1993), etc.

The Matlab function of LVQ neural network is newlvq (Mathworks, 2002).

8.1.2.4 *Adaptive resonance theory*

In competitive networks, weight matrix may not converge in many cases. How to achieve both the stability of learning process and the higher adaptability is thus important. Adaptive resonance theory (ART) is an improvement and extension of competitive learning. It was developed to solve stability/adaptability dilemma (Carpenter and Grossberg, 1987, 1990; Carpenter *et al.*, 1991; Zhang, 2010). ART is self-stable and makes self-organizing recognition on complex environmental patterns. There are three kinds of ART, i.e., ART1, ART2, and ART3, which are used in the situations of Boolean inputs, continuous inputs, and hierachical search, respectively. ART may adapt to nonstationary environments and can learn from the environment in real time. It has a stable recognition performance on learned objects and can fastly adapt to new objects never learned (Zhang, 2010).

In an ART network, there are two layers, the input layer and the output layer (Hagan *et al.*, 1996). The input layer compares input patterns to desired patterns returned from the output layer. If these patterns cannot match with each other then reset the output layer and the winner neuron and desired pattern will be cancelled, and a new round of competition is performed in the output layer. The new winner neuron of the output layer returns the input layer a desired pattern. This process continues until the input and desired patterns match with each other. If they can match mutually, then the input layer produces a new prototype pattern by combining desired and input patterns (Yang and Zhang, 2000; Zhang, 2010).

Suppose there are n neurons in the input (Boolean input) layer and m neurons in the output layer; $\dot{w}_{ij}$ is the connection weights from the input layer to the output layer, and $\ddot{w}_{ji}$ is the connection weights from the output layer to the input layer, $i = 1, 2, \ldots, n; j = 1, 2, \ldots, m$. ART1 algorithm is (Carpenter and Grossberg, 1987; Yan and Zhang, 2000; Zhang, 2010):

(1) Initialize all connection weights. Assign values $\dot{w}_{ij} = a_j, i = 1, 2, \ldots, n; j = 1, 2, \ldots, m$, where $a_m < \ldots < a_1 < 1/(\alpha + n), 0 < \alpha << 1$. $\dot{w}_{ji} = 1, j = 1, 2, \ldots, m; i = 1, 2, \ldots, n$.
(2) Input a Boolean sample $x, x = (x_1, x_2, \ldots, x_n)^T$.
(3) Calculate the weighted value for every output neuron

$$y_j = \sum_{i=1}^{n} \dot{w}_{ij} x_i, \quad j = 1, 2, \ldots, m$$

(4) Find the winner neuron K

$$K = \arg \max_j y_j$$

(5) Input and desired patterns did not match with each other, if

$$\sum_{i=1}^{n} \ddot{w}_{ji} x_i / \sum_{i=1}^{n} x_i < \beta$$

where β is the warning threshold, otherwise continue the following steps.

(6) For the winner neuron K, modify the connection weights

$$\dot{w}_{ij} = \dot{w}_{ij}, \quad j \neq K$$

$$\dot{w}_{ij} = \ddot{w}_{ij}/(\alpha + \sum_{ji}^{n} \ddot{w}_{ij}x_i), \quad j = K$$

$$\ddot{w}_{ij} = \ddot{w}_{ij}, \quad j \neq K$$

$$\ddot{w}_{ij} = \ddot{w}_{ij}x_i, \quad j = K$$

(7) Return step (2), and input a new sample.

8.2 Feedback Neural Networks

In a feedback neural network, all neurons are the same and can connect to each other (Yan and Zhang, 2000; Zhang, 2010). A feedback neural network can be produced from a two-layer feedforward network that the numbers of neurons in the input layer and the output layer are the same, if each output is directly connected to the corresponding input (Bian and Zhang, 2000). A feedback neural network may converge to a stable state. The associative memory is achieved through transition of states of neurons (Zhang, 2010). Feedback neural networks are nonlinear dynamic systems due to their feedback mechanism (Fecit, 2003). Their nonlinear behaviors, such as stability problem, attractors, chaos, unpredictability, randomness, etc., are interesting properties for self-organizing research.

8.2.1 *Elman network*

An Elman neural network contains several layers. The first layer has connection weights coming from the input, and each subsequent layer has connection weights coming from the previous layer. All layers, except for the last one, have recursive weights. The last layer is the network output (Mathworks, 2002; Zhang, 2010). Elman network is a nonlinear system. It is used to approximate any function with desired accuracy in a definite time (Zhang, 2007a, b). It is able to learn not only spatial patterns but also temporal patterns (Hagan *et al.*, 1996; Fecit, 2003). Matlab function for Elman neural network is newelm (Mathworks, 2002).

8.2.2 *Hopfield neural network*

Hopfield neural network shares the common properties of feedback or recursive networks (Hopfield, 1982, 1984; Fecit, 2003). Moreover, it has these characterisitcs (Bian and Zhang, 2000; Zhang, 2010): (1) connection weights are symmetrical ($w_{ji} = w_{ji}$) and the weight matrix is thus a symmetrical matrix; (2) no self-feedbacks of neurons ($w_{ii} = 0$) exist. Hopfield neural network is stable only isolated attractors can be achieved due to the symmetrical property.

Hopfield neural network is used to store one or more stable target vectors. These vectors may be evoked at a certain time. Given an initial condition, the network will converge to these vectors (Zhang, 2010). It is also used for optimization, linear programming, A/D transformation, and pattern recognition (Hopfield and Tank, 1985; Hagan *et al.*, 1996; Zhang, 2010).

8.2.2.1 *Discrete Hopfield neural network*

Suppose that the discrete Hopfield neural network is a n-order network, in which $W_{n\times n}$ is a symmetrical weight matrix, T is a n-dimension vector, and T_i is the threshold of neuron i. Neurons may only take the value 1 or -1. The discrete Hopfield network is represented by a group of nonlinear difference equations (Yang and Zhang, 2000; Zhang, 2010)

$$x_i(t+1) = \mathrm{sgn}\left[\sum_{j=1}^{n} w_{ij}x_j(t) - T_i\right]$$

$$i = 1, 2, \ldots, n$$

where $x_j(t)$: the state of neuron j at time t, and t: a positive integer. The network's state is $x(t) = (x_1(t), x_2(t), \ldots, x_n(t))^T, x_i(t) \in \{1, -1]$.

The discrete Hopfield network functions with two alternative rules (Bian and Zhang, 2000; Zhang, 2010): (1) asynchronous rule, only one neuron changes its state at a time, and (2) synchronous rule, some neurons change their states at a time.

If there is a time t, such that $x(t + \Delta t) = x(t)$, then the state equation is stable. A Lyapunov function (energy function) is defined to determine the stability of the network:

$$E(t)=\sum_{i=1}^{n}T_i x_i(t)-\sum_{i=1}^{n}\sum_{j=1}^{n}w_{ij}x_i(t)x_j(t)/2$$

Following the asynchronous rule, the change of energy function as x_i changes its state is

$$\Delta E(t)=-[x_i(t+1)-x_i(t)]\left[\sum_{j\neq i}w_{ij}x_j(t)-T_i\right]$$

Energy function declines when the state of network changes, and finally reaches a minimum (i.e, the equilibrium state). The nonlinear system owns multiple isolated equilibrium states, i.e., isolated attractors (Zhang, 2010). Following the synchronous rule, the attractors will be isolated attractors or oscillation cycles with the length 2.

8.2.2.2 *Continuous Hopfield neural network*

The continuous Hopfield neural network is similar to the discrete one (Hopfield, 1984). The changes of neurons' states are represented by a group of nonlinear differential equations (Yang and Zhang, 2000; Zhang, 2010):

$$\begin{aligned}C_i\,\mathrm{d}x_i/\mathrm{d}t&=\sum_{j=1}^{n}w_{ij}y_j-x_i/R_i+I_i\\ y_i&=\mathrm{g}(x_i)\\ i&=1,2,\ldots,n\end{aligned}$$

And its energy function is

$$E(t)=-\sum_{i=1}^{n}I_i y_i(t)-\sum_{i=1}^{n}\sum_{j=1}^{n}w_{ij}y_i(t)y_j(t)/2+\sum_{i=1}^{n}\int_0^{yi}\mathrm{g}^{-1}(y)\mathrm{d}y/R_i$$

If $\mathrm{g}^{-1}(\cdot)$ is a monotonous and continuous function, and $C_i>0, w_{ji}=w_{ij}$, then $\mathrm{d}E/\mathrm{d}t\leq 0$.

8.2.3 *Boltzmann machine*

Boltzmann machine is a full connected feedback neural network (Hinton, 1984; Fecit, 2003). It is similar to Hopfield network. Boltzmann machine converges to the global minimum of energy function and the network distribution will match the probability distribution of the environment. However, it is a time-cost network. In a Boltzmann machine, neurons have only two states, 1 and -1, and only one neurons changes its state at a time. All neurons are stochastic ones. Hidden neurons can be embedded in the network. There are no self-feedback connections, and all connections are bidirectional and symmetrical (Yan and Zhang, 2000; Zhang, 2010).

In Boltzmann machine, all visible neurons interact with the environment. Visible neurons are fixed at certain states, and hidden neurons change their states in response to the inputs received. Define the energy function as (Yan and Zhang, 2000; Zhang, 2010):

$$E = -\sum_i \sum_{j \neq 1} w_{ji} s_j s_i$$

The probability of state change of neuron j, from s_j to $-s_j$, is

$$P(s_j \to -s_j) = 1/[1 + \exp(-\Delta E_j/T)]$$

where the energy change ΔE_j, resulting from state change is

$$\Delta E_j = -2s_j v_j = -2s_j \sum_i w_{ji} s_i$$

The probability of state change of neuron j from $-s_j$ to s_j is $1 - P(s_j \to -s_j)$. Boltzmann machine learning aims to minimize E. Suppose the visible neurons are either input neurons or output neurons. The algorithm of Boltzmann machine is (Yan and Zhang, 2000; Zhang, 2010):

(1) Initialize connection weights, w_{ji}, by being assigned uniformly distributed random values in $[-a, a]$, where a can be set a value like 1.
(2) Fix the states of visible neurons with desired values of data pattern. Relax the network using SA, i.e., change the states of neurons following the rule:

$$s_j = 1, \text{ with the probability } P(v_j)$$
$$s_j = -1, \text{ with the probability } 1 - P(v_j)$$

where

$$v_j = \sum_{i \neq j} w_{ji} s_i$$
$$P(v_j) = 1/[1 + \exp(-2v_j/T)]$$

Perform certain iterations for every temperature T, until a given lower temperature is achieved. Finally, calculate

$$p_{ji}^+ = < s_j s_i >^+ = \sum_x \sum_y Q_{xy}^+ s_{j|xy} s_{i|xy}$$
$$i, j = 1, 2, \ldots, n; \quad i \neq j$$

where x, y: the states of visible neuron and hidden neuron respectively, $s_{j|xy}$: the state of neuron j when the state of visible neuron is x and the state of hidden neuron is y, n: total number of neurons, and

$$Q_{xy}^+ = Q_{y|x}^- Q_x^-$$

where Q_x^+: the probability of a visible neuron at state x when network's states of visible neurons are fixed, Q_x^-: the probability of a visible neuron at state x when network's states of visible neurons are free. State x may be assigned values from 1 to 2^c, and state y takes values from 1 to 2^d, where c and d are the number of visible and hidden neurons, respectively.

(3) Fix the states of input neurons only, and repeat (2) to compute

$$p_{ji}^- = < s_j s_i <^- = \sum_x \sum_y Q_{xy}^- s_{j|xy} s_{i|xy}$$
$$i, j = 1, 2, \ldots, n; \quad i \neq j$$

(4) Modify connection weights by the rule

$$w_{ji} = w_{ji} + \eta(p_{ji}^+ - p_{ji}^-)$$
$$i, j = 1, 2, \ldots, n; \quad i \neq j$$

(5) Repeat steps (2) to (4) until w_{ji} do not change, i.e., the network converges to its stable state.

A mean field approximation method has been developed to improve the performance of Boltzmann machine (Perterson and Anderson, 1987; Haykin, 1994). In this method, the stochastic neuron is replaced with a deterministic neuron. Mean field theory can enhance the training efficiency of Boltzmann machine by 1 to 2 orders of magnitude (Zhang, 2010).

8.3 Learning Theory, Architecture Determination and Mechanism Interpretability

The learning theory, architecture determination, and mechanism interpretability of neural networks are worthy to the research of self-organizing algorithms.

8.3.1 *Learning theory*

In artificial neural networks, learning theory works to address such problems as (Yan and Zhang, 2000): are the learning results able to approximate hidden mechanism as the increase of training samples (i.e., learnability)? How many samples and how long are computational time needed to learn the mechanism (i.e., complexity of learning)? How about the convergence of learning algorithm?

8.3.1.1 *Learnability*

In a supervised learning, a neural network is trained to make output y approximate desired output $\hat{y}$ by finding a specific $F(x, w)$, i.e., adjusting weights, from a given set of functions (a given architecture of neural network), i.e., $y = F(x, w)$, where $x \in R^n, y \in R^m$, and w is the weight matrix (Zhang, 2010). We are interested in analyzing if a given set of samples load enough information such that the trained neural network possesses the ideal generalization capacity. The theory on minimization of empirical risk has been proposed to address the problem (Vapnik, 1995, 1999), as described here. Given the empirical risk

$$R(w) = \int (\hat{y} - F(x, w))^2 \mathrm{d}P(x, \hat{y})$$

where $P(x, \hat{y})$ is the joint distribution of x and $\hat{y}$. Assume w_e is the weight matrix minimizing the empirical risk $R_e(w)$, and if $R_e(w)$ uniformly converges to the pratical risk $R(w)$, i.e.

$$p\{\sup_{w \in W} |R(w) - R_e(w)| > \varepsilon\} \to 0, \quad N \to \infty$$

Probabilistically, $R(w_e)$ converges to the possible minimum of $R(w)$. It is obvious that the minimization of empirical risk is possible. Nevertheless, the convergence speed should be noted because the number of training samples is limited (Yan and Zhang, 2000). The convergence speed is directly related to the *VC* dimension of the problem. *VC* dimension of a function class F is defined as the cardinality of S, the set of N points in R^n, that can be finely divided by F (Zhang, 2010). If *VC* dimension is limited, the problem will be learnable and the empirical risk converges to the minimum. For feedforward neural networks, their *VC* dimension (i.e., the capacity of multilayer neural networks) represents classification capacity. For the feedforward network with only a hidden layer, the *VC* dimension, d, is (Sontag, 1998; Zhang, 2010)

$$2n[h/2] \leq d \leq 2|W| \log_2(eN)$$

where h: the number of hidden neurons, $|W|$: the number of adjustable connection weights, N: the number of neurons, [ ]: the maximum integer, n: the input dimension, and $e = 2.7183$.

VC dimension should be reduced as lower as possible. Some methods can be used to reduce *VC* dimension: (1) constrain the number of hidden neurons; (2) constrain the number of connection weights, and (3) reduce the dimension of input space.

Amari *et al.* (1997) presented a simple criterium for network learning, i.e., the model is optimally trained when the ratio of the number of training samples to the number of the connection weights exceeds 30.

The number of hidden neurons (N) is sometimes determined by using the following rule (Nagendra and Khare, 2006), i.e., "rule of thumb": (1) N = number of input neurons + number of output neurons; (2) The maximum number of neurons in the hidden layer (Nmax) is two times the number of input layer neurons (Swingler,

1996; Berry and Linoff, 1997); and (3) $N =$ the number of the training patterns divided by five times of the number of input and output neurons (Zhang, 2010).

8.3.1.2 *Complexity of learning*

PAC learning is a better choice for neural network learning (Zhang, 2010). PAC learning only needs the approximate learning in the sense of probability at given learning error (Valiant, 1984; Anthony and Biggs, 1992; Yan and Zhang, 2000). Given a probability distribution p and the real values $0 < \varepsilon, \delta \leq 1$, if there is an algorithm A, by using the samples of C, A is able to output the hypothesis h over the polynomial time of $1/\varepsilon, 1/\delta$, and N (scale of the problem to be studied, or the complexity of $c \in C$), such that

$$P(\Delta(h,c) < \varepsilon) \geq 1 - \delta, \quad \forall c \in C$$

Then C is PAC learnable (Zhang, 2010).

Suppose the *VC* dimension of a neural network is d, and $d \to \infty$ if $n \to \infty$, where n is the input dimension. When $n \to \infty$, for any $\varepsilon > 0$, the algorithm with the number of samples less than $(d-1)/(32\varepsilon)$ is not PAC learnable (Venkatech, 1992). For a given neural network with Boolean output, the relationship between sample size (N) and error (ε) should meet the following rule (Yan and Zhang, 2000):

$$N \geq d/\varepsilon$$

where $|W|$ is the number of adjustable connection weights.

8.3.1.3 *Dynamic learning*

The neural network learning is a dynamic process. Thus, a learning rule with continuous time can be treated as a differential equation (Amari, 1990; Yan and Zhang, 2000; Zhang, 2010).

Suppose a single neuron has the input $x \in R^n$ and the weight vector is $w \in R^n$. The desired output is $\hat{y}$. The input is interrupted by the environment and represented by $p(x, \hat{y})$. Amari (1990) developed a unified learning equation as the following:

$$\tau dw(t)/dt = -w(t) + \eta(t)r(w(t), x(t), \hat{y})x(t)$$

which is equivalent to

$$\Delta w = \eta r(w, x, \hat{y})x - \lambda w \quad \text{discrete equation}$$
$$dw/dt = \eta r(w, x, \hat{y})x - \lambda w \quad \text{continuous equation}$$

where $\lambda, \eta > 0$, and $r(w, x, \hat{y})$ is the learning signal. Let $\lambda = 0$ and $r(w, x, \hat{y}) = \hat{y} - y = \hat{y} - \text{sgn}(w^{\text{T}}x)$ in the above discrete equation, then it is the perceptron learning rule; if $\lambda = 0$ and $r(w, x, \hat{y}) = \hat{y} - y = \hat{y} - w^{\text{T}}x$, then it is the Widrow-Hoff learning rule, and if $\lambda = 0$ and $r(w, x, \hat{y}) = y$, Hebb learning rule is achieved.

Various new learning rules may be developed based on the unified learning equation. Dynamic properties of learning rules can be analyzed according to the equation (Zhang, 2010).

8.3.1.4 *Generalization capacity*

The generalization capacity represents a neural network's ability to respond unknown samples, with the same mechanism as that hidden in training samples. Using the neural network to interpolate, predict, recognize, etc., requires a high generalization capacity. An over-fitted network over-fits the details (chaos) of training samples and ignores the general trends or mechanisms (Zhang, 2010). Such network thus has a little generalization capacity.

Some measures of generalization capacity are described in the following (Zhang, 2010).

(1) The general rule to describe the relationship between generalization error and training error is (Lee and Tenorio, 1993)

$$\text{GMEE} \le \varepsilon + \lambda((d/N)\ln(N/d))^{1/2}$$

where GREE: the generalization error (prediction error, or generalized minimum empirical error), ε: the training error, d: *VC* dimension, and N: the number of the training samples.

(2) The extension of *VC* dimension, i.e., regular dimension, can be used to describe generalization-training relationship (Gu and Takahashi, 1996)

$$R = \varepsilon + \sigma(d_{Re}/N))^{1/2}$$

where R: the generalization (prediction) risk, ε: the training error, d_{Re}: regular dimension, σ: the variance of loss function $R(w)$, and N: the number of the training samples.
Training error ε can be expressed as the ratio, misclassified samples *vs.* total samples.

(3) A network capacity based measure was developed to represent generalization capacity (Yan and Zhang, 2000)

$$G = \log_2 C/m$$

where C: the network capacity, i.e., the number of different mappings or functions to be realized by adjusting connection weights, m: the output dimension. A small G represents a large generation capacity.

8.3.2 *Architecture determination*

A better network architecture may maximize the generalization capacity. Compared with complex networks, a simple network needs a longer training time and has larger training error (Zhang, 2010). However, it is easier to be understood, to extract knowledge and rules, and to be realized by hardware. In addition, it has a larger generalization capacity. Therefore, the general principle is to choose the simplest architecture under the same conditions.

A method to determine architecture is to follow the rule:

$$\begin{aligned}&\text{Criterion of architecture determination}\\&\quad= \text{logarithmic likelihood function}\\&\qquad+ \lambda \times \text{ architecture complexity}\end{aligned}$$

where the first term denotes the fitting goodness of model output to input samples, the second term is to penalize the model for its complexity, and λ is a factor to adjust the constraint strength (Zhang, 2010).

Removing some trivial connection weights or neurons from a redundant network, i.e., conducting network pruning (Reed, 1993; Castellano *et al.*, 1997; Setiono, 1997; Yan and Zhang, 2000; Zhang, 2010), is also a useful method. For example, in the weight decay method, a penalty function is added in the goal function of BP algorithm, and the connection weights that do not inflict learning process can thus decline exponentially (Zhang, 2010).

The iterative pruning algorithm, proposed by Castellano *et al.* (1997), can be used to prune not only hidden neurons but also connections in feedforward neural networks. In this algorithm, network pruning and weights modifying are conducted simultaneously to have the same network outputs. The procedures of the algorithm is (Zhang, 2010):

(1) Let $k = 0$.
(2) Remove the neuron h that exerts the smallest influence on network N^k, by following the rule:

$$h = \arg \min_{i \in H} \sum_{i \in lh} w_{hi}^2 ||y_h||_2^2$$

where w_{hi}: the connection weight from the neuron h to i, H: the set of hidden neurons, I_h: the sending domain of the neuron $h, I_h = \{j \in V|(i,j) \in L\}, V$: the set of all neurons, L: the set of all connections, and y_h: the signal of the neuron h to I_h.
(3) Solve the following equation to obtain δ using conjugate gradient algorithm:

$$Y_k \delta = Z_k$$

(4) Generate a new network: $N^{k+1} = \{V^{k+1}, L^{k+1}, w^{k+1}\}$, where

$$\begin{aligned} V^{k+1} &= V^k - \{h\} \\ L^{k+1} &= L^k - \{h\} \times I_h^k \cup R_h^k \times \{h\} \\ w^{k+1} &= w_{ji}^k, \quad i \notin I_h \\ w^{k+1} &= w_{ji}^k + \delta_{ji} \quad i \in I_h \end{aligned}$$

where R_h: the receiving domain of the neuron $h, R_h = \{j \in V|(j,i) \in L\}, \delta_{ji}$: the value for modifying connection weights.
(5) Let $k = k + 1$, repeat steps (2) to (4), until the performance of network N^k declines significantly.

In addition to the methods above, a neural network can be constructed step by step from a simple and small architecture (Zhang, 2010). There are some algorithms for network constructing, for example, tiling algorithm (Mezard and Nadal, 1989), sequential addition (Marchand *et al.*, 1990), upstart algorithm (Frean, 1990), etc.

Finally, the number of hidden layers, neurons, etc., can be adjusted for network optimization (Zhang, 2007a, b; Zhang *et al.*, 2007; Zhang *et al.*, 2008a, b; Zhang, 2010).

8.3.3 *Mechanism interpretability*

One of the most useful functions of artificial neural networks is their ability in exploring the mechanism hidden in the data (i.e., mechanism interpretability), for example, detecting the relative importance of input variables of a system (Kemp *et al.*, 2007; Zhang, 2010).

So far, a lot of methods to address the problem of mechanism interpretability, such as neural interpretation diagram (Ozesmi and Ozesmi, 1999), sensitivity analysis (Lek *et al.*, 1996; Scardi, 1996, Recknagel *et al.*, 1997; Gevrey *et al.*, 2006), inference rule extraction, randomization test of significance (Scardi and Harding, 1999; Olden, 2000; Olden and Jackson, 2002; Kemp *et al.*, 2007), partial derivatives (Dimpopoulos *et al.*, 1999; Reyjol *et al.*, 2001), connection weight methods (Olden *et al.*, 2004; Zhang and Wei, 2008), have been proposed (Özesmi *et al.*, 2006; Gevrey *et al.*, 2003).

8.3.3.1 *Sensitivity analysis*

In sensitivity analysis method, the network output corresponding to each input variable is determined by assigning designed values to the variable at a time while holding the other input variables constant (Lek *et al.*, 1996; Özesmi *et al.*, 2006). The variables that are held constant are assigned their minimum, first quartile, median (or mean), third quartile, and maximum values successively. The relative importance of the variable is determined by comparing input-output relationships. Moreover, white noise can be added to each input variable and the output error should be examined (Scardi and Harding, 1999; Zhang, 2010).

8.3.3.2 *Neural interpretation diagrams*

Neural interpretation diagrams are used to explore how a neural network is weighing every input variables and how the input variables interact with each other to produce the network output (Özesmi and Özesmi, 1999; Özesmi *et al.*, 2006). They are drawn by scaling the thickness of the connection lines, shoot from the input neurons, according to the relative values of their weights, in which positive and negative signals are denoted by different colors. By scaling the thickness of the connection lines, the most important variables can be found, and the interactions between input variables can also be observed. Neural interpretation diagrams are useful for neural networks with simple architecture (Zhang, 2010).

8.3.3.3 *Randomization test of significance*

A randomization test method was proposed to assess the statistical significance of connection weights and input variables of neural networks (Olden, 2000; Olden and Jackson, 2002). In this method, a neural network uses randomized data and all connection weights are recorded. This procedure is repeated for, e.g., 1000 or 10000 times, to produce a null distribution for each connection weight. Each connection weight is then compared to the actual value to calculate the significance level (Özesmi *et al.*, 2006; Zhang, 2010). The connections with little impact on network output are removed. The independent variables significantly contributing to network prediction are thus identified.

Holdback Input Randomization (HIPR) is a network-independent randomization test (Kemp *et al.*, 2007). HIPR is a refinement of the method proposed by Scardi and Harding (1999). The procedures of HIPR are summarized as follows (Zhang, 2010):

(1) Optimize the neural network.
(2) Use the test dataset to detect the relative importance of input variables: (a) sequentially input each data point in the test dataset to the neural network but replace the values of one input variable by uniformly distributed random values in the interval (0.1, 0.9), the range over which the network was originally trained; (b) calculate the mean squared error of the neural network when the randomized test set is presented, and (c) repeat

this procedure for each input variable, and each time replace the original values with uniformly distributed random values.

8.3.3.4 *Partial derivatives*

The partial derivatives of the network output with respect to the input variables are used to determine the relative importance of input variables (Dimopoulos *et al.* , 1999; Reyjol *et al.*, 2001). By plotting the partial derivatives of the network output with respect to an input variable, the changes of network output with increasing values of the input variable are determined (Özesmi *et al.*, 2006; Zhang, 2010).

8.3.3.5 *Connection weight methods*

The connection weights are equivalent to the importance of between-neuron connections and are thus used to detect the relative importance of input variables.

The input variable relevance is proposed as a connection weight method. In this method, the relevance of an input variable is the sum square of weights for that input variable divided by the total sum square of weights for all input variables. Variables with higher relevance are more important (Özesmi and Özesmi, 1999; Zhang, 2010).

A connection weight method, proposed by Olden *et al.* (2004), is said to outperform other methods in detecting the relative importance of input variables. In this method, the sum of the product of connection weights from input neuron to hidden neurons and the connection weights from hidden neurons to output neurons for all input variables is achieved (Kemp *et al.*, 2007). The larger the sum of connection weights, the greater the importance of the variable connected to this input neuron (Zhang, 2010). The relative importance of an input variable is determined by

$$V_i = \sum_{k=1}^{N} w_{ik} w_{ko}$$

where V_i: the relative importance of the input variable i, N: the total number of hidden neurons, w_{ik}: the connection weight from the input variable i to the hidden neuron k, w_{ko}: the connection weight from the hidden neuron k to the output neuron o.

A connection weight method, IDM (Importance Detection Method), was proposed for a specific neural network ANNSSM (the artificial neural network for state space modeling). In this method, the total importance of the input variable i is calculated by the following formula (Zhang and Wei, 2008; Zhang, 2010):

$$\begin{aligned} V_i &= \sum_{k=1}^{n} \sum_{j=1}^{n} \mathrm{LW}\{k+n, j\}/(1+\exp(-IW\{j,i\})) \\ &= \sum_{j=1}^{m} w_{ij}/(1+\exp(-w'_{ji})) \\ i &= 1, 2, \ldots, n \end{aligned}$$

where n: the number of state variables, w_{ij}: the between-hidden layer connection weight, and w'_{ji}: the input-hidden layer connection weight. The relative importance of the input variable i to the output variable k is obtained by

$$\begin{aligned} V_{ki} &= \sum_{j=1}^{n} \mathrm{LW}\{k+n, j\}/(1+\exp(-\mathrm{IW}\{j,i\})) \\ k, i &= 1, 2, \ldots, n \end{aligned}$$

The larger importance value represents the greater importance of the input variable. For ANNSSM, IDM is better than the previous connection weight methods (Olden *et al.*, 2004; Zhang, 2010).

Many methods to determine relative importance of input variables implicitly assume that input variables are all unit independent or have the same unit representation (Zhang, 2010). For example, the numbers of individuals of different biological taxa are identical (Zhang and Wei, 2008). Nevertheless, input variables usually have different physical units, e.g., temperature (°C) and relative humidity (%). Sometimes temperature is mathematically a more important factor than relative humidity according to the sensitivity analysis of neural networks, but relative humidity is actually more influential than temperature if variation in relative humidity is much larger than variation in temperature (Zhang, 2010). Therefore, the range of each input variable (or training set) must be carefully determined according to the practical variation of the input variable. Moreover, sequentially removing one input variable from all input variables to test model's performance is also an alternative.

8.4 How to Create Neural Networks in Matlab

Matlab is a powerful platform for mathematical and computational methods (Fecit, 2003; Mathworks, 2002, 2007). It allows users to create artificial neural networks with self-defined architectures. A new network can be created by performing the command "network" in the command window. The neural network to be created should be assigned the following properties, with initial parameters, structures, or functions, etc. (Zhang, 2010).

(1) Architecture properties: numInputs (0; the number of inputs); numLayers (0; the number of layers); biasConnect ([ ]; bias connections); inputConnect ([ ]; input-layer connections); layerConnect ([ ]; between-layer connections); outputConnect ([ ]; layer-output connections); targetConnect ([ ]; layer-target connections); numOutputs (0; read-only; the number of outputs); numTargets (0; read-only; the number of targets); numInputDelays (0; read-only; the number of input delays); numLayerDelays (0; read-only; the number of layer delays).
(2) Subobject structures: inputs (0×1 cell of inputs); layers (0×1 cell of layers); outputs (1×0 cell containing no outputs); targets (1×0 cell containing no targets); biases (0×1 cell containing no biases); inputWeights (0×0 cell containing no input weights); layerWeights (0×0 cell containing no layer weights).
(3) Functions: adaptFcn (none; adapt function); gradientFcn (none; gradient function); initFcn (none; initialization function); performFcn (none; performance function); trainFcn (none; training function).
(4) Parameters: adaptParam (none; parameters for adapting network); gradientParam (none; parameters for gradient function); initParam (none; parameters for initialization); performParam (none; parameters for network performance); trainParam (none; parameters for training network).

(5) Weight and bias values: IW (0×0 cell containing no input weight matrices; input-layer connection weights); LW (0×0 cell containing no layer weight matrices; between-layer connection weights); b (0×1 cell containing no bias vectors; bias).

A neural network can be created, for example, using the following settings (Fecit, 2003; Zhang, 2010):

```
%Create an empty neural network
net=network;
%Architecture properties
%Set the number of inputs and network layers
net.numInputs=2;     %Number of input variables
net.numLayers=3;     %Number of layers
%Set bia connections. Bias connection matrix is a 3 × 1 matrix. We may
    set layer i has bias
% connection, i.e., net.biasConnect(i)=1. Suppose layer 1 and layer 3
    have bias
%connections in this network
net.biasConnect=[1;0;1];
%Set input-layer connections. Input-layer connection matrix is a
%3 × 2 matrix. For example, input 1 to layer 1: net.inputConnect(1,1)=1;
%input 1 to layer 2: net.inputConnect(2,1)=1;
%input 2 to layer 2: net.inputConnect(2,2)=1;
net.inputConnect=[1 0;1 1;0 0];
%Set between-layer connections.Layer connection matrix is a 3 × 3 matrix.
    The syntax:
%net.layerConnect(i,j) means a weight connection of layer j to layer i.
%Here there are weight connections of layer 1 to layer 3 and layer 2
    to layer 3
net.layerConnect=[0 0 0;0 0 0;1 1 0];
%Set output and target connections. Both output and target matrices are
    1 × 3 matrix, which
%means there is one target and 3 layers are %connected to 1 target.
net.outputConnect=[0 1 1]; %Connect layers 2 and 3 to output
net.targetConnect=[0 0 1];
%Set the layer 3 to target connection. The output of layer 3 will be
```

```
   compared to the target of
%layer 3 and yields an error signal which is used in the training.
%Subobject structures
%Some layer properties may be changed. Here layer 3 has default
   %settings excepting its
%initialization function
net.inputs{1}.range=[-5 10;-5 10];  %Set range of input 1
net.inputs{2}.range=[-5 5;-5 5;-5 5;-5 5;-5 5];  %Set range of input 2
net.layers{1}.size=8;  %There are 5 neurons in layer 1
net.layers{1}.transferFcn='tansig';  %Set transfer function
net.layers{1}.initFcn='initnw';
%Nguyen-Widrow function is the initialize function
net.layers{2}.size=5;  %There are 4 neurons in layer 2
net.layers{2}.transferFcn='logsig';
net.layers{2}.initFcn='initnw';
net.layers{3}.initFcn='initnw';
%Here delays of input or layer weights are set.
net.inputWeights{2,1}.delays=[0 1];
net.inputWeights{2,2}.delays=1;
net.layerWeights{3,3}.delays=1;
%Set some functions
net.trainFcn='trainlm';     %Training function: Levenberg-Marquardt
net.performFcn='mse';       %Performance function: mse
net.initFcn=ínitlay';
%Network initialization function: initlay. This means the network will
   initialize itself using
%layer initialization function.
```

The following codes are used to initialize, simulate and train the network, and training goals are defined:

```
net=init(net);
x={[2;3] [-3;4]; [2;-1;0;3;1;4] [1;-2;0;4;1;3]};
y={1 -1};
z=sim(net,x);
net.trainParam.epochs=1000;
net.trainParam.goal=1e-06;
net=train(net,x,y);
```

In addition to neural network architectures, various functions used in neural networks, e.g., transfer functions, topological functions, distance functions, initialization functions, training functions, etc., can also be customized, designed, and loaded by users (Zhang and Li, 2008; Zhang, 2010).

8.5 Applications of Neural Networks

Some of the applications of neural networks, specifically self-organizing neural networks, are summarized as follows.

8.5.1 *Pattern classification of HLA-DRB1 alleles, human races and populations (Zhang and Qi, 2005)*

HLA is the most complex human genome. It features 128 genes and 96 pseudogenes (Jia, 2001), in which there are 293 HLA-DRB1 alleles. These alleles mainly determine the antigenic immuning of human. A comparative study of HLA-DRB1 alleles will help trace the origin of mankind, migration, integration of history, and develop group-specific bio-engineering drugs.

In the self-organizing competitive learning neural network, 10 neurons are used, and the training was repeated 2000 times (Zhang and Qi, 2005). Data of the world's 54 human races and populations and 14 common HLA-DRB1 alleles (Jia, 2001) were used for training. Some results are abstracted as follows.

14 neurons	12 neurons	10 neurons	8 neurons	6 neurons	4 neurons	2 neurons
0302	1401	0101	0101	1104	1104	1202
0901	0101	0302	0302	0101	0101	1501
1501	0302	1402	1401	0302	1602	0701
0701	1402	1401	0901	1401	0302	1104
1402	0901	0803	1402	0901	0301	0101
0803	0803	0701	0803	1402	1401	1502
1202	0701	0301	0701	1502	0901	1602
1401	1602	1501	0301	0803	1402	0803
1502	0301	0405	1602	1501	1202	0405
0101	1104	1602	0405	0701	1501	0302
0405	1202	1202	1202	0301	0701	0301
1104	1501	0901	1501	1202	1502	0901
0301	1502	1104	1104	1602	0803	1402
1602	0405	1502	1502	0405	0405	1401

8.5.2 *Self-organizing neural networks to support the discovery of DNA-binding motifs (Mahony et al., 2006)*

Mahony *et al.* (2006) presented three self-organizing neural networks that have applicability to the motif-finding problem. Their core system is a previously described SOM-based motif-finder named SOMBRERO. The motif-finder is integrated in this work with a SOM-based method that automatically constructs generalized models for structurally related motifs and initializes SOMBRERO with relevant biological knowledge. It was shown that such a method can act as an effective structural classifier of novel motifs.

8.5.3 *Classification of motor commands using self-organizing feature map neural network (Sebelius et al., 2005)*

Sebelius *et al.* (2005) proposed a control system for an advanced prosthesis. Their system was investigated in two different biological systems: (1) the spinal withdrawal reflex system of a rat, and (2) voluntary movements in two human males: one normal subject and one subject with a traumatic hand amputation. The small-animal system was used as a model system to test different processing methods for the prosthetic control system. It was found that a modified self-organizing feature map composed of a combination of a Kohonen network and the conscience mechanism algorithm was superior in performance to the reference networks with regards to training time, low memory consumption, and simplicity in finding optimal training parameters and architecture.

8.5.4 *Mapping and fuzzy classification of macromolecular images (Pascual et al., 2000)*

Pascual *et al.* (2000) used an algorithm combining Kohonen's self-organizing feature maps and Fuzzy c-means to obtain a powerful clustering technique with the best properties inherited from both. Two different datasets obtained from the G40P helicase from *B. Subtilis* bacteriophage SPP1 were used for testing the proposed method.

8.5.5 *Investigation of impoundment impact on fish assemblages in the river (Penczak et al., 2012)*

Penczak *et al.* (2012) used a self-organizing map neural network to identify patterns in fish assemblages. They applied a new, presently available, clustering technique to obtain the higher typology levels. Ideal results were obtained using the neural network method.

8.5.6 *Pattern classification and recognition of invertebrate functional groups using self-organizing neural networks (Zhang, 2007a, b)*

Zhang (2007) made a pattern classification and recognition on sampling information using two self-organizing neural networks. Invertebrate functional groups sampled in the irrigated rice field were classified and recognized using one-dimensional self-organizing map and self-organizing competitive learning neural networks. Differences between neural networks, distance (similarity) measures, and number of neurons were compared. The results indicated that self-organizing map and self-organizing competitive learning neural networks were effective in pattern classification and recognition of sampling information. Overall the performance of one-dimensional self-organizing map neural network was better than self-organizing competitive learning neural network. The number of neurons could determine the number of classes in the classification. Different neural networks with various distance (similarity) measures led to similar classifications. Some differences, dependent upon the specific network structure, can be found.

Chapter 9

Ant Colony Optimization

9.1 Proposition of Ant Colony Optimization

Ants always find the shortest path between the nest and food. Observing the foraging behavior of ants, people found that moving ants can continue to secrete a volatile substance, i.e., pheromone, which has two main functions: (1) ants communicate with each other through it — it can direct the ants to travel along the path of higer concentration of pheromone; and (2) volatilization effect of pheromone, which makes the impact of initially long-distanced paths and not-used paths to be gradually reduced. Colony foraging process of ants is a typical process of swarm intelligence. The collective behavior of ant colony composed by a large number of ants exhibits a positive feedback, i.e., the more ants walk on a path, the greater the probability of later ants choosing the path. Ants search for food through communication of pheromone between individuals (Huang, 2013), and finally travel along the shortest path. The basic nature of this method is: (1) selection mechanism — a path with more pheromone has the greater probability being selected; (2) updation mechanism — the pheromone along a path will increase with ants' travel but at the same time decrease gradually over time; and (3) coordination mechanism — ants communicate and coordinate with each other through pheromone.

Inspired by the communication mechanism of ant foraging, Dorigo (1991) proposed the algorithm of ant colony optimization [ACO; some researchers use the name Ant Colony Algorithm (ACA) also] to solve the classic "Traveling Salesman Problem," i.e., finding a shortest path to visit n cities but each city must and can only be accessed one time (Colorni *et al.*, 1991, 1992, 1994;. Dorigo, 1992; Dorigo *et al.*,

1996, 1999; Dorigo and Gambardella, 1997). It is one of the heuristic search algorithms, following simulated annealing algorithms, genetic algorithms, emergency search algorithms, artificial neural network algorithms, and other meta-heuristic search algorithms for combinatorial optimization problems (Sreeja and Sankar, 2013; Zhang, 2013a, 2014). In solving the traveling salesman problem, the ant colony optimization designs a virtual "ants machine." It will explore the different paths, and leave the virtual pheromone that gradually disappears over time. Virtual pheromone will volatilize. Each ant randomly chooses a path to go, and it tends to choose the path with much pheromone and shorter distance. The algorithm uses the positive feedback mechanism, with the shorter path having a greater chance to be selected, and it may not be limited to the local optima due to the use of probability principles.

ACO is advantageous for solving complex optimization problems (especially discrete optimization problems). It is especially suitable for multi-point nondeterministic searching in the solution space of discrete optimization problems. However, it also has some shortcomings. For example, a perfect theoretical basis on ACO has not yet been established; as a bionic evolutionary algorithm, it lacks a rigorous mathematical basis; the search time is longer, especially when pheromone is updated each step (if the current route is not globally optimal and pheromone is but of high concentration, the algorithm may not improve the problem by itself); it will fall into the local convergence and cannot find global optimal solution if the transition probability is too large although the algorithm will thus have a faster convergence rate. Therefore, the autocatalytic function inspired by positive feedback principle tends to strengthen the solution of good performance but may lead to premature convergence. Anyway, the application of ACO proves to be successful (Garbarino *et al.*, 2013; Oakley *et al.*, 2013). The ACO algorithm have been extended for changing from one-dimensional static optimization problems to multidimensional dynamic combinatorial optimization problems, and from discrete to continuous problems (Duan *et al.*, 2004, 2007).

9.2 Principles and Characteristics of ACO

Different from conventional algorithms, ant colony optimization achieves optimization by imitating the behavior of ants. The program itself finds the best configuration based on the random running of certain rules, and thus avoids the lengthy programming and planning. That is, when the program begins to achieve the target, the optimal path is almost impossible and numerous paths may have been searched. However, the program can continuously revise the original route based on ants' pheromone principle. The whole route thus becomes shorter and shorter. The longer the time of program execution, the more likely the route closes to the optimal route. Therefore, ACO appears to be a self-learning process.

ACO algorithm is divided into adaptation phase and collaboration phase. During the adaptation phase, each candidate solution continuously adjusts its structure based on pheromone accumulated; more pheromone left on the path, the greater the probability of being chosen, and pheromone will disappear over time. In the collaboration phase, the candidate solutions exchange pheromone to produce better solutions. For a specific combinatorial optimization problem, ACO algorithm determines decision point between exploration and exploitation using pheromone; and thereafter determines the behavior and orientation of whole ant colony. The optimal solution is thus achieved by repeating the procedure (Hao, 2009).

In addition to the foraging behavior of artificial ants, there are two other mechanisms for ACO — pheromone trail volatilization mechanism and daemon actions. Forgetting is an advanced intelligent behavior. As a form of forgetting, pheromone on paths volatilizes over time, which drives ants to explore new area in the solution space in order to avoid the earlier convergence to a local optimal solution. Daemon actions include the neighborhood (local) searching process and collection of global information. Combining such constructive method and improved iterative methods will perform better.

Foraging behavior of real ants follows such rules as foraging rule, moving rule, obstacle-avoiding rule, and pheromone rule. These rules allow an ant colony to find the shortest route to the food, and allow the ant colony to show diversity (i.e., some ants will certainly have a chance of mistake to forage toward a new and wrong path) and positive feedback (that is, a shorter path can gather more ants, leaving more pheromones to attract more ants) in solving this problem. The two characteristics hold in the ACO algorithm and ACO thus shows strong properties of self-organization, evolution, and positive feedback (Duan, 2005).

The ACO algorithm is based on the following assumptions:

(1) Ants communicate through pheromone and the environment. Each ant only reacts according to its environment, and only has an impact on its local environment.
(2) Reaction of an ant on the environment is only determined by its intrinsic mode. Ants are genetic organisms, and their behaviors are thus actually adaptation responses.
(3) At the individual level, each ant makes an independent selection only according to the environment; at the population level, the single ant behaves randomly, but the ant colony can generate highly ordered group behavior through self-organization process.

Artificial ants and real ants share same features:

(1) They are a group of cooperating individuals. Like real ant colony, ACO is composed of a group of artificial ants. They find the optimal solution of the problem through the synchronous/asynchronous collaboration between artificial ants. Although a single artificial ant can construct solution of the problem, the optimal (suboptimal) solution can be achieved only through the mutual cooperation between artificial ants. This collaboration is implemented by writing/reading state variables of the problem.
(2) Both of them use pheromone track and stigmergy mechanism pheromones. Like real ants, artificial ants collaborate indirectly by changing the digital status and pheromone of the problem (in ACO, it is the pheromone). Pheromone is the only way to communicate between artificial ants in ACO. This communication method plays a crucial role in the use of group's knowledge. In addition, ACO uses the volatilization mechanism, which corresponds to the volatilization of phenomenon in real ant

colony. Volatilization mechanism enables ant colony to gradually forget the past history, so the search is less affected by past poor solution in order to better guide the direction of the subsequent search of ants.

(3) Search the shortest path and move locally. Both artificial ants and real ants have the same task, that is, they construct the shortest route from the origin (the nest) to the destination point (food source) by moving locally.
(4) Random state transition strategy. Both artificial ants and real ants achieve the transition from one state to another neighboring state based on some rules of probability transition. Probability transition rules are the functions of the problem and the environmental information. Both of them only use local information (both spatial and temporal) in the state transition process. They have not used preview strategy to predict the future state.

However, there are some differences between artificial and real ants:

(1) Artificial ants live in discrete time; they grow from a discrete state to another discrete state.
(2) Artificial ants have their own internal states, i.e., they have some ability to remember their paths toured;
(3) Amount of pheromone released by artificial ants is the function of the quality of their generated solution;
(4) The timing of artificial ants for updating pheromone depends on the particular problem. For example, most artificial ants update pheromone on the path only after they have found a solution.

Sometimes, in order to improve the efficiency of the ACO, we can incorporate forecast, local optimization, and other strategies in the algorithm.

ACO is a random search algorithm simulating swarm intelligence, which has the following characteristics (Dorigo *et al.*, 1996; Duan, 2005; Chen and Ma, 2013):

(1) It possesses a strong robustness in different conditions and environments.

(2) It has a strong nature of self-organization and evolution. Self-organization means that ACO ensures the increase of entropy of the system in the absence of external action, which reflects the dynamic changes of the system from the disorder to order states. Complex behavior of the ant colony is an intelligence generated by interaction between individuals in the colony.
(3) Each individual can only perceive the local information, and they cannot directly use the global information. An individual can change the environment, and communicates indirectly through environment.
(4) It is a probabilistic and global search method. This non-deterministic mechanism creates more opportunities to find the globally optimal solution.
(5) It is essentially a parallel search process without central control. It uses a distributed control mechanism. Simultaneous parallel computing of individuals asynchronous and indirect communication between individuals, and collaborative optimization greatly increase computing power and efficiency of the algorithm. It searches from multiple points simultaneously.
(6) Its optimization process does not depend on the strict mathematical nature of the problem itself, such as the accurate mathamatical description on the continuity, derivability, objective function, and constraint function.
(7) It is obviously a multi-agent intelligent algorithm. Emergency can be found in the algorithm.
(8) It is easily fused with other algorithms, such as genetic algorithms, artificial neural network, particle swarm optimization, artificial immune algorithms.
(9) It has a positive feedback mechanism which makes the search process to converge.
(10) It takes a longer search time and thus prone to stagnation (i.e., all individuals will likely achieve the same solution after searching for a certain time, and thus no further search can be made in the solution space).

Similar to genetic algorithms, artificial neural network, particle swarm optimization, artificial immune algorithms, ACO is a bionic optimization algorithm, they all belong to a class of intelligent optimization algorithms that simulate natural biological systems and optimize their living conditions to adapt the environment through unconscious search behavior.

However, ACO is different from those algorithms in mechanism and implementation of the algorithm. ACO uses positive feedback mechanism; it requires a longer search time and is prone to stagnation; convergence performance of ACO is sensitive to the settings of initialization parameters.

9.3 Algorithms of ACO

The earliest ACO algorithm is time-cost, easy to stagnation, and prematured. A number of researchers have improved the ACO and developed various algorithms. They can be divided into two categories, ACO with discrete domain and ACO with continuous domain. The former includes adaptive ACO, polymorphic ACO, ACO based on ant pheromone diffusion, ACO based on mixed behaviors, ACO with clustering algorithm, etc., and the latter includes continuous domain ACO based on meshing strategy, continuous domain ACO based on information distribution function, continuous domain ACO based on crossover and mutation operations, and continuous domain ACO based on immune mechanism, etc. (Duan, 2005; Ren *et al.*, 2007).

Basic steps of ACO are as follows:

(1) Initialize parameters. Let time $t = 0$, and iteration $N = 0$; set the maximum number of iterations, N_{max}, and set m ants into n elements, etc.
(2) Let $N = N + 1$.
(3) Let the index in tabu list of ant $k = 1$.
(4) Let the number of ants $k = k + 1$.
(5) An ant chooses the element j according to the probability calculated from state transition probability formula and moves forward.
(6) Modify the pointer of tabu list, i.e., an ant moves to a new element after that choice is made, and the element is moved to the tabu list of this ant individual.

(7) If the elements in the collection C are not traversed ($k < m$), then go to the step (4), otherwise go to step (8).
(8) Update the pheromone on each path according to the rule.
(9) If the termination condition is satisfied, i.e., $N \geq N_{max}$, the algorithm terminates; or else clear the tabu list and go to step (2).

9.3.1 *Ant system (AS)*

AS is a simulated evolutionary algorithm. It is a random search algorithm proposed by Coiorni and Dorigo *et al.* (1992), inspiring by the behavior of real ant colony in nature. In AS, each ant search for the solution independently in the space of candidate solution, and leaves a certain pheromone on the solution it finds. The better the solution is, the greater the pheromone on the solution left, and the greater the possibility being chosen again is. The better path has a shorter distance and is more likely updated with much pheromone. The strength of information (pheromone) left by each ant is determined by the following formula. At the initial phase, all solutions have the same pheromone. With the progress of the algorithm, the pheromone of better solution increases and the algorithm eventually converges to optimal or near-optimal solution (Colorni *et al.*, 1991, 1992, 1994; Dorigo *et al.*, 1999; Hu *et al.*, 2008).

AS algorithm originally included three algorithms, ant-cycle, ant-density and ant–guantity, among which ant–cycle was proven to be the best. Thus AS has generated from ant-cycle.

The following is the program frame of AS:

```
initialize graph;
while (termination-condition-not-satisfied);
for m = 1 to N;
while (not-every-city-visited);
select-next-city;
end;
calculate-edge-length;
end;
```

```
Update-pheromone-of-graph;
end;
```

The following are some rules for the algorithm:

$$p_{ij}(t) = \tau_{ij}(t)^{\alpha}\eta_{ij}(t)^{\beta} \Big/ \sum_{k\in\text{avail. nodes}} [\tau_{ik}(t)^{\alpha}\eta_{ik}(t)^{\beta}] \quad \text{if } j \in \text{available nodes}$$

$$= 0 \quad \text{if } j \in \text{unavailable nodes} \tag{1}$$

$$\tau_{ij}(t+1) = (1-\rho)\tau_{ij}(t) + \Delta\tau_{ij}(t) \tag{2}$$

where η_{ij}: inspired information of the edge, $\eta_{ij} = 1/d_{ij}$; α, β: weights for determining pheromone/inspired information (adjusted artificially); τ: amount of pheromone; ρ: volatilization rate of pheromone; $\Delta\tau_{ij}$: increase of pheromone, and

$$\Delta\tau_{ij}(t) = \sum_{m=1}^{N} Q/L_m \quad \text{if edge}(i,j) \in \text{ edge of ant } m \tag{3}$$

where Q is a fixed positive number, and L_m is the length of edge for ant m.

AS has been used in various areas, such as assignment problem (Colorni *et al.*, 1991), TSP problem (Colorni *et al.*, 1994), minimum spanning tree problem (Ma *et al.*, 1999), vehicle routing problem (Liu and Xu, 2004).

9.3.2 *ACO*

Due to the shortcomings of AS, Dorigo and Gambardella (1997) developed a new algorithm, ACO. Later, Dorigo *et al.* (1999) further improved AS, which emphasized heuristic information and used a variety of methods to improve the performance of the algorithm, and eventually formed the completed ACO.

ACO is different from the previous algorithms in some aspects: (1) Versatility. ACO is suitable to solve problems that can be converted into connected graphs, including many NP problems; it can also solve the problems that properties of graph structure change in the calculation process. (2) Positive feedback. It uses local solutions to construct the global solution, usually by strengthening better local solutions. (3) Negative feedback. It avoids local convergence by evaporating pheromone. (4) Indirect communications. There are no direct communications between the ants, and the transmission of information is conducted through pheromone on the arc. (5) Self-organization. ACO relies on the coordination of ants to solve the problem.

The following is the program frame of ACO:

```
initialize graph;
while (termination-condition-not-satisfied)
for m= 1 to N
while (not-every-city-visited)
read-ant-routing-table;
select-next-city;
if (online-pheromone-update)
leave-pheromone-on-the-visited-edge();
update-ant-routing-table();
end;
end;
calculate-tour-length;
if (online-delayed-pheromone-update)
for each visited-edge do
leave-pheromone-on-the-visited-edge();
update-ant-routing-table();
end;
end;
end;
update-pheromone-of-graph;
daemon-actions();
end;
```

ACO uses ant-routing-table to record the tour history of ants. When moving from one point to another point, we can always change the pheromone (online-pheromone-update). Once a solution is achieved, ants can track this solution (online-delayed-pheromone-update). Volatilization avoids the premature convergence to suboptimal solutions. Background processing conducts centralized control of algorithmic convergence. For example, select some arcs to assign special pheromone (offline-pheromone-update). It theoretically should prove that swarm intelligence is consistent with the statistical laws (Iourinski *et al.*, 2002).

So far, ACO has been applied to solve a lot of problems such as traveling salesman problem (Dorigo *et al.*, 1999), vehicle routing problem (Bullnheimer *et al.*, 1999), scheduling problems (den Besten *et al.*, 2000), etc.

9.3.3 *Elitist ant system (EAS)*

Dorigo *et al.* (1996) used elitism to improve the pheromone updation mechanism of AS. That is, enhance the importance of ants that find the optimal edges in each of the iterations, and increase additional pheromone on the edges they find. This strategy improves the ability of AS to solve large-scale problems. Ants release pheromone that is inversely proportional to their path length. The better the ants' path is, the more pheromone is left on the edges of the path, and the greater the probability with which ants select these edges in the forthcoming iterations is. The improvement is as follows:

$$\begin{aligned}
\tau_{ij}(t+1) &= \tau_{ij}(t) + \sum_{k=1}^{m} \Delta\tau_{ij}^{k}(t) + e\Delta\tau_{ij}(t) \\
\Delta\tau_{ij}(t) &= 1/C \qquad \text{if edge}(i,j) \in P \\
&= 0 \qquad\quad\ \ \text{otherwise}
\end{aligned}$$

where e: the constant to define the weight of P, and C: the length of P.

Hemmatian *et al.* (2013) used EAS to solve the optimization problem in order to reduce the weight and cost of the composite laminate sequence.

9.3.4 *Rank-based AS (ASrank)*

ASrank (Bullnheimer *et al.*, 1999) is another revision of AS (Priscila *et al.*, 2007). In ASrank, the algorithm records the best solution in the history. Once pheromone is updated, update the path contained by the globally optimal solution. The rule used to TSP problem is:

$$\tau_{ij}(t+1) = (1-\rho)\tau_{ij}(t) + \Delta\tau_{ij}(t) + \Delta\tau_{ij}^{*}(t) \tag{4}$$

where $\Delta\tau_{ij}(t)$ is achieved by the equation (3), and

$$\begin{aligned} \Delta\tau_{ij}^{*}(t) &= k/L(t) && \text{if edge}(i,j) \in P_{globally\ best} \\ &= 0 && \text{otherwise} \end{aligned} \tag{5}$$

If edge$(i, j) \in P_{globally\ best}$, add pheromone when updating pheromone each time, and the increment of pheromone is positively proportional to the size of the globally optimal solution.

In ASrank, the algorithm keeps both the best $(Q - 1)$ optimal solutions and the current globally optimal solution, and updates pheromone on the paths corresponding to these solutions. Updation rules follow the equation (4). However, the calculation in it follows this rule:

$$\Delta\tau_{ij}(t) = \sum_{k=1}^{m-1} \Delta\tau_{ij}^{k}(t) \tag{6}$$

$$\begin{aligned} \Delta\tau_{ij}^{k}(t) &= (m-k)Q/L_k && \text{if edge}(i,j) \in P_{k-th\ best\ path} \\ &= 0 && \text{otherwise} \end{aligned} \tag{7}$$

9.3.5 *ACS*

ACS has improved AS in the following aspects:

(1) ACS uses elitism strategy to update pheromone, and uses the following rule to optimal solutions achieved from each of iterations and the globally optimal solution (Dorigo and Gambardella, 1997):

$$\tau_{ij}(t+1) = (1-\rho)\tau_{ij}(t) + \rho\Delta\tau_{ij}^{best}(t) \tag{8}$$

(2) The probability for selecting the next edge is determined by a pseudo-random proportion.

$$p_{ij}(t) = \tau_{ij}(t)\eta_{ij}(t)^{\beta} / \sum_{k \in avail.\ nodes}$$

$$
\begin{aligned}
& [\tau_{ik}(t)\eta_{ik}(t)^{\beta}] && \text{if } j \in \text{available nodes} \\
&= 0 && \text{otherwise} \qquad (9)
\end{aligned}
$$

$$
\begin{aligned}
P' &= \arg\max\, [\tau_{ij}(t)\eta_{ij}(t)^{\beta}] && \text{if } q \leq q_0 \\
&= p_{ij}(t) && \text{otherwise} \qquad (10)
\end{aligned}
$$

where q is a random number, and q_0 is a variable that determines to explore a new path or to use the existing path. If $q \leq q_0$, the algorithm begins to explore a new path; otherwise, the algorithm uses the existing path information to determine the probability of selecting the next path.

(3) ACS also updates local pheromone. In addition to the global updation, ACS also updates pheromone in each of the iterations.

Vlachos *et al.* (2013) used ACS to solve the problem of maintenance and scheduling of heat generators. Jong and Lai (2013) used ACS to the scheduling and optimization of injection mold manufacturing.

9.3.6 *Max-min ant system (MMAS)*

MMAS (Stützle and Hoos, 2000) is a type of ACO to solve the TSP, QAP, and other problems. MMAS modified the method of pheromone updation in AS. It only allows to update the optimal ant (i.e., the ant constructing the shortest path in the iteration), or only the best ant so far releases pheromone. Pheromone concentration on the path is limited to the range $[\tau_{\min}, \tau_{\max}]$. In addition, the initial value of pheromone concentration is set to be $\tau_{\max}$, which helps to increase the search capability at initial stage of the algorithm. In order to avoid stagnation, the pheromone on each edge of the problem space will be reinitialized.

MMAS increases the concentration of pheromone on only the best path, thereby making better use of historical information. In order to avoid premature convergence, the algorithm limits the pheromone concentration to the range $[\tau_{\min}, \tau_{\max}]$, and the values beyond this range are forced to $\tau_{\min}$ or $\tau_{\max}$, which effectively avoids the divergence of the algorithm. The pheromone concentration on each path is set to be the $\tau_{\max}$, so that it can fully perform optimization.

After all of the ants have completed a path search, the pheromone is updated through volatilization. Reserve of new pheromone is:

$$\tau_{ij}(t+1) = \tau_{ij}(t) + \Delta\tau_{ij}^{best}(t)$$
$$\Delta\tau_{ij}^{best}(t) = 1/C$$

Tang *et al.* (2013) used MMAS to solve vehicle routing problem; Jin *et al.* (2011) used MMAS solve the path planning of logistics.

9.3.7 *Ant-solver*

Ant-Solver is an AS-type algorithm to specifically address Constraint Satisfaction Problem (CSP) (Solnon, 2002). It simplifies AS, in which the ant tour in AS is replaced by a process of randomly constructing the solution of CSP, and pheromone is updated using the results of construction process. The pheromone updation rule is:

$$\tau_{ij}(t+1) = (1-\rho)\tau_{ij}(t) + \sum_{Ak \in best\,of\,iteration} \Delta\tau_{ij}(A_k)$$

9.3.8 *Multiple ant colony system (MACS)*

MACS is expanded from the basic AS (Liu *et al.*, 2013). MACS contains multiple and independent ant colonies. In MACS, not only ants in a colony but also colonies collaborate and interact with each other through pheromone, thereby improving the efficiency of the algorithm and the probability of finding the optimal solution. Mutual cooperation between colonies can use a variety of strategies (Liu *et al.*, 2013)

$$s = \arg \max_{i \notin mhk} [\tau_h(i)^\alpha \eta(i)^\beta] \qquad \text{if } q \leq q_0$$

where q is a random value in [0, 1], $0 \leq q_0 \leq 1$; $\tau_h(i)$: pheromone that ant colony h deposits on the i-th area to be divided; $\eta(i)$: heuristic information, i.e., the proportion of the area i to be divided *vs.* total area to be divided; α, β: parameters that determine the relative importance of pheromone and heuristic information; m_{hk}: set of areas selected by k-th ant in h-th colony $(1 \leq k \leq m, 1 \leq h \leq G)$. If $q \leq q_0$, what area is selected is determined by the maximum of $[\tau_h(i)^\alpha \eta(i)^\beta]$. If $q > q_0$, what area is selected is determined by the following rule:

$$p_{hk}(i) = \underset{i \notin mhk}{[\tau_h(i)^\alpha \, \eta(i)^\beta]} \Big/ \sum_{j \notin mhk} [\tau_h(j)^\alpha \eta(j)^\beta] \qquad \text{if } i \notin m_{hk}$$

$$= 0 \qquad \text{otherwise}$$

where $p_{hk}(i)$ is the probability of selecting i-th area by k-th ant in h-th colony. The rule for local updation is:

$$\tau_h(i) = (1 - \rho)\tau_h(i) + \rho\tau_0$$

where τ_0 *is* the initial value of pheromone, $0 \leq \rho \leq 1$. The rule for global updation is:

$$\tau_h(i) = (1 - \alpha)\tau_h(i) + \alpha\Delta\tau_h(i)$$

where α is the initial parameter, and $\Delta\tau_h(i)$ is the utilization of raw materials of the optimal ant in h-th colony. The rule for updation between colonies is:

$$\tau_h(i) = \tau_h(i) + \lambda\Delta\tau_{best}(i)$$

where λ is the initial parameter, and $\Delta\tau_{best}(i)$ is the utilization of raw materials of the optimal colony.

9.3.9 *Adaptive AS*

Adaptive AS is different from AS in decision and selection rules and global updation rules (Wang, 2013; Zhou and Liu, 2013). The algorithm is as follows:

(1) Define initialize rules and initialize the artificial ant colony Ω.
(2) Set the maximum number of iterations or other termination conditions, and start iteration.
(3) Each ant gradually constructs a solution, i.e., generates a feasible path, based on the state transition rule. Each ant should independently complete a search process and generate a feasible solution. Suppose at current time t, the ant k locates in the node

i, it will go to next node j with the probability $p_{ijk}(t)$, following the state transition rule:

$$
\begin{aligned}
j = &\ \arg\max\, [\tau_{iu}(t)^{\alpha}\eta_{iu}(t)^{\beta}] && u \in J_k(i), q \le q_0 \\
= &\ J && q > q_0
\end{aligned}
$$

where q is a random value in [0, 1], and

$$
\begin{aligned}
p_{ijk}(t) &= [\tau_{ij}(t)^{\alpha}\eta_{ij}(t)^{\beta}] \Big/ \sum_{u \in Jk(i)} [\tau_{iu}(t)^{\alpha}\eta_{iu}(t)^{\beta}] && j \in J_k(i) \\
&= 0 && j \notin J_k(i)
\end{aligned}
$$

(4) In each of the iterations, the ant k updates pheromone on every edge it walked, using the rule for local updation:

$$\tau_{ij}(t+1) = (1-\varepsilon)\tau_{ij}(t) + \varepsilon\tau_0$$

where ε is the volatilization coefficient, $0 \le \varepsilon \le 1$; τ_0 is the initial value of pheromone, and:

$$\tau_0 = 1/(mL_{mm})$$

where L_{mm} is the path length, and m is the number of nodes for constraint conditions. If m ants have finished their search, go to (5); otherwise go to (3).

(5) Knowing that m ants have generated m solutions, i.e., they all finished a tour, choose a shortest path as the optimal solution and update pheromode on every edge of the path with the global updation rule:

$$\tau_{ij}(t, t+n) = (1 - \rho(t, t+n))\tau_{ij}(t) + \rho(t, t+n))\Delta\tau_{ij}(t, t+n)$$

where $\rho(t, t+n)$ is the volatilization coefficient, $0 \le \rho(t, t+n)$ ≤ 1, and

$$
\begin{aligned}
\Delta\tau_{ij}(t, t+n) &= 1/L_b && (i, j) \in V_{best} \\
&= 0 && (i, j) \notin V_{best}
\end{aligned}
$$

where L_b is the optimal solution and V_{best} is the globally optimal path.

The volatilization coefficient is adjusted by using adaptive adjustment strategy:

$$\rho(t, t+n) = \psi p(t) \qquad \psi p(t) > \rho_{min}$$
$$= \rho_{min} \qquad \text{otherwise}$$

where ψ is the constraint for volatilization coefficient, $0 < \psi < 1$, and

$$\psi = 2\psi_0/[1 + \exp(\varphi h/h_{max})]$$

where φ: positive constant, and h: iterations so far.

(6) If the termination condition is met, print results; otherwise go to (2).

9.3.10 *Best-Worst AS*

Given the low efficiency and poor quality of AS, the Best-Worst AS enhances the quality of search process, so that the search of ants is more concentrated in the domain around the best path found by the current search, which avoids premature convergence and also shortens the search time. Search rule is:

$$\tau(r, s) = (1 - \rho)\tau(r, s) - \varepsilon L_{worst}/L_{best}$$

where L_{worst} and L_{best} are path length of the worst and best paths respectively in the current iteration, and $\tau(r, s)$ is the amount of pheromone between cities r and s.

9.3.11 *Stochastic perturbation ACO*

Stochastic Perturbation ACO is divided into two parts, coarse tuning and fine tuning. During the process of coarse tuning, pheromone concentration and transition probability are set in advance, and then some parameters, determining impact degree of path choice and updation degree of pheromone on the path, are adjusted. In the process of fine tuning, the deterministic choice leads to the choice of the path with maximum transition coefficient, and the stochastic choice leads to the certain stochasticity in computing transition coefficient. The two processes jointly make the algorithm with a stronger capability of global search.

In addition to the algorithms above, Pourtakdoust and Nobahari (2004) proposed an ACO searching only under the guidance of the pheromone to solve continuous optimization problem. Dreo and Siarry (2002) developed the algorithm, Continuous Interating Ant Colony, for solving multi-objective continuous optimization problem.

9.4 Applications of ACO

ACO has been used in various areas, in particular, in solving many combinatorial problems (Wang, 2014):

(1) TSP
ACO solution for a TSP problem is generally described as: (1) a colony of ants depart randomly from the starting point, encounter food, catch food, and return along the same route; (2) ants leave pheromone on their roads of a round trip; (3) pheromone will gradually volatilize over time, which is generally described by a negative exponential function; (4) for an ant starting to depart the nest, the probability of choosing a path is proportional to the concentration of pheromone on the path.

(2) Vehicle Routing Problem (VRP)
Various types of evolutionary algorithms, such as genetic algorithms, and tabu search have been used in VRP. ACO solution for TSP problems can also be used to solve VRP.

(3) JSP (Job-Shop Scheduling Problem)
Northwestern University in USA has used ACO for painting trucks in assembly line. They take groups of workers and each group paints one color only. When a group's assignment is particularly tense, another group can assist that group. It proved to greatly improve the overall productivity.

(4) DcMsT (Degree-constrained Minimum Spanning Tree)
If the number of vertices in a tree is restricted, the minimum spanning tree problem becomes a DcMST. DcMST refers to problems finding to find a spanning tree with constrained vertex degree and minimum total weight in all of the spanning trees. There are many examples such as: plumbing, circuit design,

communication systems, computer networks, etc. Using ACO to solve these optimization problems has achieved ideal results, especially when the degree constraints are more stringent. In this case, ants move based on the construction rules of the spanning tree and their transition probabilities are similar to that in TSP (Lin and Ye, 2013).

(5) Multi-objective Shortest Path Problem

(6) Multi-objective Optimal Tree Problem
Multi-objective optimal tree problem can be derived from the classic concept of the minimum spanning tree. Circuit layout problem, spanning tree problem of Min-Max vertex degree, etc., are examples of the problem.

(7) Termite-inspired Robots
Inspired by termites, scientists at Harvard University have created robots that use a few rules and environmental informations to build castle-like structures and pyramids (Gibney, 2014). The robots work independently, and independently acting robots constitute a robust system. Each travels along a grid and can move, climb a step and lift and put down bricks. They use sensors to detect other robots and existing bricks, and react to these stimuli according to a few simple rules. The template for each three-dimensional structure is translated into a specific set of traffic rules and combined with fixed laws of robot behavior. Combining traffic rules and robot behavior decomposes a complex engineering problem into two sets of information working together.

(8) Dual DNA Sequence Alignments
ACO has been successfully used to align dual DNA sequences (Chen and Chen, 2006).

For two DNA sequences, S = CGGATC, and T = CGAATTC, the alignment matrix, derived from Ant Colony Pairwise Alignment Model, is shown in Table 1.

The main steps are as follows (Chen and Chen, 2006):

(1) Initialize the parameters, i.e., initial pheromone, orientation weights, initial iterations, time counting.

Table 1. Pairwise sequence alignment matrix based on Ant Colony Pairwise Alignment Model.

	⊸	C	G	A	A	T	T	C
⊸								
C								
G								
G								
A								
T								
C								

(2) According to the two DNA sequences and the given rule, generate the match matrix, in which positive values are the original ones and negative values take the reciprocal of the absolute values of the original values. That is, the value is 2 if characters are equal and there are no spaces; it is 1 if characters are not equal and there are no spaces; it is 0.5 if there are some spaces.

(3) Take the unique point corresponding to the two spaces at the upper left of the similarity matrix as the starting point of each ant, and set the starting point into the tabu list. Each ant searches starting from the starting point, and set the points searched into the tabu list until the endpoint. If maximum iterations are achieved, or the optimum does not change in N iterations, terminate the algorithm.

(4) There are three possible moving directions for ants, downward, right and lower right. First, the program generates a random value. If the random value is less than the fixed value p_0, then select to move to lower right; if it is the random greater than p_0, then calculate the transition probabilities of the three moving directions, p_k, and select a direction to move using roulette wheel selection. Finally, record the migration path and alignment score of each ant.

(5) If an ant selects to move downward, insert a space "-" in the corresponding position of the sequence T; if it moves to the right side, insert a space "-" in the corresponding position of the sequence S. When an ant has reached the far right, it can

only move vertically downward. When it comes to the bottom, it can only move horizontally to the right side.

(6) A round of iteration ends if all ants reach the endpoint. Update the pheromone matrix according to the migration path and alignment score of each ant, and record the maximum and mean scores.

(7) Go to step (3).

According to Chen and Chen (2006), the implementation framework is:

```
input the DNA sequences, S0,S1,...,SN−1
output the optimal and approximately optimal scores for alignment of DNA
sequences S0,S1,...,SN−1
initialize the parameters
for (iter=1; iter<=MaxIter; iter++)  //iterations
for (k=1; k<=N; k++)    {   //N ants are uniformly assigned to N sequences
for (l=1;l<=|Sk|;l++)  {   //character cycling in sequence Sk
for (n=1;n<=N;n++)      {   //sequence cycling
if (n!=k) {
for (m=loc(k,l,n);m<=loc(k,l,n)+h-1;m++)
calculation of p(k,l,n,m)
calculate m such that p(k,l,n,m) reaches the maximum value
if (Sn(m)==Sk(l)) select Sn(m)
else choose the space with a certain probability, and choose Sn(loc(k,l,n))
   with the
remaining probability
} //if
} //n
} //l
insert spaces, and align characters that ants did not select in order to to
   get the full alignment
calculate the fitness function value of this alignment according to the
   given rule
update pheromone according to the given rule
} //k
if the alignment score is higher than all previous score, reserve the
```

```
alignment and
score
if (iter>start) regulate pheromone
adjust the parameters a, b, c according to the given rules
} //iter
```

In the future, the following aspects of ACO should be strengthened: (1) mining the theoretical basis of ACO; (2) improving ACO algorithms; (3) integrating ACO and other optimization algorithms; and (4) more applications.

Chapter 10

Fish and Particle Swarm Optimization

10.1 Artificial Fish Swarm Algorithm (AFSA)

Artificial Fish Swarm Algorithm (AFSA) is an optimization algorithm on the basis of simulating the behaviors of fish swarm (Li, 2003; Li *et al.*, 2004). AFSA has been widely used in combinatorial optimization, parameter estimation, optimization of artificial neural networks, power network layout, scheduling system, and solving nonlinear equation group, etc. (Ma and Liu, 2013; Zhang *et al.*, 2013). So far, AFSA has been expanded to solve multi-dimensional and dynamic combinatorial optimization problems from the original one-dimensional static problems. AFSA features such advantages as: (1) excellent performance to avoid local optimization; (2) fast track to drifting of minimum; and (3) only the domain of satisfied solutions is achieved.

From the view of agent-based modeling, a fish individual in AFSA can be treated as an agent/component.

10.1.1 *Algorithmic procedure*

Suppose the state of a fish individual is $X = (x_1, x_2, \ldots, x_n), Y = f(x)$ is the food density (the value of objective function, i.e., fitness value) at the current foraging location of the fish individual, $d_{ij} = |X_i - X_j|$ is the between-individual distance, Visual is the perceivable distance of fish individual, step is the moving step of fish individual (usually a random step), and $\delta(0 < \delta < 1)$ is the crowing factor.

The procedure of AFSA is as follows:

(1) Foraging behavior. The present state of a fish individual is X_i. Randomly select a state X_j within its perceivable range; if $Y_i < Y_j$, then move a step at the direction of X_j; otherwise, randomly select a new state X_j, and determine if the condition to move ahead is met. If certain times of trials are conducted and the condition is still not met, the algorithm moves ahead one step randomly.
(2) Swarming behavior. Suppose the present state of a fish individual is X_i. Find the number of its adjacent individuals, n_i, within its perceiving range ($d_{ij} <$ Visual), and the central position Y_c; if $Y_c/n_i = \delta Y_i$ and $Y_i < Y_c$, which represents much food in the center and not crowded individuals, then the algorithm moves ahead a step along the center's direction; otherwise continue foraging. Continue foraging if $n_i = 0$.
(3) Tracking behavior. Suppose the present state of a fish individual is X_i. Find its adjacent individual X_j with the greatest fitness value within its perceiving range ($d_{ij} <$ Visual), and the central position Y_c; if $Y_j/n_i = \delta Y_i$ and $Y_i < Y_c$, which represents much food around X_j, and not crowded individuals around it, then the algorithm moves ahead a step along X_j; otherwise continue foraging. Continue foraging if $n_i = 0$.
(4) Random behavior (default behavior). Randomly select a state in the perceivable scope and move ahead a step at this direction.
(5) Update bulletin board. Set a bulletin board to record the state X_{best} of the historically best individual in fish swarm. Each fish individual checks its own state during an iteration process. Change X_{best} to X_i, if $f(X_i) < f(X_{best})$.

The convergence of AFSA was theoretically proven using Markov chain (Huang *et al.*, 2012). The states of all individuals in the fish swarm are mapped onto a discrete space, and calculate the transition probability of each individual from a state to another state, P_{ij}. Current state of an individual, X_i, corresponds to a state in Markov chain. The state is proven to be stable, thus the state can converge to an optimal point.

However, the convergence of AFSA is fast during its early stage, but it is easy to locally converge in later iterations. A variety of revised algorithms are thus developed to improve the efficacy of AFSA.

10.1.2 *Parametrical-based revisions of AFSA*

Based on AFSA, Wang and Shi (2010) proposed a global and parametrical-based version of AFSA. It was proven that a small step length may lead to better performance of searching solution. A shorter step length will enhance the random foraging and eliminate the possibility of local convergence.

In a revised AFSA, the parameters can be dynamically adjusted (Wang *et al.*, 2009a, b). The adjustment of parameters ensures the large scale and global search in the early stage, and ensures the fine and local search in the late stage. The adjustment of step length is performed as follows:

$$\begin{aligned} \boldsymbol{\alpha} &= \exp(30(t/T_{max})^3 \\ visual &= visual \times \boldsymbol{\alpha} + visual_{min} \\ step &= step \times \boldsymbol{\alpha} + step_{min} \end{aligned}$$

Xiao *et al.* (2006) used an adaptive method to automatically adjust such parameters as step length, perceivable range, and crowing factor δ, etc., in order to ensure fine and local search in the late stage. In their algorithm, the parameters change as follows:

$$\begin{aligned} k &= (f(t) - f(t,n))/f(t,n) \\ \delta &= D(f(t), f(t-n), f(t-2n)) \\ step &= f(step), g = f(g)(k \leq \theta, \delta \leq \Phi) \\ step &= step, g = g(k > 0, \delta > \Phi) \end{aligned}$$

where δ and Φ can be set in advance. The algorithm proved to be superior to adaptive particle swarm optimization, simulated annealing particle swarm algorithm, and original AFSA.

10.1.3 *Intelligence fused revisions of AFSA*

Zhang *et al.* (2006) introduced a mutation operator in AFSA. In their algorithm, a threshold, beststep, was predefined to mark the optimum value on the bulletin board. If the optimum value on the bulletin board did not change or less changed within a certain number of iterations, then the present best individual should be kept and other individuals will mutate at a certain probability. The mutation method is: produce the random values, $\gamma \in (0, 1)$, for each dimension of each individual that need to perform mutation operation. If $\gamma \leq P_m$, then perform mutation to the dimension, otherwise the dimension does not change. After the completion of mutation, compare with the best on the bulletin board and revise the best individual on it.

To further improve the algorithm, Zhang *et al.* (2006) used the simulated annealing (SA) in the late stage of AFSA. The convergence probability of nearly one of SA at low temperatures guarantees the fine local search for the solution.

Liu and Zhou (2009) embedded AFSA in the framework of cultural algorithm and thus presented a new hybrid global optimization algorithm. The algorithm contains two spaces, population space and knowledge space. Fish swarm learns from its evolution and generates a knowledge space. Knowledge space and population space co-evolve in the algorithm. The main steps include:

(1) Initialize N fish individuals in the population space, and choose 30% of them to construct the knowledge space;
(2) Use the best individual in population space to replace the worst individual in knowledge space, and the knowledge evolves.
(3) Perform n generations of evolution in population space, under the guidance of knowledge space. Use the better individuals in knowledge space to replace the poor individuals (with small fitness values) in population space. Here we have:

$$n = 15 + endstep - k/endstep \times 200$$

where *endstep*: the default maximum number of iterations, and k: the current iteration.
(4) Fish swarm executes tracking operation and swarming operation. The default is foraging operation.
(5) Judge whether the algorithm terminates or not.

In the PSO-AFSA hybrid algorithm, Wang *et al.* (2010) divided the population to two sub-populations. A population evolves in PSO way, another one evolves in AFSA way. Two populations exchange information by sharing the optimum individual.

Qu *et al.* (2011) introduced the rules of particle swarm evolution and evolutionary strategies to AFSA and proposed a new algorithm of better performance.

10.2 Particle Swarm Optimization (PSO)

Early in 1975, Wilson proposed a theory on swarm. In a swarm, each individual can share other individuals' discovery and experience to escape from predators and acquire food. It established the theoretical foundation of particle swarm optimization (PSO).

For a bird swarm, its distribution is wider in space. Each individual can position itself in the swarm. Every individual perceives the flight movements of surrounding neighbors to adjust its own trajectory, finally makes the whole swarm to be controlled by a central system. According to this phenomenon, we can think that every bird must perceive three aspects: its own position, the positions of two or three neighbors, and the trajectory of entire swarm. Based on this fact, Reynolds (1987) proposed a distributed behavioral model. Further, Kennedy and Eberhart (1995) put forward PSO to deal with optimization problems of continuous functions (Yang and Li, 2004). Similar to other evolutionary algorithms, PSO searches the optimal solution by cooperation and competition between individuals. However, evolutionary operators such as crossover, mutation, selection, are not needed in PSO. Every individual is treated as a particle in a higher dimensional space. Each particle moves at a certain speed in the solution space, and towards its historically best position and the historically best position of neighbors, in order to evolve as a candidate solution.

From the view of agent-based modeling, a bird individual/particle is a component or agent.

10.2.1 *Algorithmic procedure*

The location and velocity of a particle i in the D-dimensional space are $x_i = (x_{i1}, x_{i2}, \ldots, x_{iD}), v_i = (v_{i1}, v_{i2}, \ldots, v_{iD})$, respectively, where $x_i \in (a_i, b_i), i = 1, 2, \ldots, D; a_i, b_i \in R$. x_i and v_i are initialized with random values at the start of the iterations (Imran *et al.*, 2013). Local and global extremums are obtained by comparing extremums of particles. A new global extremum is achieved at each iteration. The updating rules for location and velocity are as follows:

$$v_{id}(t+1) = v_{id}(t) + c_1 r_1(t)(pbest_{id}(t) - x_{id}(t)) + c_2 r_2(t)(gbest_d(t) - x_{id}(t))$$

$$x_{id}(t+1) = x_{id}(t) + v_{id}(t+1)$$

where c_1, c_2: learning factors (accelerating constants), and generally $c_1 = c_2 = 2$. The two parameters adjust the maximal step moving to the global best individual and the local best individual respectively. $r_1, r_2 \in r\,and\,[0, 1]$. *pbest* and *gbest*: extremums of an individual and entire swarm respectively. $v_{id}(t+1)$ denotes "inertia" or "momentum" that reflects the movement habit of the particles. $v_{id}(t)$ denotes cognition term, representing particle's memory on its own historical experience. $c_1 r_1(t)(pbest_{id}(t) - x_{id}(t)) + c_2 r_2(t)(gbest_d(t) - x_{id}(t))$ denotes the knowledge sharing and collaboration in the swarm, which means the trend of the particle moving to global and local extremums. The detailed steps of PSO are as follows (Xie *et al.*, 2003; Yang and Li, 2004):

(1) Initialize the location and velocity of each particle randomly. The pbest coordinates of each particle are set to be its current location; calculate the fitness value of extreme point of each individual. The best in all of pbest is the global extremum; record the ID number of the best particle. gbest is set to be the location of the best particle.
(2) Evaluate fitness. Calculate each particle's fitness value, and compare with the original extremum of the particle. If the present fitness is better than the original, then pbest is set to be the present particle's location, and update the particle's extremum. If the best in all of pbest is better than the current global extremum, gbest is set to be the location of the best particle, and record the ID number of the best particle, and update the global extremum.
(3) Update location and velocity of each particle according to the rules above.

(4) Test whether the requirements for termination are met. If the current iterations reach the maximum, or a predetermined value/minimum error requirement is achieved, the algorithm terminate, otherwise return step (2) to continue.

PSO has two versions, global version and local version (Yang and Li, 2004). The global version converges fast but may easily converge to the local optimum. The local version features several attractors to avoid local convergence. For a particle i, define a particle set in a neighborhood of size l:

$$N_i = \{\text{pbest}_{i-l}, \text{pbest}_{i-l+1}, \ldots, \text{pbest}_{i-1}, \text{pbest}_i, \text{pbest}_{i+1}, \ldots, \text{pbest}_{i+l+1}, \text{pbest}_{i+l}\}$$

Select the best from N_i, and set its location as lbest to replace gbest. Other steps are the same with the global version. Local version converges slower than global version, but not easily converges to a local optimum. It is suggested that the global version is used for coarse search and the local version is for fine search (Yang and Li, 2004).

It is noted that PSO is more applicable to the problems of larger swarm size, and the problems with the area of more complex, non-differential, and multiple peaks (Ling *et al.*, 2008).

10.2.2 *Some revisions of PSO*

10.2.2.1 *Cluster analysis*

Kennedy (2000) proposed a PSO revision based on cluster analysis. The procedure is as follows:

(1) Choose a few particles as the center of the different clusters, for example, set the location of C as a cluster's center.
(2) Calculate the distances of all particles with C, and all nearest particles to C are defined as a cluster.
(3) Calculate the center of each cluster, and compare to obtain the global center pgbest or local center plbest.
(4) Return step (2) to continue, until the location of the center is stable.

10.2.2.2 *Selection rule*

Angeline (1998) put forward a PSO revision based on an additional selection criteria. His algorithm is superior to conventional PSO in solving simple problems. Two main steps are included as follows:

(1) Based on the individual's current location, compare its fitness value with k other individuals.
(2) For the $k + 1$ individuals, put the individuals with the greatest fitness value to the top of the swarm.

10.2.2.3 *Adaptive PSO*

These methods introduce the adaptive selection of parameters in PSO, for example, introducing random term in inertia weight (Eberhart and Shi, 2001), adjusting inertia weight using fuzzy logic (Shi and Eberhart, 2001), using an independent secondary PSO to optimize the parameters of the original PSO (Doctor *et al.*, 2004), and automatically determine all parameters including the number of particles and the topological structure of particles (Zhang *et al.*, 2006).

Chapter 11

Synergy, Coevolution, and Evolutionary Algorithms

11.1 Synergy and Biological Evolution

11.1.1 *Synergy*

As described in the previous chapter, synergy means that in an open and complex system — made of a large number of subsystems, far from equilibrium, and having matter and energy exchange with outside the system — its subsystems interact in a nonlinear way to produce coherent and synergistic effects and thus make the system generate a self-organization structure in space/time with certain functions. Due to both fierce competition and close synergy between populations in the same environment populations, synergetics therefore can and has been used in ecology and biology (Haken, 1978, 1983; Bjumenefld, 1985). Between-species ecological interactions can be divided into three cases: competition, predation/parasitization, and symbiosis. Each of the three interactions must maintain a dynamic equilibrium in order to adapt the environment. Applying synergetics to biological morphology, the morphogenesis is interpreted such that the diffusion and reaction of certain chemical substances form a so-called morphological source field. Morphological source field governs genes to control cell differentiation, and thus forms the organism.

11.1.2 *Synergy and competition*

In his book, *The Origin of Species*, Darwin (1911) first used the terminology "competition," and considered competition as part of the struggle for existence. Since Darwin, many biologists believed that the struggle for existence had promoted biological evolution under natural selection. Integrating various definitions of biological competition (Begon *et al.*, 1994; Grime 1997), we may consider the competition as the between-individual or between-population negative interaction due to resource constraints. In ecology and biology, synergy refers to mutual collaboration and mutual adaptation either between organisms or between organisms and the environment. Compared to competition, synergy enables organisms to exist and reproduce at the minimum cost (i.e., maximum fitness). Synergy and competition can mutually transform under certain conditions, which helps to enhance the self-organization, orderly, and biodiversity of ecosystems. Competitors try to reduce niche overlap and adapt to the evolution, which is called disruptive coevolution (Wilson and Lee, 1994). Generally, biologists and ecologists, tend to use coadaptation, coevolution, etc., to represent the terminology "synergy" (Williams, 1992).

11.1.3 *Coevolution*

11.1.3.1 *Coevolution*

The terminology "coevolution" was first proposed by Ehrlieh and Raven (1964) in their research on mutual influence between plants and herbivorous insects in the evolution. Hereafter, Roughgarden (1979) defined coevolution as the simultaneous evolution of multiple interactive species. Jazen (1980) defined coevolution as the traits of a species evolving as a reaction to the evolution of the traits of another species, and vice versa. The definition considers the specialty, mutuality, and simultaneity, and is thus called pairwise coevolution. However, the pairwise coevolution is only found in certain pairs of symbiotic, parasitic or competitive species, for example, mimicry in butterflies, insect, and flowering plants/pollinating insects, etc. Contrary to pairwise coevolution, the diffuse coevolution is most widely used and is usually equivalent to the evolution. Diffuse coevolution is defined as a mutual evolution that the traits of one or more species evolve under the influence of the traits of other multiple species

(Futuyma and Slatkin, 1983). Different from pairwise coevolution, the diffuse coevolution does not meet simultaneity.

In a broad sense, coevolution refers to the mutual dependence between species and the environment in the evolution. Compared to Darwin's theory, coevolution stresses the between-species interactions and the adaptation of species to their biological environments rather than physical environments.

Coevolution occurs at the molecular//cellular/individual/population/ecosystem levels. However, at the population level is the focus.

Organisms have certain ecological investment strategies in the growth process. A large number of experiments have shown that each of the energy investment strategies is a trade-off among ecological costs (Stinchcombe, 2002; Strauss *et al.*, 2002), i.e., an increase of an ecological cost means the decrease of another ecological cost. For example, a great energy cost in reproduction and growth, which enhances the organism's competitiveness, may lead to not enough energy to withstand adverse environment, and the organism's ability to escape predators will be reduced. Pursuiting excessive energy cost will lead to degradation of the entire species. As a consequence, coevolution/coadaptation becomes a realistic evolutionary strategy.

Establishment of biological communities and ecosystems is the result of coevolution of species-species and species-environment. Interactions and coevolution of animals, plants, and microorganisms maintain the stability and functions of the ecosystems.

11.1.3.2 *Research methodology of coevolution*

There are various types of synergies/coevolution, e.g., the large scale synergy, threshold effect, functional complementation, integrated environmental conditions, risk/cost ratio, information sharing, swarm intelligence, symbiosis, and complex divisions of works in biological systems. Most synergies can be measured and quantified, and most are directly concerned with the survival and reproduction (Corning, 2014). The following briefly describes some research methodology of coevolution.

(1) Make a long-term observation to some of the traits possibly caused by coevolution between species to investigate the relationship between genetic variation, for example, Pimentel's study on housefly-parasitoid relationship, and Barertt's study on pathogen virulence and plant resistance (Futuyma and Slatkin, 1983). This method is appropriate for two species with pairwise coevolution.
(2) Use historical fossils and written material to analyze the long-term evolution of species, and to speculate the possibility of producing coevolution. This is a method often used by paleontologists. These data can be used to analyze the reasons for massive extinction of species. We can also reconstruct the ancient climate change through some biological residual material (such as phytoliths).
(3) Use phylogenetic methods to study the long-term evolutionary process of species. This method is mainly used to research coevolution between insects and plants.
(4) Make in-depth research on the physical structure, morphology, as well as some behaviors of species to reveal potential coevolution. This helps to understand the mechanism of behavioral responses between coevolved species.
(5) Study of the mutual adaptive behaviors between geographical populations and biological habitats in different environments helps to explain the coevolution between organisms and the environment. For example, similar geographical environments or habitats will usually produce communities with similar species composition and structure.
(6) Develop ecological models to approach between-species coevolution and draw biological interpretations. Interspecific competition models, predator-prey models, etc, are most often used.
(7) Explore the changes of different trophic levels between species in the community. This helps to reveal the mechanism of coevolution in maintaining the community stability.

Corning (2014) summarized four categories of experiments to study the biological synergy: (1) removal test. Remove some parts from a system and observe the changes of the system; (2) biological econometric method. Under different situations, calculate the cost/benefit of changing an adaptive trait; (3) comparative experiments. This can accurately measure synergies; and (4) classical ecological experiments, namely multi-factor control experiments, which consider the interactions between various factors.

11.2 Evolutionary Programming (EP)

Evolutionary programming, evolutionary strategies, and genetic algorithms emphasize a distinct aspect of natural selection. Evolutionary programming (EP) focuses on behavioral change at the level of the species. Foegl *et al.* (1966) firstly used Evolutionary Programming (EP) to generate finite automata. Since then, EP was substantially revised by Yao *et al.* (1999), Iwamatsu (2002), Ji *et al.*, (2004), Zhang and Lu (2008), and many others. The target of EP is to design a finite automata of the best predictive ability. The basic procedure of EP is as follows (Eberhart and Shi, 2007):

(1) Initialize population.
(2) Calculate fitness values of individuals.
(3) Mutating to produce offspring.
(4) Evaluate individuals and their offspring.
(5) Select individuals from parents and their offspring to construct a new population.
(6) Repeat steps from (2) to (5), until the termination condition is met.

In view of agent-based modeling, individuals in evolutionary algorithms can be treated as agents/components.

11.3 Evolutionary Strategies (ES)

ES is an evolutionary algorithm independently developed by Rechenberg (1965, 1971, 1973) and Schwefel (1965, 1975), inspired by biological mutation. ES stresses the behavioral changes at the level of the individual. ES has been further improved by Schewefel (1981),

Beyer (2001), Zhou and Li (2003), Arnold and Beyer (2006), and many other researchers. ES procedure is as follows:

(1) Construct fitness (adaptation) function, $f(p)$.
(2) Randomly select individuals, $p_i \in R^n, i = 1, 2, \ldots, N$, to produce initial population.
(3) The element p_{ij} of parent p_i reproduces offspring, p'_i, following the rule, $p'_{ij} = p_{ij} + N(0, \sigma)$, where $N(0, \sigma)$ is a random value of normal distribution.
(4) Compare $f(p)$ and $f(p')$, and select N individuals of larger fitness values from $2N$ individuals to evolve to next generation.
(5) Repeat steps from (3) to (4), until the termination condition is met.

11.4 Differential Evolution Algorithm (DEA)

Differential Evolution Algorithm (DEA) is a global optimization algorithm based on continuous space and real valued encoding (Storn and Price, 1997). It is a population-based random search algorithm. Different from conventional evolutionary algorithms, DEA uses the difference-based mutation. Through difference mutation, the target individual $x_i^{(g)}$ in generation g reproduces the offspring individual, $v_i^{(g+1)}$, and the weighted difference values of two target individuals are added to the third individual. DEA procedure is as follows (Liu and Gu, 2013):

(1) Produce an initial population. Randomly produce N individuals following the rule $x_{ij}(0) = x_{ij}^L + rand(0,1)(x_{ij}^U - x_{ij}^L)$ where $rand(0,\ 1)$ is a random value of uniform distribution in $[0,\ 1]$.
(2) Mutation. Randomly select three individuals, X_{p1}, X_{p2}, X_{p3}, where $p1 \neq p2 \neq p3 \neq i$, from the Population, and let $h_{ij}(g) = x_{p1} + F(x_{p2j} - x_{p3j})$.
(3) Crossover. Let

$$\begin{aligned} v_{ij}(g+1) &= h_{ij}(g), rand(0,1) \leq \text{CR}, \text{ or } j = rand(1,n) \\ &= x_{ij}(g), rand(0,1) > \text{CR}, \text{ or } j \neq rand(1,n) \end{aligned}$$

where CR is the crossover probability.

(4) Selection. Compare the vectors $v_i(g+1)$ and $x_i(g)$, and let

$$\begin{aligned} x_i(g+1) &= v_i(g+1), \qquad f(v_i(g+1)) < f(x_i(g)) \\ &= v_i(g), \qquad\quad\;\; f(v_i(g+1)) \geq f(x_i(g)) \end{aligned}$$

(5) Repeat steps from (2) to (4), until the termination condition is met.

11.5 Quantum-inspired Evolutionary Algorithm (QEA)

Combining quantum theory and evolutionary algorithms, Narayanan *et al.* (1966) proposed the concept of quantum-inspired genetic algorithm. Han and Kim (2000) developed a genetic quantum algorithm. It was further expanded as the Quantum-inspired Evolutionary Algorithm (QEA) (Han *et al.*, 2001). The procedure of QEA is (Wei *et al.*, 2011):

(1) Initialize the evolution time ($t = 0$), and the population $Q(t)$.
(2) Produce $P(t)$ from $Q(t)$.
(3) Perform the crossover and mutation operations to produce $P(t)$.
(4) Evaluate the fitness of $P(t)$, and preserve the optimal solution.
(5) Termination judge. If the termination condition is met, print the optimal individual, and the algorithm terminates; otherwise continue.
(6) Update $Q(t)$, let $t = t + 1$, and go to step (2).

QEA has been successfully used in the optimization of artificial neural networks, engineering, and industrial production.

11.6 Genetic Algorithms (GAs)

GAs are stochastic global search methods that mimics the metaphor of natural biological evolution (Chipperfield *et al.*, 2014). They are substantially evolutionary algorithms. According to John Holland, GAs are adaptive plans. In an adaptive plan, structures in a search space are selected and continuously modified by some rules according to the quality of their performance in the past. In a GA, a set of structures is a population and each structure is genotype/individual. A genotype/individual is the expressions of features in the structures of the problem domain. Adaptive operators (genetic operations) include selection, crossover, and mutation. GAs operate on a population using the principle of survival of the fittest, in order to produce better and better approximations to a solution. Genetic operations are used in each generation to mimic natural adaptation, which leads to the evolution of individuals that are better suited to their environment than their parent individuals (Chipperfield *et al.*, 2014).

11.6.1 *Simple GA (SGA)*

Consider a optimization problem as the following (Wang and Cao, 2002):

$$\max f(x) = x\sin(10\pi x) + 2, x \in [-1, 2]$$

The simple genetic algorithm (SGA) to solve above problem is as follows (Goldberg, 1989a, 1989b; Wang and Cao, 2002; Chipperfield *et al.*, 2014):

(1) Encoding. The real variable x here is phenotype. We try to find a mapping from phenotype to genotype, i.e., encoding. The x is usually expressed as a binary string (an individual) and the string length is determined by desired accuracy. For example, x with six decimal of accuracy requires a binary string of 22 bits.
(2) Produce an initial population. An individual is a binary string. Set crossover probability P_c, mutation probability P_m, and termination rules; produce an initial population $P(0)$ of N individuals being generated randomly , and let $t = 0$.
(3) Calculate fitness of all individuals in population $P(t)$. In this example, we may directly use the object function, $f(x)$, as the fitness function, i.e., $f(s) = f(x)$, where s is the binary string of x. The individual with the greatest $f(s)$ is the most optimal individual.

(4) Genetic operation (population evolution). Three genetic operations, selection, crossover (recombination), and mutation are performed.

(a) Selection (parents). Suppose the following two parent individuals have been selected by roulette wheel selection. First the single point crossover is conducted:

$$s_2 = < 00000\ 01110000000010000 >$$

$$s_3 = < 11100\ 00000111111000101 >$$

In total of $M/2$ pairs of parents are selected from $P(t)$, where $M \geq N$.

(b) Crossover. Select a crossover point randomly, i.e., the point between bit 5th and bit 6th, as indicated above. The crossover operation is performed to produce two new offspring strings:

$$s_2' = < 00000\ 00000111111000101 >$$

$$s_3' = < 11100\ 01110000000010000 >$$

And the fitness values of the two individuals are $f(s_2') = 1.940865$, and $f(s_3') = 3.459245$, respectively. Obviously the fitness value of offspring s_3' is higher than its two parents.

Crossover is performed to $M/2$ pairs of parents selected, with the probability of P_c, and thus resulting M intermediate individuals.

The crossover operation is not necessarily performed on all strings in the population. Instead, it is applied with a probability when the pairs are chosen for breeding (Chipperfield *et al.*, 2014).

(c) Mutation. Perform mutation to M intermediate individuals respectively, with the probability P_m, and produce M candidate individuals. Mutation causes the change of individual genetic representation based on some probabilistic rule. For example, mutating the fifth bit of s_3' leads to the new string:

$$s_3' =< 11101\ 00000111111000101 >$$

Mutation is generally considered to be an operator that ensures that the probability of searching a particular subspace of the problem space is never zero (Chipperfield *et al.*, 2014).

It tends to inhibit the possibility of converging to a local optimum.

(d) Selection (offspring). After above operations have been performed, the objective function can be evaluated and a fitness value may be assigned to each M candidate individual. Select N individuals from M candidate individuals according to their fitness values, and produce a new population $P(t+1)$.

The process continues generation by generation. By doing this, the average performance of individuals in the population is expected to increase. Highly fit individuals are preserved and have a high probability of being selected for mating and the less fit individuals die out (Chipperfield *et al.*, 2014).

The algorithm terminates when some criteria are satisfied, that is, a certain number of generations, a mean deviation in the population, or when a particular point in the search space is encountered; go to the next step; or else $t = t+1$, go to step (3)/(4).

(5) Simulation result. For the example above, suppose the initial population size 50, crossover probability 0.25 and mutation probability 0.01. The optimum individual after 89 generations is $s_{max} =< 1101001111110011001111 >$, which corresponds to $x_{max} = 1.850549$ and max $f(x_{max}) = 3.850274$.

A pseudo-code of the SGA (Chipperfield *et al.*, 2014) is indicated bellow. Here $p(t)$ is the population at time t (the initial population is $p(0)$):

```
procedure GA {
t=0;
initialize p(t);
evaluate p(t);
while not-terminated {
t=t+1;
select p(t) from p(t-1);
reproduce pairs in p(t);
evaluate p(t); }}
```

From the above algorithm, it can be seen that GAs differ from conventional search and optimization methods. The most significant differences include (Chipperfield *et al.*, 2014): (1) GAs search a population in parallel; (2) GAs do not require derivative information or other auxiliary knowledge — only the objective function and corresponding fitness values affect the directions of search; (3) GAs use probabilistic transition rules rather than deterministic ones; and (4) GAs work on an encoding of the parameter set, rather than the parameter set itself.

So far, numerous revisions of SGA have been developed. GAs have also been used in various areas, for example, Feng and Zhang (2004) used GA to improve BP algorithm. Function optimization is the classic applications of GAs. For some optimization problems of non-linear, multi-model, and multi-objective functions, GAs perform better than conventional methods. GAs have been successfully used to solve the traveling salesman problem, packing, layout optimization, graphic delineation, and other optimization problems with NP difficulty. In addition, they are used in automatic control, robotics, and image processing.

11.6.2 *Encoding, genetic operations, and fitness functions*

11.6.2.1 *Encoding*

Encoding is not necessarily binary encoding. However, encoding should meet the requirements: completeness, closure, compactness, scalability, multiplicity, flexibility, modularity, redundancy, and complexity (Wang and Cao, 2002).

11.6.2.2 *Fitness functions and fitness scaling*

Without the use of any external information, GA uses fitness function only in the evolutionary search. Fitness function determines the convergence of GA. Generally, the fitness function is transformed from objective function $f(x)$. Fitness scaling refers to the transformation of range of objective function. A better fitness function should be single-valued, continuous, non-negative, and maximized. It is uniform, widely applicable, and have less calculation cost. Several often used fitness functions are (Wang and Cao, 2002):

(1) Fit $(f(x)) = f(x)$ for maximizing problem, or Fit $(f(x)) = -f(x)$ for minimizing problem.
(2) Fit $(f(x)) = c_{max} - f(x)$, if $f(x) < c_{max}$
    $=0$, otherwise
    for minimizing problem, or
    Fit $(f(x)) = f(x) - c_{min}$, if $f(x) > c_{min}$
    $=0$, otherwise
    for maximizing problem.
(3) Fit $(f(x)) = 1/(1 + c + f(x))$, if $c \geq 0$, and $c + f(x) \geq 0$ for minimizing problem, or
    Fit $(f(x)) = 1/(1 + c - f(x))$, if $c \geq 0$, and $c - f(x) \geq 0$ for maximizing problem.

Fitness scaling includes some methods as follows (Wang and Cao, 2002)

(1) Linear transformation: $f' = \alpha f + \beta$
(2) Power function transformation: $f' = f^k$
(3) Exponential function transformation: $f' = \mathrm{e}^{-\alpha f}$

11.6.2.3 *Genetic operations*

(1) Selection
    Each individual in the population has a probability to be selected. The used assignment of selection probability includes (Wang and Cao, 2002): (a) proportional fitness assignment, e.g., the selection probability of an individual i is $p_i = f_i / \sum_{j=1}^{m} f_j$, where f_i is the fitness value of individual i; (b) rank-based fitness assignment, i.e., for nonlinear rank, $\mathrm{Fit(Pos)} = NX^{Pos-1} / \sum_{i=1}^{N} X^{i-1}$, where N: population size, *Pos*: rank of the individual, and X is the root of

$$(Y - 1)X^{N-1} + YX^{N-2} + \ldots + YX + Y = 0, Y \in [1, N - 2]$$

For linear rank, $\text{Fit}(\text{Pos}) = 2 - Y + 2(Y - 1)(Pos - 1)/(N - 1)$, where $Y \in [1, 2]$.

Selection methods include roulette wheel selection, stochastic universal sampling, local selection, and tournament selection.

(2) Crossover (recombination)

Crossover is the process of exchanging the parts of two parents' structure to produce new offspring. Two methods, real value crossover and binary crossover (single point crossover: crossover occurs in a single point; multi-point crossover: crossover occurs randomly at several points; uniform crossover: each point is the potential crossover point, and crossover occurs uniformly at each point), can be used. In addition, partially matched crossover, ordered crossover, cycle crossover, shuffle crossover, and crossover with reduced surrogate, etc., are used in crossover operations.

(3) Mutation

Mutation itself is a local random search, which ensures the convergence of GA and diversity of population. However, the mutation with mutation probability greater than 0.5 may lead to complete random search. Two methods, real value mutation and binary mutation (for each individual, the value of variable is randomly changed) are usually used. Other mutation methods include displacement, duplication, addition, and deletion, etc.

11.6.3 *Some mathematical principles for GAs*

A schema refers to the structure with the same characteristic bits in various gene strings. It represents the same configuration in different strings (Wang and Cao, 2002). It is a subset of a string. The number of determined (known) bits in the schema H is called the schema order of $H, O(H)$. Schema order reflects the between-schema difference in certainty. A greater schema order means the higher certainty of the schema. For a schema H, the distance between the first known bit and the last known bit is called the defining length of $H, \delta(H)$.

11.6.3.1 *Schema theorem*

Schema Theorem (Holland, 1975): Under the action of genetic operations such as selection, crossover, and mutation, the schemas with the lower order, short defining length, and higher average fitness than that of population will exponentially grow in the offspring generations.

Schema theorem is the fundamental theory of GAs. It guarantees the exponential growth of better schemas (i.e., more optimal solutions). However, schema theorem is currently valid for binary encoding only. In addition, it does not prove the convergence of GAs.

The schemas with the lower order, short defining length and higher average fitness than that of population are called building blocks (Wang and Cao, 2002). They combine to reproduce individuals with even higher fitness through genetic operations, and thus produce more optimal solutions. Building blocks hypothesis holds that a GA can eventually approximate the global optimal solution through the combination of building blocks under the operation of genetic operators.

11.6.3.2 *Walsh schema transformation (Goldberg 1989a, 1989b)*

Walsh functions are defined as $\Psi_j(X) = (-1)^\eta, \eta = \sum_{i=0}^{l-1} j_i x_i$, where $X = (x_{l-1}x_{l-2...}x_0)$, and $x_i = \{0,1\}, i = 0,1,\dots,l-1$. The base functions of Walsh functions are orthogonal functions and meet the condition:

$$\sum_{X=0}^{2\wedge l-1} \Psi_i(X)\Psi_j(X) = 2^l \quad \text{if } i = j$$
$$= 0 \quad \text{if } i \neq j$$

Any function $f(X)$, with the domain of X of $\{0,1,\dots,2^l-1\}$, can be expanded and expressed as the Walsh polynomial:

$$f(X) = \sum_{X=0}^{2\wedge l-1} \omega_j \Psi_j(X)$$

where the Walsh coefficient $\omega_j = (\sum_{X=0}^{2\wedge l-1} f(X)\Psi_j(X))/2^l$. The average fitness of a schema H can be calculated following the steps (Wang and Cao, 2002):

(1) calculate $f(X)$;
(2) calculate Walsh coefficient using the Walsh transformation $f(H) = (\sum_{X\in H} f(X))/|H| = \sum_{j=0}^{2\wedge l-1} \omega_j \sum_{X\in H} \Psi_j(X)/|H|$; and
(3) Calculate the value of average fitness of H based on Walsh coefficient and Walsh transformation table.

Chapter 12

Synergy: Correlation Analysis

Correlation analysis helps to exploit the synergic mechanism. This chapter describes some general correlation measures and their partial correlations.

12.1 Pearson Linear Correlation and Partial Linear Correlation Analysis

In a series of earlier studies (Zhang, 2007, 2011d, 2012a, 2012b), methodology for constructing ecological networks by correlation analysis of community sampling data have been proposed. I mentioned that a statistically significant Pearson linear correlation means an indirect or direct interaction between two taxa, and a statistically significant partial (net or pure) correlation based on Pearson linear correlation means a candidate direct interaction between two taxa. For studies of ecological communities and ecosystems, there are generally two types of interactions — direct interactions and indirect interactions. Direct interactions refer to predation, parasitism, competition, amensalism, mutualism, protocooperation, commensalism, etc. An interaction means a dependency relationship in state changes of two taxa (direct interaction). Conversely, a seeming dependency relationship in state changes of two taxa does not necessarily mean an interaction (indirect interaction). People always use Pearson linear correlation to find interactions (Goh *et al.*, 2000; Pazos and Valencia, 2001; Tu, 2006). However, some undeterministic interactions may be found and some candidate interactions may be missed when using this method, as pointed out by Zhang (2011d). In the present chapter, we discuss the error of predicting interactions with correlation analysis (Zhang and Li, 2015).

12.1.1 *Methods*

A network is globally linear network, quasi-linear network, or non-linear network. Furthermore, a network changed in a local domain (short time, small extent) can be approximated as a linear network (Zhang, 2011d, 2012a, 2012b), i.e., in the local domain, all between-node (or -taxon, -component, etc) changes are treated as linear ones, and linear correlation measures can thus be used. Pearson linear correlation is a measure to reflect the linear dependence between two taxa. A statistically significant Pearson linear correlation represents a direct or indirect linear interaction between two taxa. Partial (net or pure) linear correlation is based on Pearson linear correlation. It has eliminated the indirect effects produced by the remaining taxa. A statistically significant partial linear correlation represents a candidate direct linear interaction between two taxa (Zhang, 2007, 2011, 2012a, 2012b). In present study, we treated the linear interactions, predicted by partial linear correlation, as candidate direct interactions.

Data of various biological networks were obtained from that of Zhang (2011d). These biological networks are different in countries, years, seasons, types of taxa, and number of taxa. Therefore, we expect the wide representativeness of conclusions drawn from them.

12.1.2 *Results*

12.1.2.1 *Relationship between partial linear correlation and Pearson linear correlation*

Partial linear correlation (y) is approximately half of the Pearson linear correlation (x), as indicated by:

$$y = -0.0064 + 0.4785x, r^2 = 0.173, p < 0.00001, n = 1447$$

This relationship means that indirect interactions in a network increase mean interaction strength of taxa in the network.

12.1.2.2 *Error estimation*

Our results are listed in Table 1. In all predicted candidate interactions by partial linear correlation, about 34.35% (x, $0 \sim 100\%$) of them are not successfully detected by linear correlation. In all predicted interactions by Pearson linear correlation, 50.58% (y, $0 \sim 100\%$) of them are undeterministic interactions, i.e., not successfully detected by partial linear correlation. In all predicted interactions by Pearson linear correlation, 49.42% (z, $0 \sim 100\%$) of them are candidate interactions, i.e., successfully detected by partial linear correlation.

It is found that the Pearson linear correlations between N and x, y, z are -0.49 ($p = 0.07$), 0.48 ($p = 0.18$) and -0.48 ($p = 0.18$), respectively. The proportion of missed (x), mis-predicted (y) and precisely predicted (z) candidate direct interactions by Pearson linear correlation analysis decreases, increases, and decreases with the number of taxa respectively. However, statistically there is no significant linear dependency between number of taxa (N) and these percentage indices.

12.1.2.3 *Precisely predicted (z) candidate direct interactions by Pearson linear correlation analysis*

The results showed that the precisely predicted (z) candidate direct interactions by Pearson linear correlation analysis are not necessarily those with the highest Pearson linear correlations. For example, the predicted interactions by Pearson linear correlation analysis (and Pearson linear correlations in parentheses) of network CN-06Sep (taxon: family) are as follows, of which italic interactions are candidate direct interactions:

(1,2)(0.3962)
(2,3)(0.4021)
(3,4)(0.3877)
(4,10)(0.3531) *(4,16)(0.4379)* *(4,18)(0.4954)*
(5,12)(0.4095) *(5,13)(0.6502)*
(6,17)(0.43)

Table 1. Comparison of results of Pearson linear correlation (PLC) and partial linear correlation (Partial PLC).

Network ID (data set)	No. taxa (N)	No partial PLC> PLC	No. absolute partial PLC >absolute PLC(NAP)	NAP/N(%)	No. SS yes partial PLC but SS not PLC (SSN)	$x = SSN/(SSN+SYY)$ (%)	No. SS Yes PLC but SS not partial PLC SPN	$y = SPN/SPN+SYY)$ (%)	No. SS Yes PLC & SS yes partial PLC (SYY)	$z = SYY/(SPN+SSY)$ (%)
CN-06sep	4	0	0	0.00	1	**100**	1	**50.00**	1	**50.00**
CN-06sep	4	2	3	50.00	2	**100**	0	**0.00**	2	**100.00**
CN-06Oct	4	2	3	50.00	2	**100**	1	**50.00**	1	**50.00**
PH-Mar	21	99	126	60.00	0	**0**	6	**66.67**	3	**33.33**
PH-Apr	20	57	106	55.79	4	**33.33**	14	**63.64**	8	**36.36**
PH-Sep	21	70	122	58.10	1	**25**	10	**76.92**	3	**23.08**
PH-Oct	21	98	100	47.62	0	**0**	8	**72.73**	3	**27.27**
PH-Mar	7	10	12	57.14	1	**50**	0	**0.00**	1	**100.00**
PH-Apr	7	4	6	28.57	0	**0**	5	**71.43**	2	**28.57**
PH-Sep	7	5	7	33.33	0	**0**	5	**83.33**	1	**16.67**
PH-Oct	7	11	13	61.90	0	**0**	0	**0.00**	2	**100.00**
CN-06Sep	23	120	174	68.77	3	**25**	10	**52.63**	9	**47.37**
CN-06Oct	23	116	127	50.20	1	**14.29**	10	**62.50**	6	**37.50**
CN-06Oct	27	168	248	70.66	5	**33.33**	14	**58.33**	10	**41.67**
Mean						**34.35**		**50.58**		**49.42**
$\pm$(p **≤0.05**)						**77.76**		**58.19**		**58.19**

No. SS Yes Partial PLC but SS Not PLC: Total No. of statistically significant Partial PLC ($p \le 0.01$) but statistically not significant PLC ($p \le 0.01$); No. SS Yes PLC but SS Not Partial PLC: Total No. of statistically significant PLC ($p \le 0.01$) but statistically not significant Partial PLC ($p \le 0.01$); No. SS Yes PLC & SS Yes Partial PLC: Total No. of both statistically significant PLC ($p \le 0.01$) and statistically significant Partial PLC ($p \le 0.01$).

(8,11)(0.4017) (8,12)(0.5605)
(10,12)(0.3611) (10,18)(0.4883)
(11,12)(0.3784) (11,18)(0.451)
(12,13)(0.4703)
(14,22)(0.4201)
(16,21)(0.3541)
(17,20)(0.43)

Therefore we should not try to choose a portion [e.g., 49.42% (z), as calculated above] of predicted interactions with the greatest Pearson linear correlations as candidate direct interactions.

The proportions of missed (x), mis-predicted (y), and precisely predicted (z) candidate direct interactions by Pearson linear correlation analysis have significant biological meaning, for example, if we want to detect candidate direct interactions between species by Pearson linear correlation analysis, a false conclusion may sometimes be drawn (y). For example, Tu (2006) found that some true interactions (i.e., the interactions confirmed by experiments; A candidate direct interaction is a true interaction if it is confirmed by experiments) between proteins that have no statistically significant Pearson linear correlation. In medical science (biological control, biodiversity conservation), assume there is a true direct interaction between two proteins (a predator and a prey, two species), A and B. We want to develop a medicine (release A, release A) to affect A or B (to control B, to balance B) directly, and hope that the measure will make A affect (control, balance) B. However, we know that A may likely not affect (control, balance) B due to the proportion x (medicine failure, failure of biological control, failure of balancing plan).

According to Tu (2006), the proportions of missed (x) and precisely predicted (z) true direct interactions by Pearson linear correlation analysis in small-cell lung cancer (SCLC; $N = 19$ protein sequences) were 25% and 75%, respectively. For non-small-cell lung cancer (NSCLC; $N = 92$ protein sequences), the proportions of missed (x) and precisely predicted (z) true direct interactions by Pearson linear correlation analysis were 42.5% and 57.5%, respectively. The results (means of SCLC and NSCLC: $x = 33.8\%$,

$z = 66.3\%$) are of better coincident with our conclusions ($x \approx 34.35\%$; $y \approx 49.42\%$). The results of Tu indicated that true direct interactions by Pearson linear correlation analysis are not necessarily those with the highest Pearson linear correlations, which also confirmed our conclusion.

In ecological networks (systems), the phenomena above are popular. For example, it is well known that ladybird *Coccinella septempunctata* is the natural enemy of cotton aphid *Aphis gossypii* Glover, and thus there is a true (direct) interaction between the two species. However, in some cotton fields, the lady bird lacks foods and mainly feeds on the cotton aphid, and the Pearson linear correlation between them is thus likely significant (the interaction belongs to z). In other cotton fields, the ladybird has many available foods and the Pearson linear correlation between them is thus likely insignificant (the interaction belongs to x). In this situation, the ladybird, as a natural enemy of cotton aphid, is ineffective.

In the present study, we used the significance level $p < 0.01$. To avoid missing candidate interactions as much as possible, the significance level can be adjusted to a reasonable value, for example, $p < 0.05$.

In general, we suggest jointly using Pearson linear correlation and partial linear correlation to analyze various interactions. Candidate direct interactions detected by both linear correlation measures should be the most focused interactions, which have the most significant biological meaning, seconded by those interactions detected by only partial linear correlation and by Pearson linear correlation.

12.2 Partial Correlation of Some Correlation Measures

To eliminate indirect between-attribute interactions of other attributes not being tested, partial linear correlation of Pearson linear correlation measure has been widely used. However for most general correlation measures, partial correlation measures are not available. In the present chapter, I describe some correlation measures and partial correlation measures for some correlation measures (Zhang, 2015b).

12.2.1 *Methods*

12.2.1.1 *Correlation measures*

There are a variety of correlation measures (Zhang, 2007c, 2011d, 2012a, 2012b; Zhang *et al.*, 2014b). I proposed three correlation measures, revised Dice coefficient, overlap coefficient, and proportion correlation, and use more correlation measures to derive their partial correlations, in exception of Pearson linear correlation which has been affiliated with partial linear correlation (Zhang, 2011d, 2012a; Zhang and Li, 2015), among which Jaccard correlation, revised Dice coefficient, overlap coefficient, and point correlation are for binary attributes (nodes, taxa, variables, etc), and Spearman rank correlation and proportion correlation are for interval value attributes (nodes, taxa, variables, etc.).

First, assume there are m attributes (nodes, taxa, variables, etc) and n samples (components, members, etc), and the raw data is a matrix (x_{ij}), $i = 1, 2, \ldots, m; j = 1, 2, \ldots, n$.

Jaccard correlation is:

$$r_{ij} = (e - (c + b))/(e + c + b) \quad i, j = 1, 2, \ldots, m$$

where $-1 \leq r_{ij} \leq 1, c$ is the number of sample pairs of 1 for attribute i but not for $j; b$ is the number of sample pairs of 1 for attribute j but not for $i; e$ is the number of sample pairs of 1 for both attribute i and attribute j.

Revised Dice coefficient is:

$$r_{ij} = 4^{*}e/(b + c) - 1 \qquad i, j = 1, 2, \ldots, m$$

where $-1 \leq r_{ij} \leq 1, b$ is the number of sample of 1 for attribute $i; c$ is the number of sample of 1 for attribute $j; e$ is the number of sample pairs of 1 for both attribute i and attribute j.

Point correlation is (Zhang, 2007c)

$$r_{ij} = (ad - bc)/((a + b)(c + d)(a + c)(b + d))^{1/2} \qquad i, j = 1, 2, \ldots, m$$

where $-1 \leq r_{ij} \leq 1$, both attribute i and attribute j take values 0 or 1. a is number of both attribute i and attribute j take value 0, b is number of attribute i takes 0 and attribute j takes 1, c is number of attribute i takes 1 and attribute j takes 0, and d is number of both attribute i and attribute j take value 1.

Overlap coefficient is:

$$r_{ij} = 2e/(e+b) - 1 \qquad i,j = 1,2,\ldots,m$$

where $-1 \leq r_{ij} \leq 1, b$ is the number of sample pairs with different values; is the number of sample pairs of 1 for both attribute i and attribute j.

Proportion correlation is:

$$r_{ij} = 2\sum_{k=1}^{n}(p_{ik}p_{jk}) \Big/ \left(\sum_{k=1}^{n} p_{ik}^2 \sum_{k=1}^{n} p_{jk}^2\right)^{1/2} - 1 \qquad i,j = 1,2,\ldots,m$$

where $-1 \leq r_{ij} \leq 1$, and $p_{il} = x_{il,}/\sum_{k=1}^{n} x_{ik}$, $p_{jl} = x_{jl,}/\sum_{k=1}^{n} x_{jk}$, $l = 1,2,\ldots,n$. It is used specifically to the situations of $r_{ij} \geq 0, i = 1,2,\ldots,m; j = 1,2,\ldots,n$.

Spearman rank correlation is (Spearman, 1904; Schoenly and Zhang, 1999):

$$r_{ij} = 1 - 6^{*}\sum d^2 \Big/ [n(n^2-1)] \qquad i,j = 1,2,\ldots,m$$

where $-1 \leq r_{ij} \leq 1, d = r(i) - r(j)$, and $r(i)$ and $r(j)$ are rank of an element in the attribute i and attribute j, from the smaller to the larger values in n samples.

For the correlation measures, calculate $t = |r_{ij}|/[(1-r_{ij}^2)/(n-2)]^{1/2}$, and if $t > t_\alpha(n-2)$, the correlation (either positive or negative correlation) between attributes i and j is statistically significant.

12.2.1.2 *Partial correlations*

The partial correlation between two attributes eliminates the indirect effects produced by the remaining attributes. Let $R = (r_{ij})$, where

r_{ij} are correlations calculated above. Calculate

$$\sum_{k=1}^{m} r_{ik} r_{kj}^{-} = 0 \qquad i,j = 1,2,\ldots,m; i \neq j$$

$$\sum_{k=1}^{m} r_{ik} r_{ki}^{-} = 1 \qquad i = 1,2,\ldots,m$$

That is, calculate $R^{-} = (r_{ij}^{-})$ such that $RR^{-} = I$ and I is the unit matrix. Partial correlation between attributes i and j is

$$parr_{ij} = -r_{ij}^{-}/(r_{ii}^{-*} r_{jj}^{-})^{1/2} \qquad i,j = 1,2,\ldots,m; i \neq j$$

where $-1 \leq parr_{ij} \leq 1$. It should be noted that a singular or ill-condition matrix, R, will lead to the wrong results.

For the partial correlation measures, calculate $t = |parr_{ij}|/[(1 - parr_{ij}^{2})/(n - m)]^{1/2}$, and if $t > t_{\alpha}(n - m)$, the partial correlation (either positive or negative correlation) between attributes i and j is statistically significant, and there is a direct interaction and interdependence between attributes i and j in terms of the correlation measure being used.

12.2.2 *Matlab codes*

The Matlab codes of the full algorithm above and of finding interactions are listed as follows, of which the Pearson linear correlation measure (for interval value attributes) is included also:

```
measure=input('Input a number for correlation measure (1: Pearson linear
   correlation; 2: Spearman rank
correlation; 3: Proportion correlation; 4: Point correlation; 5: Jaccard
   correlation; 6: Revised Dice coefficient; 7:
Overlap coefficient)')
sig=input('Input significance level(e.g., 0.01)')
%X is m*n raw data matrix. m: number of attributes (taxa, etc); n: number of
   samples.
dim=size(X);
m=dim(1); n=dim(2);
r=zeros(m);
```

```
switch measure
     case 1
          for i=1:m-1; for j=i+1:m; r(i,j)=linearcorre(X(i,:)',X(j,:)');
            end;end;
     case 2
          for i=1:m-1; for j=i+1:m; r(i,j)=spearman(X(i,:)',X(j,:)');end;end;
     case 3
          for i=1:m-1; for j=i+1:m;
r(i,j)=propcorre(X(i,:)',X(j,:)');end;end;
     case 4
          for i=1:m-1; for j=i+1:m; r(i,j)=pointcorre(X(i,:)',X(j,:)');
            end;end;
     case 5
          for i=1:m-1; for j=i+1:m; r(i,j)=jaccard(X(i,:)',X(j,:)');end;end;
     case 6
          for i=1:m-1; for j=i+1:m; r(i,j)=reviseddice(X(i,:)',X(j,:)');
            end;end;
     case 7
          for i=1:m-1; for j=i+1:m; r(i,j)=overlap(X(i,:)',X(j,:)');end;end;
end
for i=1:m-1; for j=i+1:m; r(j,i)=r(i,j);end;end;
for i=1:m; r(i,i)=1;end;
disp('Correlation matrix')
r
tvalues=abs(r)./sqrt((1-r.^2)/(n-2));
alpha=(1-tcdf(tvalues,n-2))*2;
sigmat=alpha<sig;
sigmat=sigmat.*r-eye(m);
sigmatr=sigmat;
disp('Pairs with statistically significant correlation')
if (sigmat~=ones(m))
[pairx,pairy,rvalues]=find(sigmat);
temp1=pairx; temp2=pairy;
pairxs=pairx(temp1<temp2);
pairys=pairy(temp1<temp2);
rvaluess=rvalues(temp1<temp2);
```

```
PairsAndCorrelations=[pairxs pairys rvaluess]
else
disp('No significant pairs')
end
inversr=inv(r);
for i=1:m-1; for j=i+1:m;
parr(i,j)=-inversr(i,j)/sqrt(inversr(i,i)*inversr(j,j));end;end;
for i=1:m-1; for j=i+1:m; parr(j,i)=parr(i,j);end;end;
for i=1:m; parr(i,i)=1;end;
disp('Partial correlation matrix')
parr
if (n>m)
tvalues=abs(parr)./sqrt((1-parr.^2)/(n-m));
alpha=(1-tcdf(tvalues,n-m))*2;
sigmat=alpha<sig;
sigmat=sigmat.*parr-eye(m);
sigmatparr=sigmat;
disp('Pairs with statistically significant partial correlation')
if (sigmat~=ones(m))
[pairx,pairy,rvalues]=find(sigmat);
temp1=pairx; temp2=pairy;
pairxs=pairx(temp1<temp2);
pairys=pairy(temp1<temp2);
rvaluess=rvalues(temp1<temp2);
PairsAndPartialCorrelations=[pairxs pairys rvaluess]
else
disp('No significant pairs')
end
else
disp('The number of samples is not enough to support the statistic test of
  partial correlations')
end
x=sigmatparr & (~sigmatr);
y=(~sigmatparr) & sigmatr;
z=sigmatparr & sigmatr;
for i=1:3;
```

```
switch i
    case 1
        mat=x; s='Significant partial correlation but insignificant linear
          correlation';
    case 2
        mat=y; s='Significant linear correlation but insignificant partial
          correlation';
    case 3
        mat=z; s='Significant both partial correlation and correlation';
end;
[pairx,pairy]=find(mat);
temp1=pairx; temp2=pairy;
pairxs=pairx(temp1<temp2);
pairys=pairy(temp1<temp2);
disp([s])
SignificantPairs=[pairxs pairys]
end;
```

Seven M function files, linearcorre.m, spearman.m, propcorre.m, pointcorre.m, jaccard.m, reviseddice.m, and overlap.m are as follows:

```
function linearcor=linearcorre(x,y)     %x and y: two column vectors to be
tested.
m=max(size(x));
if(m~=max(size(y)))
  error('Array sizes do not match.');
end
if((min(size(x))~=1)|(min(size(y))~=1))
  error('Both x and y are vectors');
end
xbar=mean(x);
ybar=mean(y);
aa=sum((x-xbar).*(y-ybar));
bb=sum((x-xbar).^2);
cc=sum((y-ybar).^2);
linearcor=aa/sqrt(bb*cc);
```

```
function spearm=spearman(x,y)           %x and y: two column vectors to be
tested.
if(max(size(x))~=max(size(y)))
  error('Array sizes do not match.');
end
if((min(size(x))~=1)|(min(size(y))~=1))
  error('Both x and y are vectors');
end
n=max(size(x));
for i=1:n
rx(i)=0;ry(i)=0;xx(i)=0;yy(i)=0;
end
for j=1:n
nx=1;ny=1;
for i=1:n
if (x(i)<x(j)) nx=nx+1; end
if (y(i)<y(j)) ny=ny+1; end
end
rx(j)=nx;
ry(j)=ny;
end
for j=1:n
if (rx(j)==(n+1)) continue; end
nx=rx(j);
ntie=-1;
for i=1:n
if(rx(i)~=nx) continue; end
ntie=ntie+1;
xx(i)=rx(i);
rx(i)=0;
end
for i=1:n
if (rx(i)~=0) continue; end
xx(i)=xx(i)+(ntie*0.5);
rx(i)=n+1;
end
```

```
end
for j=1:n
if (ry(j)==(n+1)) continue; end
ny=ry(j);
ntie=-1;
for i=1:n
if (ry(i)~=ny) continue; end
ntie=ntie+1;
yy(i)=ry(i);
ry(i)=0;
end
for i=1:n
if(ry(i)~=0) continue; end
yy(i)=yy(i)+ntie*0.5;
ry(i)=n+1;
end
end
rs=0;
rs=sum((xx-yy).^2);
spearm=1-((6*rs)/(n*(n^2-1)));

function prop=propcorre(x,y)      %x and y: two column vectors to be tested.
m=max(size(x));
if(m~=max(size(y)))
  error('Array sizes do not match.');
end
if((min(size(x))~=1)|(min(size(y))~=1))
error('Both x and y are vectors');
end
xp=x/sum(x);
yp=y/sum(y);
aa=sum(xp.*yp);
bb=sum(xp.^2);
cc=sum(yp.^2);
prop=2*aa/sqrt(bb*cc)-1;
```

```
function pointcor=pointcorre(x,y)     %x and y: two column vectors to be
                                          tested.
if(max(size(x))~=max(size(y)))
 error('Array sizes do not match.');
end
if((min(size(x))~=1)|(min(size(y))~=1))
error('Both x and y are vectors');
end
aa=sum((x==0) & (y==0));
bb=sum((x==0) & (y~=0));
cc=sum((x~=0) & (y==0));
dd=sum((x~=0) & (y~=0));
pointcor=(aa*dd-bb*cc)/sqrt((aa+bb)*(cc+dd)*(aa+cc)*(bb+dd));

function jac=jaccard(x,y)       %x and y: two column vectors to be tested.
if(max(size(x))~=max(size(y)))
 error('Array sizes do not match.');
end
if((min(size(x))~=1)|(min(size(y))~=1))
 error('Both x and y are vectors');
end
bb=sum((x==0) & (y~=0));
cc=sum((x~=0) & (y==0));
dd=sum((x~=0) & (y~=0));
jac=(dd-(cc+bb))/(dd+cc+bb);

function dic=reviseddice(x,y)     %x and y: two column vectors to be tested.
if(max(size(x))~=max(size(y)))
 error('Array sizes do not match.');
end
if((min(size(x))~=1)|(min(size(y))~=1))
 error('Both x and y are vectors');
end
dd=sum((x~=0) & (y~=0));
bb=sum(x~=0);
cc=sum(y~=0);
```

```
dic=4*dd/(bb+cc)-1;


function overl=overlap(x,y)        %x and y: two column vectors to be tested.
if(max(size(x))~=max(size(y)))
  error('Array sizes do not match.');
end
if((min(size(x))~=1)|(min(size(y))~=1))
  error('Both x and y are vectors');
end
bb=sum(x~=y);
dd=sum((x~=0) & (y~=0));
overl=2*dd/(dd+bb)-1;
```

12.3 Jointly Using Linear Correlation and Rank Correlation to Find Interactions

In earlier studies (Zhang and Li, 2015; Zhang, 2015b), I pointed out that a network changed in a local domain (a short time, a small extent) can be approximated as a linear network, i.e., in the local domain, all between-node (or -taxon, -component, etc) changes are treated as linear ones, and Pearson linear correlation measure can be used. For a little wider domain, the quasi-linear measure, Spearman rank correlation can also be used. Therefore, we may jointly use Pearson linear correlation measure and Spearman rank correlation measure, and their partial correlations to find interactions.

Here I defined (Zhang, 2015a):

A: Interactions detected by statistically significant partial correlations for Pearson linear correlation measure;

B: Interactions detected by statistically significant partial correlations for Spearman rank correlation measure;

C: Interactions detected by statistically significant Pearson linear correlations;

D: Interactions detected by statistically significant Spearman rank correlations.

When finding interactions, some principles should be followed: (1) Pearson linear correlation measure is more reliable than Spearman rank correlation measure; (2) partial correlation measure is more reliable than correlation measure; (3) jointly use of Pearson linear correlation measure and Spearman rank correlation measure is more reliable than the single use of two measures; and (4) jointly use partial correlation measure and correlation measure is more reliable than the single use.

According to the above principles and the rules of set operation, I defined reliability levels (basically, from the set of most reliable interactions, L1, to the sets of most unreliable interactions, L61 and L62) as follows:

L1: A∩B∩C∩D;	(candidate direct interactions)
L21: A∩C; L22: B∩D;	(candidate direct interactions)
L3: A∩B;	(candidate direct interactions)
L41: A; L42: B;	(candidate direct interactions)
L5: C∩D;	(indirect/direct interactions)
L61: C; L62: D.	(indirect/direct interactions)

The Matlab codes of the full algorithm are listed as follows:

```
% X is the m*n raw data matrix. m: number of attributes (taxa, proteins,
    genes, etc); n: number of samples.
sig=input('Input significance level(e.g., 0.01)')
dim=size(X);
m=dim(1); n=dim(2);
if (n<=m)
disp('The number of samples is not enough to support the statistic test of
    partial correlations')
pause
end;
r=corr(X');
disp('Pearson linear correlation matrix')
r
for i=1:m-1; for j=i+1:m; spr(i,j)=spearman(X(i,:)',X(j,:)');end;end;
for i=1:m-1; for j=i+1:m; spr(j,i)=spr(i,j);end;end;
for i=1:m; spr(i,i)=1;end;
disp('Spearman rank correlation matrix')
```

```
spr
for k=1:2;
if(k==1) rr=r; else rr=spr; end;
inverse=inv(rr);
for i=1:m-1; for j=i+1:m; parr(i,j)=-inverse(i,j)/sqrt(inverse(i,i)*
    inverse(j,j));end;end;
for i=1:m-1; for j=i+1:m; parr(j,i)=parr(i,j);end;end;
for i=1:m; parr(i,i)=1;end;
if (k==1)
disp('Partial correlation matrix for Pearson linear correlation')
peparr=parr;
peparr
else
disp('Partial correlation matrix for Spearman rank correlation')
spparr=parr;
spparr
end;
end;
for k=1:2;
if(k==1)
tvalues=abs(r)./sqrt((1-r.^2)/(n-2));
alpha=(1-tcdf(tvalues,n-2))*2;
sigmat=alpha<sig;
sigmatr=sigmat.*r;
partvalues=abs(peparr)./sqrt((1-peparr.^2)/(n-m));
paralpha=(1-tcdf(partvalues,n-m))*2;
parsigmat=paralpha<sig;
parsigmatr=parsigmat.*peparr;
else
tvalues=abs(spr)./sqrt((1-spr.^2)/(n-2));
alpha=(1-tcdf(tvalues,n-2))*2;
spsigmat=alpha<sig;
spsigmatr=spsigmat.*spr;
partvalues=abs(spparr)./sqrt((1-spparr.^2)/(n-m));
paralpha=(1-tcdf(partvalues,n-m))*2;
spparsigmat=paralpha<sig;
```

```
spparsigmatr=spparsigmat.*spparr;
end;
end;
L1=(parsigmat & spparsigmat & sigmat & spsigmat);
                                    % candidate direct interactions
L21=(parsigmat & sigmat); L22=(spparsigmat & spsigmat);
                                    % candidate direct interactions
L3=(parsigmat & spparsigmat);       % candidate direct interactions
L41=parsigmat; L42=spparsigmat;     % candidate direct interactions
L5=(sigmat & spsigmat);             % indirect/direct interactions
L61=sigmat; L62=spsigmat;           % indirect/direct interactions
for i=1:9;
switch (i)
    case 1
        mat=L1; s='L1';
    case 2
        mat=L21; s='L21';
    case 3
        mat=L22; s='L22';
    case 4
        mat=L3; s='L3';
    case 5
        mat=L41; s='L41';
    case 6
        mat=L42; s='L42';
    case 7
        mat=L5; s='L5';
    case 8
        mat=L61; s='L61';
    case 9
        mat=L62; s='L62';
end;
disp(['Interactions (the interactions found in higher levels are also
  listed) with statistically significance for
reliability level: ' s])
[pairx,pairy]=find(mat);
```

```
temp1=pairx; temp2=pairy;
pairxs=pairx(temp1<temp2);
pairys=pairy(temp1<temp2);
InteractionPairs=[pairxs pairys]
end;
```

The M function file, spearman.m, is listed as above.

Chapter 13

Community Succession and Assembly

There are no control and organizing centres in the evolution of community succession and assembly (Whittaker, 1979). It is a self-organizing process (Zhang, 2014b). A biological community is a self-organizing system, generated from repeating species invasions, selection, adaptation, and optimization. According to Young *et al.* (2001), community succession refers to the regular and predictable process of replacement of species composition in the environment such that all species have been eliminated or disturbed. Succession likely leads to the restoration of the community to the previous non-disturbed state. Community assembly is the process that species grow and interact to establish a community. Community assembly stresses the change of community over a single phase (Warren *et al.*, 2003). Community assembly may likely lead to a diverse and stable community.

So far, some mechanisms governing community succession and assembly have been confirmed. For example, a community can have multiple distinct steady distributions or alternative steady states. They represent different species assemblages occurring at possible similar conditions. History of community succession determines which steady state will occur (Hraber and Milne, 1997). Understanding the mechanism of community succession and assembly is the focus of ecologists (May, 1973; Case, 1990; Pimm, 1991; Cohen *et al.*, 1993; Hraber and Milne, 1997; Zhang, 2012b).

This chapter describes advances, methods, and theories in community succession and assembly. Details can be found in Zhang (2012a, 2013a, 2013b, 2013c, 2014a, 2014b, 2014c, 2014d).

13.1 Research Advances of Community Succession and Assembly

Theory on community assembly provides basic rules for species assembly (Chase, 2003; Zhang, 2012c, 2014d). Community assembly rule was first proposed by Diamond in 1975. He held that community assembly is the process that species in the regional species bank join the local community through the multiple-layer filtering of the environment and biological interactions (Diamond, 1975; Wang *et al.*, 2014; Zhang, 2014b). Wilson and Roxburgh (1994) maintained that the rules for plant community assembly are a series of potential rules restricting the presence or increase of species. So far, many rules have been proposed. However, the most accepted rule is species co-occurrence hypothesis proposed by Diamond (1975). About the community assembly mechanism, Fukami (2010) defined the mechanism of community assembly as a construction and conservation process of local community through sequential arrival of species from the external species bank and increase/extinction of species in the community. Ecological niche theory and neutral theory are two important interpretations of the community assembly mechanism. Methods used in community assembly include establishing research plots, indoor simulation, etc. In addition, methods of community succession can be used in community assembly research.

Community succession is one of the most important topics in ecology. It was first discussed by Clements in 1916. Community succession is suggested to be studied at landscape level (spatial heterogeneity and disturbance, etc.) (Margalef, 1968). The rules and directions of community succession have been disputed among ecologists (Connel and Slatyer, 1977). Major theories on community succession include relay floristic hypothesis, initial floristic composition hypothesis, C-S hypothesis, life history strategy hypothesis, resource ratio hypothesis, Odum-Magelef's ecosystem development hypothesis, McMahon system concept model, shifting mosaic steady state hypothesis, and scale and hierarchical system hypothesis, etc. Research of community succession can be performed in two ways: a long period of survey in the same research plot, or multi-site sampling in a large area to replace time costing survey in the same plot.

The theory of community assembly holds that ecological drift and niche differentiation lead to the formation of the community. However, the theory of community succession maintains that the formation of a community is predictable. It is a process towards a certain direction. Young (2001) suggested that both try to understand the mechanism of community formation; they emphasize the history of community formation; both recognize that the community tends to be stable over time, and both recognize the role of interspecific interactions, in particular competition.

13.1.1 *Theories and methods on community assembly*

13.1.1.1 *Concept of community assembly*

Community assembly can be divided into two categories, habitat filtering and micro-environment/resource blocking. Micro-environment/resource herein generally can restrict similar species in the community (Cornwell and Ackerly, 2009). Community assembly might be constrained by local (abiotic characteristics of habitat) or regional (composition of species pool, habitat isolation) factors (Bossuyt *et al.*, 2005).

13.1.1.2 *Theories on community assembly*

(1) *Ecological niche theory*

As early as in the 1895, Warming pointed out that the similar phenotypes shared the same habitat. Niche theory holds that the similar habitats in different regions will finally harbor plants with similar physiological characteristics. And plant phenotypes may match their environments in a predictable way. The theory emphasizes that the differences in habitat or resource requirements of species and in environmental influences greatly affect the outcome of interspecific interactions (Chase and Leibold, 2003).

Competitive exclusion principle, proposed by Gauss, has become one of the core paradigms in community ecology. It points out that two species with the similar ecological niche requirements will be impossible to live together. Competition between species with similar fundamental eco-niches constrains their long term co-existence (Maherali and Klironomos, 2007). Differentiation in habitat, food spectrum, activity time, or other biological habits will occur if they live in the same environment. It means that some species will become extinct if a species successfully joins the community (Weiher and Keddy, 1999). The species belonging to the same family might not survive in the same habitat. The community with more niche resources may accommodate more species (Bossuyt *et al.*, 2005).

(2) *Neutral theory*

Ecological niche theory is unable to explain the extreme biodiversity in rain forests. Thus the neutral theory, i.e., neutral mutation and random genetic drift, originated from molecular evolution, was used to describe community assembly. This theory states that most molecular mutations contributing to the evolution are neutral or nearly neutral in the sense of natural selection; the evolution of neutral mutations is a process of random genetic drift: they may be fixed in the population or disappear. It assumes that all species and individuals at the same trophic level are completely equivalent. Random walk and richness of species in the community are dependent upon the dynamic equilibrium between species formation, extinction, and immigration. In this situation, the relative abundance of every species coincides with logarithmic series distribution (for assembly communities that speciation by point mutations) or polynomial distribution (pervasion constrained local community or randomly species formed assembly community). In general, neutral theory emphasizes the equivalence of individuals at the same trophic level rather than species differences in ecological niches.

The unified neutral theory was proposed by adding speciation rate in the neutral theory (Chisholm and Burgman, 2004). It assumes that all individuals in the community are equivalent. Nevertheless, neglection of interspecific differences cannot explain species differences in environmental response and invasion capacity.

(3) *Incorporation of niche theory and neutral theory*

Assembly and construction is probably the outcome of joined action of random ecological drift and niche differentiation. As such, niche theory and neutral theory can be incorporated to interpret the mechanism of community assembly (Chase, 2005; Gravel *et al.*, 2006). Harte (2004) suggested treating neutral theory as the null model of community assembly. Gravel (2006) put forward the neutral-niche continuum hypothesis. It predicts that when the community saturates due to niche complementary, the immigration of propagules will preserve the redundancy of species in the community, and ecological drift can increase species richness (Purves and Pacala, 2005; Leibold and McPeek, 2006). Random niche theory, proposed by Tilman in 2004, holds that the community assembly is in essence a process of continuous species invasion.

(4) *Assembly history*

Assembly history is particularly important to small communities, probably due to the priority effect of earlier arrived species (Fukami, 2004). Assembly history is also significant to community structure.

13.1.1.3 *Research methods of community assembly*

(1) *Agent-based modeling*

Zhang (2012a, 2013a, 2013c, 2014a, 2014c) pointed out that agent-based modeling (ABM) should be used as the core methodology for community assembly, and methodological details for constructing communities using ABM were discussed. The mechanisms represented by both neutral theory and niche theory are seamlessly incorporated in agent-based modelling: species are the parental agents and individuals are offspring agents.

(2) *Phylogenetic analysis*

Using phylogenetic analysis method to approach assembly rules of local forest communities is a focus in research. Ge *et al.* (2012) utilized three fragments (rbcL, matK and psbA-trnH) commonly used in plant barcodes, the plant DNA barcodes and Phylomatic method to build the phylogenetic relationship of 183 woody plant species in a forest community of Dinghushan. The results revealed that closely related species coexisted in low-elevation habitats and habitat filtering may play a leading role; the species being phylogenetically distant coexisted in high-elevation habitats and competitive exclusion may play a leading role.

(3) *Network analysis*

Ecosystems are highly self-organizing and self-developing network systems. An ecological network is a network of interactions for material, energy flows between different compartments. Structural network analysis has become an effective tool to study structural and internal changes in ecosystems (Bailey, 2004; Fath, 2004; Fath *et al.*, 2007; Zhang, 2012a).

Some researchers have built the predation structures of some communities using network analysis, in which primary producers, herbivores, and carnivores were included but they lacked decomposers and detritivores. Therefore, the networks lacked matter circulation. These networks include cascade model (Cohen and Newman, 1985) and habitat model (Williams and Martinez, 2000). Some other networks include energy flows and functional groups and thus are relatively perfect models (Fath *et al.*, 2007).

For those networks without full knowledge on interspecific associations, in particular the large scale food webs, Zhang (2011d, 2012b, 2014b) used interspecific correlation measures to construct networks with indirect or direct interactions, and proposed the concept, theory and methodology of statistic networks. Moreover, Zhang (2012b, 2014b) found that many species in a community are redundant species and they are important for robustness of the community.

(4) *Species-area relationship analysis*

Using species-area relationship analysis, Gong *et al.* (2011) found that the importance of the same species was different at different phases for different communities.

(5) *Spatial pattern analysis*

Spatial distribution patterns may reflect the situation of interspecific interactions (Li *et al.*, 2014).

(6) *Pedigree structure analysis*

With the rapid development of techniques in molecular biology, the methods of systematic evolution have been applied to community ecology studies. Pedigree structure analysis is used in the community assembly for two reasons: (1) community assembly is treated as a process of random ecological drift of equivalent individuals (Hubbell, 2001), and the null hypothesis of this theory is that the assembled community is composed of species (the number of species is equal to the observed richness of the community) randomly selected from species pool, which is in essence similar to the neutral theory; and (2) the aggregation or divergence of pedigree structure means the habitat filtering or competitive exclusion, which emphasizes the non-equivalence of different species and thus acknowledges niche theory. Therefore, pedigree structure analysis can be used to analyze major factors of community assembly by analyzing phylogenetic linkages between species and to explore the effect of historical factors on the community (Webb *et al.*, 2002). Pedigree structure analysis is considered a combined solution of neutral theory and niche theory (Cavender-Bares *et al.*, 2009).

13.1.2 *Theories and methods on community succession*

13.1.2.1 *Concept of community succession*

Community succession is a phenomenon or process by which an ecological community undergoes more or less orderly and predictable changes following a disturbance or initial colonization of new habitat (Connell and Slatyer, 1977). It is the observed process of change in the species structure of an ecological community over time. The community begins with relatively few pioneering plants and animals and develops through increasing complexity until it becomes stable or self-perpetuating as a climax community (Wikipedia: http://en.wikipedia.org/wiki/Ecological_succession; Fig. 2). In general, succession is composed of three phases: (1) invasion and colonization. Some pioneer species invade and colonize the barren land, change the environment, and thus lay basis for successive invasions of other species. (2) Competition and equilibrium. Some species survive in the competition and some species extinct. The coexisting species reach an equilibrium state in utilizing resources. (3) Relatively stable state. Species composition and abundance are relatively fixed. Food web is complex and stable. However, so far there is no widely recognized pattern and theory on community succession (Lebrija-Trejos *et al.*, 2010).

(1) *Types of community succession*

There are two kinds of successions, i.e., primary succession and secondary succession. Succession that begins in new habitats which are not influenced by pre-existing communities is primary succession, whereas succession that follows disruption of a pre-existing community is secondary succession. Succession can be classified as century succession, long-term succession, and short-term succession. It can also be identified as autotrophic succession, heterotrophic succession, and relatively stable climax.

(2) *Influential factors of community succession*

External natural factors dominating community succession include fire, drought, CO_2, storm, volcano eruption, etc.

13.1.2.2 *Theories on community succession*

(1) *Promotion theory*

Promotion theory was first proposed by Clements in 1916. The theory held that the plant community is a super-organism. This organism is highly integrated. Through succession, a community can develop into a single climax community. Development of community is orderly and successive. It develops from a pioneer community into a complex climax community. Succession is powered mainly by the interactions between organisms and between organisms and the environment. The development of the present phase will promote the development of the next phase. The succession is directional and predictable. Species composition and environmental characteristics are predictable. The winning species in the competition becomes the dominant species, but after a certain time, the new species with stronger competition capacity will become the new dominant species.

(2) *Initial floristic composition theory*

The theory argued that succession is highly heterogeneous (Egler, 1954). It will certainly not proceed towards a climax community. Succession is a continuum and it is a species replacing process rather than community replacing process. Succession process is determined by the initial floristic composition at the earlier phase of succession. The species in the initial floristic composition randomly occurs. Succession is unpredictable.

(3) *Resource-ratio hypothesis*

Tilman (1985) maintained that the change in resource-ratio leads to the change of competion relationship and thus results in community succession.

(4) *Tolerance hypothesis*

Connell and Slatyer (1977) suggested that succession occurs through facilitation, inhibition, and tolerance between populations. They thought that the pioneer species are not important and any species can start succession. Species replacement reduces as the resources decreases and more tolerable species to resource deficiency will replace the sensitive species.

(5) *Life history strategy hypothesis*

Grime (1988) thought that the pattern of species strategies in the succession is predictable. He argued that the secondary succession is predictable. R-strategy species (which are adaptable to the environment with abundant resources temporarily) are pioneer species, C-strategy species (which are adaptable to the environment with abundant resources eternally, and are thus called competitive species) occur in mid-succession, and S-strategy species (which are adaptable to the environment lacking of resources, and are thus called tolerance species) are dominant in the climax community.

(6) *Mosaic theory*

Bormann and Likens (1979) proposed that community succession is free of external disturbance. It should be a sponteneuous and self-controlled process.

13.1.2.3 *Research methods of community succession*

(1) *Long-term observation of fixed plots*

In this method, fixed plots are observed for dozens or even hundreds of years. Species composition, environmental factors, and other data are recorded in detail.

(2) *Space-time replacement*

The community succession in different regions will be similar to each other if the environmental conditions of these regions are similar (Ladislav, 2010). Thus, we can record the full process of succession of a community over different regions.

(3) *Mathematical and GIS/GPS/RS methods*

A typical mathematical method used in community succession is state transition probability models, including Markov chain (Zhang and Wei, 2009). GIS, GPS, and RS are powerful tools for community succession research.

13.1.3 *Unified theory and methodology on community assembly and succession*

Continuing my past propositions, I proposed the unified theory and methodology on community assembly and succession. Community assembly and succession is a process of self-organization; it follows the major principles and mechanisms of self-organization (Zhang, 2013a, 2013c, 2014c). Multi-steady states are available for community succession. The community assembly and succession is unpredictable. Agent-based modeling can be used to describe the dynamics of community assembly and succession. Full details on self-organization and agent-based modelling, can be found in Zhang (2012a, 2013a, 2013c, 2014a, 2014c).

13.2 Food Web Models

A biological community is also a food web. The trophic structure of a stable food web (climax community) follows a certain pattern, which can be described by some models, e.g., cascade model, niche model, and inverse niche model, etc. (Zhang, 2012a; Zhang *et al.*, 2014a). Some rules are expected to be produced from these models and used in the self-organizing simulation of community succession and assembly.

13.2.1 *Cascade model*

The cascade model was first proposed by Cohen and Newman (1985). All species are arranged as a cascade structure so that the top species can feed on the lower species. The adjacent matrix $A_{s\times s}$, built from the food web, is thus a strict upper triangular matrix ($a_{ij} = 0, i > j$), i.e., it is a strict trophic hierarchy or cascade (Zhang, 2012a). Therefore, the food web has no cycle. The species with ID No.1 are only fed by other species and they do not feed on any other species, etc. The species with ID No. S may feed on any other species. Furthermore, this model assumes that there is a positive real value

$c(c = 2CS^2/(S-1)$, and C is the connectance, $\mathrm{C} = L/S^2)$, so that for $S \geq c$, all elements $a_{ij}, i \leq j$, follow the Bernoulli distribution with the parameter $p = c/S$. Thus some properties of the food web can be derived.

Suppose $p = c/S, q = 1-p$, and $c \geq 0, S \geq c$. Topological structure of a food web can be derived by (Cohen and Newman, 1985):

$$
\begin{aligned}
E(T) &= E(B) = (1-q^S)/p \\
E(I) &= S\left[1 - 2(1-q^S)/c + q^{S-1}\right] \\
E(L) &= pS(S-1)/2 = c(S-1)/2 \\
E(L_{BI}) &= E(L_{IT}) = (S-1)(1+q^{S-1}) - (1+q)(1-q^{S-1})/p \\
E(L_{BT}) &= (1-q^{S-1})/p - (S-1)q^{S-1} \\
E(L_{II}) &= pS(S-1)/2 - (S-1)(2+q^{S-1}) + (1-q^{S-1})(1+2q)/p
\end{aligned}
$$

From the first three formulae, we can calculate the proportions of number of top species T, number of intermediate species I, and number of basal species B, divided by total number of species. And from the last three formulae, we may calculate the proportions of number of top species — basal species links BT, number of top species–intermediate species links IT, number of intermediate species–intermediate species links II, and number of intermediate species–basal species links BI, divided by total number of species (Zhang *et al.*, 2014a).

13.2.2 *Niche model*

The niche model was proposed by Williams and Martinez (2000). In this model, each species i is assigned a niche value, n_i, which is a random value in [0,1]. Species i feeds on all species in an interval $[a, a+r] \subset [0,1]$, where $r_i = n_i^* x_i$, and x_i is a random value of β-distribution (Zhang, 2012a). The value $c_i = a + r/2$, and $c_i \in (r_i/2, n_i)$. It assumes a contiguous trophic niche for the consumer.

The niche model assigns each species a randomly drawn "niche value." The species are then constrained to consume all prey species within one range of values whose randomly chosen centre is less than the consumer's niche value. A variant of niche model was thus needed (Williams and Martinez, 2000).

According to the variant of niche model, there are two steps in the construction of food web. First, randomly assign species $i = 1, 2, \ldots, s$, to the niche nutrients in the interval [0, 1]. Second, randomly assign species i into the interval [0, 1]; all species within the interval are considered to be preys of species i. The food web matrix is thus constructed. Trophic similarity of a pair of species (s_{ij}) is the number of predators and prey shared in common divided by the pair's total number of predators and prey. Mean maximum similarity of a web can be calculated by averaging all species' largest similarity index (Williams and Martinez, 2000).

The following are Java codes, nicheModel (Zhang, 2012a), for the variant of niche model of Williams and Martinez (2000):

```
/*Niche model of food web generation (Williams and Martinez, 2000).v: number of species. d[1-v][1-v]: food web matrix, where if species j feeds on species i, the element (i,j) is 1, or else 0.*/
public class nicheModel {
public static void main(String[] args){
int v;
if (args.length!=1)
System.out.println("You must input the number of species. For example, you may type the following in the command window: java nicheModel 20, where 20 is the number of species.");
v=Integer.valueOf(args[0]).intValue();
nicheModel(v);
nicheModelVariant(v); }
public static int[][] nicheModelVariant(int v) {
int i,j,k,s;
double a,b,c;
int d[][]=new int[v+1][v+1];
double x[]=new double[v+1];
for(i=1;i<=v;i++)
x[i]=Math.random();
for(i=1;i<=v;i++) {
```

```
a=x[i]*Math.random();
b=x[i]*Math.random();
if (b<a) {
c=a;
a=b;
b=c; }
for(j=1;j<=v;j++) {
if (i==j) continue;
if ((x[j]>=a) & (x[j]<=b))
d[i][j]=1; }}
System.out.println("Variant model of food web generation (Williams and
Martinez,2000).");
print(v,d);
speciestype(v,d);
tropsimil(v,d);
return d; }
public static int[][] nicheModel(int v) {
int i,j;
double a,b,c,r;
int d[][]=new int[v+1][v+1];
double x[]=new double[v+1];
for(i=1;i<=v;i++)
x[i]=Math.random();
for(i=1;i<=v;i++) {
r=range(x[i]);
c=Math.random();
c=x[i]*c+r/2*(1-c);
a=c-r/2;
b=c+r/2;
for(j=1;j<=v;j++) {
if (i==j) continue;
if ((x[j]>=a) & (x[j]<=b))
d[i][j]=1; }}
System.out.println("Food web generation (Williams and Martinez,2000).");
print(v,d);
speciestype(v,d);
```

```
tropsimil(v,d);
return d; }
public static double range(double x) {
int i,max=1000;
double s,y,xx,r,mean,beta;
s=0;
for(i=1;i<=max;i++) {
y=Math.random();
s+=y; }
mean=s/max;
beta=1/mean-1;
s=0;
for(i=1;i<=max;i++) {
xx=1-Math.pow(1-Math.random(),1/beta);
s+=xx; }
mean=s/max;
r=mean*x;
return r; }
public static void speciestype(int v, int d[][]) {
int i,j,s;
int a[]=new int[v+1];
int b[]=new int[v+1];
for(i=1;i<=v;i++) {
a[i]=b[i]=0;
for(j=1;j<=v;j++) {
a[i]+=d[j][i];
b[i]+=d[i][j]; }}
s=0;
System.out.print("Top species:");
for(i=1;i<=v;i++)
if (a[i]==0) {
System.out.print(i+" ");
s++; }
System.out.println();
System.out.println("Total number:"+s);
System.out.println();
```

```
s=0;
System.out.print("Intermediate species:");
for(i=1;i<=v;i++)
if ((a[i]!=0) & (b[i]!=0)) {
System.out.print(i+" ");
s++;}
System.out.println();
System.out.println("Total number:"+s);
System.out.println();
s=0;
System.out.print("Basal species:");
for(i=1;i<=v;i++)
if (b[i]==0) {
System.out.print(i+" ");
s++;}
System.out.println();
System.out.println("Total number:"+s);
System.out.println(); }
public static double[][] tropsimil(int v, int d[][]) {
int i,j,k,c,cc;
double max,sum,mxsim;
double s[][]=new double[v+1][v+1];
mxsim=sum=0;
for(i=1;i<=v;i++) {
max=0;
for(j=1;j<=v;j++) {
if (i==j) {
s[i][j]=1;
continue; }
c=cc=0;
for(k=1;k<=v;k++) {
if ((d[i][k]==1) & (d[j][k]==1)) c++;
if ((d[i][k]==1)|(d[j][k]==1)) cc++; }
for(k=1;k<=v;k++) {
if ((d[k][i]==1) & (d[k][j]==1)) c++;
if ((d[k][i]==1)|(d[k][j]==1)) cc++; }
```

```
s[i][j]=c*1.0/cc;
if (s[i][j]>max) max=s[i][j]; }
sum+=max; }
mxsim=sum/v;
System.out.println("Trophic similarity of paired species:");
for(i=1;i<=v;i++) {
for(j=1;j<=v;j++)
System.out.print(s[i][j]+" ");
System.out.println(); }
System.out.println();
System.out.println("Maximum similarity of the web:"+mxsim);
System.out.println();
return s; }
public static void print(int v, int d[][]) {
int i,j,k,s;
double c;
System.out.println("There are "+v+" species in the food web.");
System.out.println("Food web matrix (if species j feeds on species i,
the element (i,j) is 1, or else 0):");
for(i=1;i<=v;i++) {
for(j=1;j<=v;j++)
System.out.print(d[j][i]+" ");
System.out.println(); }
System.out.println();
s=0;
System.out.println("Number of predators per species (in-degree):");
for(i=1;i<=v;i++) {
k=0;
for(j=1;j<=v;j++) {
k+=d[j][i];
s+=d[j][i]; }
System.out.print(k+" "); }
System.out.println("\n");
System.out.println("Number of preys per species (out-degree):");
for(i=1;i<=v;i++) {
k=0;
```

```
for(j=1;j<=v;j++)
k+=d[i][j];
System.out.print(k+" "); }
System.out.println("\n");
System.out.println("Total degree of every species:");
for(i=1;i<=v;i++) {
k=0;
for(j=1;j<=v;j++)
k+=d[i][j];
for(j=1;j<=v;j++)
k+=d[j][i];
System.out.print(k+" "); }
System.out.println("\n");
System.out.println("There are "+s+" links.");
c=s/Math.pow(v,2);
System.out.println("Directed connectance (C)="+c);
System.out.println("\n"); }
}
```

13.2.3 *Inverse niche model*

The inverse niche model was revised from the niche model, by Warren *et al.* (2010). It was developed to include parasites in the food web. In this model, there are two sub-webs, predator-prey links and parasite-host links. Rules of predator-prey links are the same with the niche model. For the parasite-host links: (1) corresponding to a value n_k, the range for parasitization of the parasite k is $r_k = (1 - n_k)^* y_k$, where y_k is a random value of β-distribution, and (2) the median value $c_k = a + r_k/2$, and $c_k \in (n_k, 1 - r_k/2)$. The small n_k means the wider range of parasitization r_k.

13.2.4 *MaxEnt model*

A MaxEnt is based on the maximum-entropy principle, which can be used to predict degree distribution of ecological network (Williams, 2010; Zhang, 2012a). MaxEnt model includes prior knowledge of the number of basal species B and does not attempt to predict the fraction of basal species. Similarly, two consumer distributions are considered, the "all-species consumer distribution" and the "restricted consumer distribution." The "all-species consumer distribution" is defined as the distribution of the number of consumers of each species, including the top species, which have no consumers. The "restricted consumer distribution" is defined as the distribution of the number of consumers of the resource species, includes prior knowledge of the number of top species T, and does not attempt to predict the fraction of top species (Williams. 2010).

In the all-species consumer distribution, the resource species or consumers of each species can range from 0 to S(S: number of species; L: links), and the mean number of links per species is L/S. In the restricted resource distribution, the number of links per consumer varies within 1 and S, and the mean number of links per consumer is $L/(S-B)$. In the restricted consumer distribution, the number of links from each resource varies within 1 and S, and the mean number of links from each resource is $L/(S-T)$.

First, a discrete distribution should be found based on some values, for example $\{x_1, \ldots, x_n\}$. The mean μ (i.e., indegree or outdegree of a node) maximizes $H = -\sum_i p_i \ln p_i$, and satisfies some conditions. The maximum entropy distribution is:

$$p_i = P(X = x_i) = Ce^{\lambda xi}, \quad \text{for } i = 1, \ldots, n.$$

Williams (2010) used the model $p_i = P(X = x_i) = x_i^\lambda$, i.e., using power law distribution but not exponential distribution. Set these conditons (Jaynes, 1957), $\sum_i P_i = 1$ and $\sum_i X_i P_i = \mu$. C and λ can thus be obtained by using Lagrange multipliers. Williams (2010) assumes that the number of consumers and resources are independent on each node of food web. For T top species without consumers, the number of links is from maximum entropy resource distribution. Similarly, For B basal species, the number of links is from maximum entropy consumer distribution. And for $S-B-T$ intermediate species, the number of links is the sum of the numbers from consumer and resource distributions.

In addition, Simpson index can also be used as the entropy index:

$$H = 1 - \sum p_i^2$$

As a supplement, some other degree distributions are also described as follows.

Bararasi (2009):

$$f(r) = r^{-\beta}$$

Zhang (2011d):

$$f(r) = \alpha(r-a)^{\beta}(b-r)^{\eta}, a \leq r \leq b;$$
$$f(r) = 0, r > a \quad \text{or} \quad r < b;$$
$$\alpha > 0, \beta \geq 0, \eta \geq 0.$$

where α: scale parameter; β, γ: shape parameters; a, b: position parameters.

The following are Java codes, maxEnt, for maximum entropy algorithm (Zhang, 2012a):

```
/*Maximum entropy algorithm for network generation. In the basal species
connection number file (maxent1) and top species connection number file
(maxent2), the numbers of basal and top species connections of the network
are given respectively.*/
public class maxEnt {
static double x[]=new double[10];
public static void main(String[] args){
int i,j,vtb,vt,vb,nf1,nf2,typedis,typeent;
double miuin,miuout;
if (args.length!=11) System.out.println("You must input the names of
basal and top species connection number tables in the database, the
different number of connections in basal species connection number table,
the different number of connections in top species connection number table,
the expected mean number of basal species connections, the expected mean
number of top species connections, the type of frequency distribution of
degrees (1: exponential distri (p(x)=ce^(dx)); 2: power law distri
(p(x)=x^(-c)); 3: Zhang's distri (p(x)=a(x-b)^c*(d-x)^e)), the type of
entropy expression (1: Shannon-Wiener index; 2: Simpson index), the number
of total species, the number of top species, and the number of basal species.
For example, you may type the following in the command window: java maxEnt
maxent1 maxent2 4 7 5 8 1 2 15 3 6, where maxent1 and maxent2 are the names
```

```
of basal and top species connection number tables respectively, there are 4
and 7 different numbers of basal and top species connections in the two
tables respectively, there are totally 15 species in which there are 3 top
species and 6 basal species, 1 means exponential distribution, 2 means
Simpson index, the expected mean numbers of basal and top species connections
are 5 and 8 respectively.");
    String tablename1=args[0];
    String tablename2=args[1];
    nf1=Integer.valueOf(args[2]).intValue();
    nf2=Integer.valueOf(args[3]).intValue();
    miuin=Double.valueOf(args[4]).doubleValue();
    miuout=Double.valueOf(args[5]).doubleValue();
    typedis=Integer.valueOf(args[6]).intValue();
    typeent=Integer.valueOf(args[7]).intValue();
    vtb=Integer.valueOf(args[8]).intValue();
    vt=Integer.valueOf(args[9]).intValue();
    vb=Integer.valueOf(args[10]).intValue();
    readDatabase readdata1=new readDatabase("dataBase",tablename1,nf1);
    readDatabase readdata2=new readDatabase("dataBase",tablename2,nf2);
    double ff1[]=new double[nf1+1];
    double ff2[]=new double[nf2+1];
    for(j=1;j<=nf1;j++)
    ff1[j]=(Integer.valueOf(readdata1.data[1][j])).intValue();
    for(j=1;j<=nf2;j++)
    ff2[j]=(Integer.valueOf(readdata2.data[1][j])).intValue();
    maxent(typedis,typeent,nf1,nf2,miuin,miuout,vtb,vt,vb,ff1,ff2); }
    public static void maxent(int typedis, int typeent, int nf1, int nf2,
double miuin, double miuout, int v, int vt, int vb, double ff1[], double
ff2[]) {
    int i,j,nf,k=0;
    double ran,mis,mos;
    double xi[]=new double[nf1+nf2+1];
    double p[]=new double[nf1+nf2+1];
    double u[]=new double[nf1+nf2+2];
    int t[]=new int[vt+1];
    int b[]=new int[vb+1];
```

```
nf=nf1;
for(i=1;i<=nf;i++)
xi[i]=ff1[i];
System.out.println("Degree distribution of basal species:");
simplex(typedis,typeent,nf,miuin,xi);
for(i=1;i<=nf;i++)
if (typedis==1) p[i]=x[1]*Math.exp(x[2]*xi[i]);
else if (typedis==2) p[i]=Math.pow(xi[i],-x[1]);
else if (typedis==3)
p[i]=x[1]*Math.pow(xi[i]-x[2],x[3])*Math.pow(x[4]-xi[i],x[5]);
u[0]=0;
for(i=1;i<=nf;i++)
u[i]=u[i-1]+p[i];
for(i=1;i<=vb;i++) {
ran=Math.random()*u[nf];
for(j=1;j<=nf;j++)
if ((ran>=u[j-1]) & (ran<u[j])) {
k=j;
break; }
b[i]=(int)xi[k]; }
nf=nf2;
for(i=1;i<=nf;i++)
xi[i]=ff2[i];
System.out.println("Degree distribution of top species:");
simplex(typedis,typeent,nf,miuout,xi);
for(i=1;i<=nf;i++)
if (typedis==1) p[i]=x[1]*Math.exp(x[2]*xi[i]);
else if (typedis==2) p[i]=Math.pow(xi[i],-x[1]);
else if (typedis==3)
p[i]=x[1]*Math.pow(xi[i]-x[2],x[3])*Math.pow(x[4]-xi[i],x[5]);
u[0]=0;
for(i=1;i<=nf;i++)
u[i]=u[i-1]+p[i];
for(i=1;i<=vt;i++) {
ran=Math.random()*u[nf];
for(j=2;j<=nf;j++)
```

```
    if ((ran>=u[j-1]) & (ran<u[j])) {
    k=j;
    break; }
    t[i]=(int)xi[k]; }
    mis=mos=0;
    for(i=1;i<=vt;i++)
    mis+=t[i];
    for(i=1;i<=vb;i++)
    mos+=b[i];
    System.out.println("Food web generation (Williams, 2010; Bararbas, 2009;
Zhang, 2011d)");
    System.out.println("Indegree of "+vb+" basal species:");
    for(i=1;i<=vb;i++)
    System.out.print(b[i]+" ");
    System.out.println();
    System.out.println("Total degree of "+(v-vt-vb)+" intermediate species:
"+(mos+mis));
    System.out.println("Outdegree of "+vt+" top species:");
    for(i=1;i<=vt;i++)
    System.out.print(t[i]+" ");
    System.out.println(); }
    public static double[] simplex(int typedis, int typeent, int nf, double
    miu, double xi[]) {
    double dt=0,fm=0,pr,t,gj,al,be,ga,sq,tk,d1,d2,fh,fh2,fl,ub,ua,ui,uj;
    int i,j,n=2,max,it,n1,n2,n3,n4,n5,h1,l1;
    max=50000;
    pr=1e-05;
    if (typedis==1) n=2;
    if (typedis==2) n=1;
    if (typedis==3) n=5;
    double xl[][]=new double[n+7][n+2];
    double d[][]=new double[n+3][n+2];
    double f[]=new double[n+7];
    t=0.01;
    al=0.01;
    be=0.01;
```

```
ga=0.01;
n1=n2=n3=n4=n5=n+1;
sq=Math.sqrt(n1);
tk=t/(n*Math.sqrt(2));
d1=tk*(sq+n-1);
d2=tk*(sq-1);
for(i=1;i<=n;i++) {
if (typedis==1) {
x[1]=Math.random();
x[2]=-1+Math.random(); }
else if (typedis==2) x[1]=1+Math.random();
if (typedis==3) {
x[1]=Math.random();
x[2]=Math.random();
x[3]=2+Math.random();
x[4]=xi[nf]+Math.random();
x[5]=2+Math.random(); }}
gj=objectivefun(typedis,typeent,nf,miu,xi);
for(j=1;j<=n;j++) d[1][j]=0;
for(i=2;i<=n1;i++) {
for(j=1;j<=n;j++) d[i][j]=d2;
d[i][i-1]=d1; }
for(i=1;i<=n1;i++)
for(j=1;j<=n;j++) xl[i][j]=x[j]+d[i][j];
for(i=2;i<=n1;i++) {
for(j=1;j<=n;j++) x[j]=x|[i][j];
f[i]=objectivefun(typedis,typeent,nf,miu,xi); }
it=0;
point1: while (it>=0) {
fh=fl=gj;
h1=l1=1;
for(i=2;i<=n1;i++)
if (f[i]<=fh) {
if (f[i]>=fl) continue;
else fl=f[i];
|1=i;
```

```
continue; }
else {
fh=f[i];
h1=i;
if (f[i]>=fl) continue;
else fl=f[i];
|1=i;
continue; }
for(j=1;j<=n;j++) {
xl[n2][j]=0;
for(i=1;i<=n1;i++) xl[n2][j]+=xl[i][j];
x|[n2][j]=(xl[n2][j]-xl[h1][j])/n;
x|[n3][j]=(1+al)*xL[n2][j]-a|*xl[h1][j];
x[j]=x|[n3][j]; }
ub=objectivefun(typedis,typeent,nf,miu,xi);
if (ub<fl) {
for(j=1;j<=n;j++) {
x|[n4][j]=(1-al)*x|[n2][j]+ga*x|[n3][j];
x[j]=x|[n4][j]; }
ua=objectivefun(typedis,typeent,nf,miu,xi);
if (ua>=fl) {
for(j=1;j<=n;j++) x|[h1][j]=xl[n3][j];
f[h1]=ub;
for(j=1;j<=n;j++) x[j]=xl[n2][j];
ui=objectivefun(typedis,typeent,nf,miu,xi);
dt=0;
for(i=1;i<=n1;i++) dt+=Math.pow(f[i]-f[l1],2);
dt=Math.sqrt(dt);
if ((it>=max)|(dt<pr)) {
for(j=1;j<=n;j++) x[j]=x|[l1][j];
fm=f[|1];
break point1; }
else {
it++;
continue point1; }}
else {
```

```
for(j=1;j<=n;j++) xl[h1][j]=xl[n4][j];
f[h1]=ua;
for(j=1;j<=n;j++) x[j]=x|[n2][j];
ui=objectivefun(typedis,typeent,nf,miu,xi);
dt=0;
for(i=1;i<=n1;i++) dt+=Math.pow(f[i]-f[|1],2);
dt=Math.sqrt(dt);
if ((it>=max)|(dt<pr)) {
for(j=1;j<=n;j++) x[j]=x|[l1][j];
fm=f[|1];
break point1; }
else {
it++;
continue point1; }}}
fh2=gj;
if (h1==1) fh2=f[2];
for(i=2;i<=n2;i++) {
if ((i==h1)|(f[i]<=fh2)) continue;
else fh2=f[i]; }
if (ub<=fh2) {
for(j=1;j<=n;j++) x|[h1][j]=x|[n3][j];
f[h1]=ub;
for(j=1;j<=n;j++) x[j]=xl[n2][j];
ui=objectivefun(typedis,typeent,nf,miu,xi);
dt=0;
for(i=1;i<=n1;i++) dt+=Math.pow(f[i]-f[|1],2);
dt=Math.sqrt(dt);
if ((it>=max)|(dt<pr)) {
for(j=1;j<=n;j++) x[j]=x|[|1][j];
fm=f[|1];
break point1; }
else {
it++;
continue point1; }}
if (ub<=fh)
for(j=1;j<=n;j++) x|[h1][j]=x|[n3][j];
```

```
for(j=1;j<=n;j++) {
xl[n5][j]=be*xl[h1][j]+(1-be)*xl[n2][j];
x[j]=xl[n5][j]; }
uj=objectivefun(typedis,typeent,nf,miu,xi);
if (uj<=fh) {
for(j=1;j<=n;j++) xl[h1][j]=xl[n5][j];
f[h1]=uj;
for(j=1;j<=n;j++) x[j]=xl[n2][j];
ui=objectivefun(typedis,typeent,nf,miu,xi);
dt=0;
for(i=1;i<=n1;i++) dt+=Math.pow(f[i]-f[l1],2);
dt=Math.sqrt(dt);
if ((it>=max)|(dt<pr)) {
for(j=1;j<=n;j++) x[j]=xl[l1][j];
fm=f[l1];
break point1; }
else {
it++;
continue point1; }}
for(i=1;i<=n1;i++) {
for(j=1;j<=n;j++) {
xl[i][j]=t*(xl[i][j]+xl[l1][j]);
x[j]=xl[i][j]; }
f[i]=objectivefun(typedis,typeent,nf,miu,xi); }
for(j=1;j<=n;j++) x[j]=xl[n2][j];
ui=objectivefun(typedis,typeent,nf,miu,xi);
dt=0;
for(i=1;i<=n1;i++) dt+=Math.pow(f[i]-f[l1],2);
dt=Math.sqrt(dt);
if ((it>=max)|(dt<pr)) {
for(j=1;j<=n;j++) x[j]=xl[l1][j];
fm=f[l1];
break point1; }
else {
it++;
continue point1; }}
```

```
if (typedis==1) System.out.println("p(x)="+x[1]+"exp("+x[2]+"x)");
if (typedis==2) System.out.println("p(x)="+"x^"+(-x[1]));
if (typedis==3)
System.out.println("p(x)="+x[1]+"(x-"+x[2]+")^"+x[3]+"("+x[4]+ "-x)^
"+x[5]);
return x; }
public static double obj(int term, int nf, double p[], double xi[]) {
double s=0;
for(int j=1;j<=nf;j++)
switch (term) {
  case 1: s+=p[j]; break;
  case 2: s+=p[j]*xi[j]; break;
  case 3: s+=Math.log(p[j])*p[j]; break;
  case 4: s+=Math.pow(p[j],2); break; }
return s; }
public static double objectivefun(int typedis, int typeent, int nf,
double miu, double xi[]) {
double f=0;
double p[]=new double[nf+1];
for(int i=1;i<=nf;i++)
if (typedis==1) p[i]=x[1]*Math.exp(x[2]*xi[i]);
else if (typedis==2) p[i]=Math.pow(xi[i],-x[1]);
else if (typedis==3)
p[i]=x[1]*Math.pow(xi[i]-x[2],x[3])*Math.pow(x[4]-xi[i],x[5]);
switch (typeent) {
case 1:f=obj(3,nf,p,xi)+Math.abs(obj(1,nf,p,xi)-1)+Math.abs(obj
(2,nf,p,xi)-miu);
        break;
case 2:f=obj(4,nf,p,xi)-1+Math.abs(obj(1,nf,p,xi)-1)+Math.abs(obj
(2,nf,p,xi)-miu);
        break; }
return f; }
}
```

13.3 Agent-based Modeling of Community Succession and Assembly

As mentioned earlier, community succession and assembly is a self-organizing process generated from multiple species invasions, selection, adaptation, and optimization (May, 1973; Cohen *et al.*, 1993; Hraber and Milne, 1997; Wang *et al.*, 2009; Nedorezov, 2012; Zhang, 2012a, 2012c, 2013a, 2013b, 2014c, 2014d). Classical explanation of ecological processes assumed that communities were balanced systems which rarely experience disturbances. Recovery from disturbance was expected to proceed in an orderly and linear way towards a stable state of a uniquely adapted species assemblage, i.e., a climax community. Species diversity, productivity, and stability were assumed to increase with time to a maximum at climax. One hypothesis, the intermediate disturbance hypothesis suggests that species diversity is highest, not at a stable endpoint but rather at intermediate levels of disturbance (i.e., frequency and intensity). External disturbances, for example species invasion, might produce unpredictable and significant influence. Up till now, some processes or mechanisms governing community succession have been confirmed (Case, 1990; Hraber and Milne, 1997; Zhang, 2012a, 2014d): (1) niche partition/Competitive exclusion. One of two species that utilizes similar resources would be replaced by another due to their competitive interaction. This resulted from adaptive evolution that selectively utilizes resources, or the competitive exclusion between species. (2) Multiple attractors. A community could have multiple distinct steady distributions or alternative steady states. They represent different species assemblages that occurred at possibly similar conditions. History of community succession determines which steady state will occur. (3) Spatial range. Biological communities possess four orderliness, number of species, number of individuals per species, space occupied by each species, and space occupied by each individual. Spatial heterogeneity, like resource aggregation or resource gradient, may reduce competition or predation effect by providing local refugee or fine adaptive mechanism. Environmental variation is indispensable to species richness, which results in different succession rules. However, spatial heterogeneity of species and individuals is always ignored. (4) Open systems. Communities are self-adaptive systems. They can respond to continuous fluctuations of the environment and population.

So far, to develop and use assembly models for describing communities is a major tool in community ecology (Morton *et al.*, 1996). However, assembly models of multispecies ecosystems with trophic structure have been less developed, starting from early works (Pimm and Lawton, 1978; Pimm, 1979; 1980; Lockwood *et al.* 1997; Bastolla *et al.*, 2001; Bonabeau, 2002).

Communities are complex systems. They coincide with the characteristics of agents. Hence, Agent-based Modeling (ABM) can be used to modeling spatial-temporal dynamics of ecosystems and biological communities (Jennings, 2000; Mellouli *et al.*, 2003). So far there are only a few studies on how ABM is used in ecology. Hraber and Milne (1997) developed an ABM model on the basis of a self-adaptive system, Echo. It can be used to simulate the dynamic process of species assemblage. In this model, there are behaviors like predation, competition and mating between individuals of different species or the same species. Different species and individuals have different genotypes and hence possess different fecundity and survival capacity. Fecundity and survival capacity might be enhanced by learning process. Genotypes of species or individuals with greater fecundity and survival capacity are more easily multiplied. Spatial heterogeneity, however, is not considered in the model. Topping *et al.* (2003) proposed an ABM model, ALMaSS, which was used to simulate the growth and spread of multiple species in the heterogeneous environment. This model does not consider the self-adaptive learning process of individuals, and there are only a few between-species interaction types. The ABM model of Savage and Askenazi (1999), Arborscapes, can be used to describe competition, growth, and spread processes of multiple tree species. Each individual in the model possesses some biological attributes and behavioral rules, and some disturbances like logging are also considered. As for between-species interactions, however, this model only includes competition.

And self-adaptive learning process of individuals is ignored in the model.

The following is a framework of ABM for community succession and assembly. Details can be found in Zhang (2012a, 2014d), etc.

ABM framework of community succession and assembly consists of three parts, obtaining dynamic data on community succession and assembly, ABM, analyzing mechanism of species establishment. Major procedures include:

(1) Define agents, and specify agents' behaviors.
(2) Identify relationship between agents, and construct interaction types between agents.
(3) Choose the platforms and environments for ABM, and set the strategies of ABM.
(4) Obtain necessary data for ABM. In the experiment and investigation of community dynamics, obtain spatial-temporal data of every species in the community. Use artificial recapture method to simulate species invasion and diffusion. In addition, obtain some data from references and internet.
(5) Test the patterns of agents' behaviors and system's behaviors.
(6) Run ABM model, and analyze the output from the standpoint of linking the micro-scale behaviors of the agents to the macro-scale behaviors of the system.
(7) Analyze the mechanism of community succession and assembly using ABM model.

Some definitions and methods for ABM of community succession and assembly are described as follows.

13.3.1 *Agents and behaviors*

ABM can be based on existing modeling platforms, like Swarm, Echo, NetLogo (NetLogo, 2004). Other platforms or methods can also be used, such as ALMaSS, Arborscapes, etc. (Repast, 2004). Java is a pure object-oriented, distributed, robust, structure-neutral, platform-independent, and dynamic programming language. We may perform systematic modeling and program using Java. Computation-extensive objects and methods can be realized as DLL (Dynamic Link Library). The available modeling environments include Windows and Linux operation systems. Modeling languages are UML, Java (JBuilder), Delphi (Borland Delphi), and C (Visual C++).

In a community, functional groups, species, individuals, etc., can be treated as agents at various levels. Several types of agents can be defined as follows (Zhang, 2012a, 2014d):

(1) Species agents. Species agents include predator agents, parasitoid agents, neutral agents, herbivore agents; or include agents that individuals of different species are just labeled with between-species coordination [positive or negative coordination, or neutral interaction (non coordination), magnitude of coordination]. Between-species coordination can be derived from sampling species assemblages, expressed as partial correlation, coordination coefficient (Zhang, 2011d, 2012b), etc.
(2) Space agents. They are represented by two-dimensional cells.
(3) Functional agents. They include interactive agents (user-model interactions), inductive agents [user induction of spatial dynamics, by such mechanisms as the change of distribution of plant resources (changing landscape structure)], data collection and analysis agents. Of these agents, some agents may be designed as Java classes or DLLs using Java, Delphi, or C++.

In the ABM model, the invasive species agent is treated as a common species agent in the community. Location and proportion of invasive species agent and frequency of invasive events are given fixed spatial or temporal probabilities. The community is initialized with random specified number and location of individuals of every species, or initialized by investigated community data. The model proceeds on a monthly or annually or daily step basis.

The model includes plant resource input (herbivorous species agents must interact with plant resources). Herbivorous species agents are given an initial supply of each plant resource in its plant resource reservoir. A species agent may acquire resources from the environment or form interactions with other species agents. Once enough resources are gathered, species agents can reproduce themselves.

Genetically-mediated behavior determines whether two species agents can interact. The species agent genome has two main regions, each with several attributes that code for a particular interaction. The *tag* region of the genome codes for attributes which are visible to other species agents. The *conditions* region represents attributes representing internal states, known only to the species agent itself. Matching a condition attribute against a tag attribute allows the interaction coded by those attributes to occur. Whether two species agents will interact is determined by comparing tag and condition attributes for that interaction. In the model, matching alleles in tag and condition attributes allows the interaction coded by those attributes. Tests for interactions are conducted sequentially: first for predation and competition, then for mutualism, and finally for mating.

Each species agent possesses some attributes: (1) a set of fixed, species-specific life history attributes, which include longevity, fecundity, age level, and others; (2) a time-relevant state (age, etc.). Species agents are autonomous and have adaptive behaviors, i.e., they adapt their behaviors (growth, feeding, habitat selection, mate choice, etc.) according to their state and the environment, to seek higher fitness by using learning algorithms, e.g., ANNs (BP, Hebbian learning, etc.) (Zhang and Barrion, 2006; Zhang, 2007b, 2010; Zhang *et al.*, 2007; Zhang and Wei, 2009; Zhang and Zhang, 2008; Zhang *et al.*, 2008a, 2008b).

Modeling adaptive system dynamics via individualistic mechanisms of adaptation is a fundamentally different approach than modeling with differential equations.

There are many behavioral models for agents. They include if-then rule and threshold model, artificial neural network and genetic algorithm rules, differential or difference equations (Zhang and Gu, 2001), optimization rules, multivariable decision-making, etc.

The landscape dynamics depend on species agent interactions (competition, mutualism, predation, etc.) and dynamic landscape structure (plant resources distribution, landscape structure, etc.).

System behavior depends on species agent interactions coupled with exogenous species agent. The spatial landscape of the model is a two-dimensional cell grid, in which each cell may be simultaneously occupied by many species (invasive species, indigenous species).

A space agent has major attributes including: (1) plant resources availability; (2) number of individual agents of each species; (3) landscape structure and space available for species, etc. Landscape structure is driven by weather and other factors. Species agents migrate between grid cells according to the states of adjacent spatial cells.

In general, the state transition method and differential/ difference equations can be used to model the dynamics of species agents and/or landscape structure (Zhang and Gu, 2001).

13.3.2 *Model objects*

On ABM platform, agents are implemented as objects. The model is expected to include these objects: Species, GridCell, Individual, Invasion, LandStruc, etc. The simplest and most common object is the Individual, the record of the state of an individual of species, including its life history attributes. Few methods are implemented in the Individual class as the Species object is responsible for the actual state transitions that an individual may undergo. For example, an individual object is sent a method and then relays the message to its species object with a reference to itself as an argument. The GridCell object provides the space in which the simulation is taking place, a simple square grid where each cell represents the area required by some individuals. The GridCell have:

(1) A reference from every occupied cell in the grid to the individual currently residing at that location.
(2) A separate list of all the individuals currently residing within its boundaries.
(3) Current landscape structure and food availability. LandStruc generates dynamic landscape structure based on state transition method (or difference equation method), or just load it from GIS.

The Species object is the most complex participant in the model. Its role is to record all of the attributes of a species and to execute the actual simulation step on behalf of every individual belonging to the species. Attributes that differ across species include age levels, longevity, fecundity, trophic level, etc. The Invasion object is responsible for species invasion events in the model. Species invasions occur in space with uniform, random or aggregated distributions at pulse way. Other objects are designed to make user-model interactions, user induction, and data collection and analysis, etc.

13.3.3 *Visualization tools*

In addition to defining agents and a schedule of events, the model should provide visual tools for the observation of the model on a time step basis. Windows will graphically track the abundance and spatial distribution of each species including invasive species, etc. A probe feature allows the user to query any cell on species, age, and attributes of the individual at the site. These window functions can be crucial in evaluating model behavior. Windows allow modification of parameters, which include size and structure of the landscape, number of species, frequency of invasion, etc.

13.3.4 *Pattern analysis*

A pattern is anything above random variation and thus indicates some kind of internal organization. Pattern-oriented ABM starts with identifying a variety of observed patterns, at different scales and at both individual and system levels, that characterize the system's dynamics and mechanisms. These patterns, along with the problem being addressed and conceptual models of the system, provide the basis for designing and testing our ABM. To analyze spatial and temporal pattern dynamics generated by the model, output can be linked to one of many pattern analysis packages. We may also write pattern analysis algorithms (spatial distribution patterns, topological structures, etc.) into the code. The model output may be linked to Fragstats (a comprehensive pattern analysis program). Its faster version is appropriate for the evaluation of the structure of cell-based models, generates metrics for area, patch density, size and variability, edge, shape, core area, diversity, contagion, interspersion, and nearest neighbor values. On the other hand, we may design pattern analysis algorithms [spatial distribution pattern, topological structure (shape, size, mosaic density, boundary, connectedness, etc.)].

Identifying the critical invasion strength for community may help develop ways to manage landscapes in maintaining a particular ecological function, or to make comparisons between different communities. The critical invasion strength for community can be determined by the method of spatial phase transitions.

13.3.5 *Parameterization*

A major problem of ABM of real systems is parameterization. Parameters are acquired from community investigation or experiments or internet. Many parameters would be uncertain or even unknown. Consequently, model results are uncertain and predictions and insights from the model become questionable. Sensitivity analysis with limited available parameter values will provide a partial solution.

13.3.6 *Model application*

The completed model will be used to approach patterns and mechanisms of community succession.

Usually, the steps of the ABM have to be repeated several times because this will lead to new theories, additional patterns, or modification of the entire ABM.

13.4 Identification of Landscape Patch Groups

Ecological processes at the landscape level are self-organizing processes. A landscape is composed of many patches (components, as called in self-organization). This section introduces a method to identify landscape patch groups. Rules for clustering patches may be produced from procudures in this method. More details can be found in Qi (2003) and Zhang *et al.* (2004, 2014b).

For most organisms, patch structure determines their feeding, survival, and prosperity (Bell *et al.*, 1991; Zhang *et al.*, 2004; Zhang *et al.*, 2014b). As a membrane or barrier or channel, the boundary between two patches affects the movement and migration of organisms (Forman, 1995; Zhang and Schoenly, 1999b; Qi, 2003; Qi and Zhang, 2002; Zhang, 2012a, 2013a, 2013c; Zhang *et al.*, 2013; Zhang, 2014c). Relationship between patch boundaries and biotic/abiotic factors has been confirmed in many studies (Gillison and Brewer, 1985; Hansen *et al.*, 1988; Holland *et al.*, 1991). It was found that homogeneous and continuous patches with less natural barriers were conducive to the spread of organisms. The number and shape of patches determines ecosystems at the landscape level.

Assume there are m patch indices and n sampling sites (patches, or components). Now we have the $n \times m$ matrix (y_{ij}), and the algorithm is as follows.

13.4.1 *Choose distance measures*

Distance measures are used to measure the similarity between patches (components). A larger distance value means a lower similarity between two patches. We chose four categories, a total of 14 types of distance measures here (Qi, 2003; Zhang, 2007c; Zhang *et al.*, 2004; Zhang, 2012a; Zhang and Fang, 1982; Krebs, 1989).

(1) Minkowski distance measures. They include Euclidean distance, Manhattan distance, and Chebyshov distance:

$$z_{ij} = \left(\sum_{k=1}^{m}(y_{ik} - y_{jk}^2/m\right)^{1/2}$$

$$z_{ij} = \sum_{k=1}^{m}|y_{ik} - y_{jk}|/m$$

$$z_{ij} = max_k|y_{ik} - y_{jk}| \qquad i, j = 1, 2, \ldots, n$$

(2) Correlation based measures. They include Pearson correlation and angular cosine:

$$z_{ij} = 1 - \sum_{k=1}^{m}((y_{ik} - y_{ib})(y_{jk} - y_{jb}))\bigg/$$

$$\left(\sum_{k=1}^{m}(y_{ik}-y_{ib})^2\sum_{k=1}^{m}(y_{jk}-y_{jb})^2\right)^{1/2}$$

$$z_{ij}=1-\sum_{k=1}^{m}(y_{ik}y_{jk})\Bigg/\left(\sum_{k=1}^{m}y_{ik}^2\sum_{k=1}^{m}y_{jk}^2\right)^{1/2}\qquad i,j=1,2,\ldots,n$$

The two categories of indices are used on the indices with real values.

(3) Nominal distance measures. They include linkage coefficient, and three co-linkage coefficients:

$$z_{ij}=1-(x^2/(x^2+n_{..}))^{1/2}$$
$$z_{ij}=1-(x^2/(n_{..}\max(p-1,q-1)))^{1/2}$$
$$z_{ij}=1-(x^2/(n_{..}min(p-1,q-1)))^{1/2}$$
$$z_{ij}=1-(x^2/(n_{..}((p-1)(q-1))^{1/2}))^{1/2}$$

where

$$x^2=n_{..}\left(\sum_{i=1}^{p}\sum_{j=1}^{q}n_{ij}^2/(n_{i.}n_{.j})-1\right)$$

$$n_{..}=\sum_{i=1}^{p}n_{i.},\quad n_{i.}=\sum_{j=1}^{q}n_{ij},\quad n_{.j}=\sum_{i=1}^{p}n_{ij}$$

where there are p nominal values, i.e., $t_1,t_2,\ldots,t_p$, for sampling site i and q nominal values, i.e., $r_1,r_2,\ldots,r_q$, for sampling site j. Assume that n_{kl} is the number of sampling site i takes value t_k and sampling site j takes value $r_l, k=1,2,\ldots,p; l=1,2,\ldots,q$.

(4) Boolean distance measures, including point correlation, quadratic correlation, two angular cosine functions, and Jaccard coefficient:

$$z_{ij}=1-(ad-bc)/((a+b)(c+d)(a+c)(b+d))^{1/2}$$
$$z_{ij}=1-sin((a+d-(b+c))/(a+b+c+d)^*3.1415926/2)$$
$$z_{ij}=1-(a^*a/((a+b)(a+c)))^{1/2}$$
$$z_{ij}=1-(a^*a^*d^*d/((a+b)(a+c)(b+d)(c+d)))^{1/2}$$
$$z_{ij}=(b+c)/(b+c+d)\qquad i,j=1,2,\ldots,n$$

where both sampling site i and sampling site j take values 0 or 1. a is the number of both sampling site i and sampling site j take value $0, b$ is the number of sampling site i takes 0 and sampling site j takes $1, c$ is the number of sampling site i takes 1 and sampling site j takes 0, and d is the number of both sampling site i and sampling site j take value 1.

13.4.2 *Randomization test*

If $min\ y_{ij} < 0$, let $y_{ij} = y_{ij} - min\ y_{ij}, i = 1, 2, \ldots, n; j = 1, 2, \ldots, m$. z_{ij} is the decimal number of y_{ij}, and $d_{ij} = 10^{Zij}$. Let

$$y_{ij} = y_{ij}\, max_j d_{ij} \qquad i = 1, 2, \ldots, n$$

Through these transformations, all of the values in sampling data become integers which are equivalent to numbers of individuals. The randomization test used in present study is based on the idea that, if no difference exists, then the distribution of individuals in sampling sites i and j will be a result of allocating the mixed sampling site values at random into two sampling site of size equal to those of the original sampling site (Solow, 1993; Manly, 1997; Zhang and Schoenly, 2001; Zhang *et al.*, 2004; Zhang, 2007c). The randomization test procedure is described as the follows. Assume the two sampling sites to be tested are i and $j, i = 1, 2, \ldots, n-1; j > i$. The $s_{ij} = \sum_{k=1}^{m} y_{ik} + \sum_{k=1}^{m} y_{jk}$ individuals of the combined sampling site are randomly reallocated into two randomized sampling sites with $\sum_{k=1}^{m} y_{ik}$ and $\sum_{k=1}^{m} y_{jk}$ labeled individuals. Calculate the expected absolute distance (similarity) between the two randomized sampling sites and compare whether it is not less than the absolute distance (similarity) between the true sampling sites i and j. Repeat the simulation many times, calculate the number of the expected are not less than the absolute distance (similarity) between the sampling site i and j, take the percentage as the p value. The p value is used to make statistical test. The threshold p value for test may be defined as 0.05, 0.01, etc. If the calculated p value is less than (in the case of similarity, larger than) p threshold, then the difference between sampling sites i and j is statistically significant.

13.4.3 *Ordered regionalization of patches (components)*

This procedure aims to conduct ordered regionalization of patches according to their similarity (Fang, 1979; Zhang, 1993). Patches with significant difference belong to different groups (clusters). In the ordered cluster analysis, the patches in the same group are geographically adjacent, and the lines between any two patches that belong to different groups are not intersected.

Assume the two-dimensional coordinate of each sampling site (patch) is $(y_{i11}, y_{i12}), i = 1, 2, \ldots, n$. And for sampling sites i and $j, i = 1, 2, \ldots, n-1; j > i$, calculate

$$g_1(i,j) = \{k, i | y_{k12} > (y_{j12} - y_{i12})(y_{k11} - y_{i11})/(y_{j11} - y_{i11}) + y_{i12}\}$$
$$g_2(i,j) = \{k, j | y_{k12} < (y_{j12} - y_{i12})(y_{k11} - y_{i11})/(y_{j11} - y_{i11}) + y_{i12}\}$$
$$G_1(i,j) = \{k, j | y_{k12} > (y_{j12} - y_{i12})(y_{k11} - y_{i11})/(y_{j11} - y_{i11}) + y_{i12}\}$$
$$G_2(i,j) = \{k, i | y_{k12} < (y_{j12} - y_{i12})(y_{k11} - y_{i11})/(y_{j11} - y_{i11}) + y_{i12}\}$$

Thus there will be $n(n-1)$ possible classifications. In each classification, for the sampling site $i(i_1 \leq i \leq i_2 - 1)$ in sub-group 1 and the sampling site $j(i_2 \leq j \leq i_3 - 1)$ in sub-group 2, calculate the p value, p_{ij}, in the randomization test, and choose the classification that meets the condition:

$$min\ pbar = \left(\sum_{i=i1}^{i2-1} \sum_{j=i2}^{i3-1} p_{ij} \right) \Big/ ((i_2 - i_1)(i_3 - i_2))$$

Suppose there have been k groups, obtain two sub-groups for each group, and choose a group to make sub-classification. If $k = n$, then stop the procedure. Therefore a cluster graph with different statistical significance levels can be achieved.

If all values of *max min pbar* are larger than the given threshold p value, then the landscape is continuous and homogeneous. Or else there are several patches with significant differences.

The algorithm is implemented as a Java program, TwoDimBoundDetector, which can be found at http://www.iaees.org/ publications/software/index.asp. In data file, the first two columns are two-dimensional coordinates of sampling sites and the remaining columns are for every index.

13.5 Network Evolution

In a sense, community succession and assembly is a process of network evolution (Zhang, 2012c). Change of the number of nodes and dynamics of connection structure are the core of network evolution, which are equivalent to change of species richness and between-species interations in the biological community.

As mentioned in Zhang (2012a), Barabasi and Albert (1999) presented a method to generate scale-free random graphs. Suppose that there are n vertices. For every pair of vertices i and j, an edge is generated with probability, p. When a vertex is connected and added, all edges are then generated in the way similar to the generation of random graph. New vertex traverses all existing vertices, but is connected and added with a fixed probability. However the probability values are different. The drawback of this algorithm is that the vertex connected and added earlier tends to more likely be the attractor, which is different from the real world. To this end, they gave a fitness value for each vertex. The rate of a vertex for accepting a new connection is positively proportional to its fitness value. At the start of the calculation, each vertex is assigned a uniform distributed fitness value. When the vertex is assigned a new edge, it increases the fitness value. Thus, with the increase of the connected edges the fitness value grows and therefore a positive feedback loop is formed.

The following are Java codes, randGraphScaFr, for generating scale-free random graph (Zhang, 2012a):

```
/*Barabasi-Albert algorithm to generate a scale-free random graph.*/
/*v: number of vertices; m: number of edges to be added; p: probability
for generating edges; d[1-v][1-v]: adjacency matrix of the generated
scale-free random graph.*/
public class randGraphScaFr {
public static void main(String[] args){
int v,e;
double p;
```

```
if (args.length!=3)
System.out.println("You must input the number of vertice, the number of
edges and the probability. For example, you may type the following in
the command window: java randGraphScaFr 20 200 0.3, where 20 is the
number of vertices and 200 is the number of edges, 0.3 is the
probability.");
v=Integer.valueOf(args[0]).intValue();
e=Integer.valueOf(args[1]).intValue();
p=Double.valueOf(args[2]).doubleValue();
randGraph(v,e,p); }
public static int[][] randGraph(int v, int e, double p) {
int i,j,k,a1,a2;
double b;
int d[][]=new int[v+1][v+1];
double s[]=new double[v+1];
s[0]=0;
for(i=1;i<=v;i++)
s[i]=s[i-1]+Math.random();
for(k=1;k<=e;k++) {
b=s[v]*Math.random();
for(i=0;i<=v-1;i++)
if ((b>=s[i]) & (b<s[i+1])) {
s[i+1]+=Math.random();
do
a1=(int)Math.round(v*Math.random()+0.5);
while (((i+1)==a1)|(d[i+1][a1]==1));
if (Math.random()<=p) {
d[i+1][a1]=1;
d[a1][i+1]=1; }
else {
d[i+1][a1]=0;
d[a1][i+1]=0; }
break; }}
System.out.println("Scale-free random graph (between-vertex edges):");
for(i=1;i<=v-1;i++)
for(j=i+1;j<=v;j++)
```

```
if (d[i][j]==1) System.out.println(i+"-"+j);
return d; }
}
```

Zhang (2012d) presented a method to simulate the network evolution or community succession/assembly as the following. This method can be improved by including learning and adaptation rules.

Suppose there are totally v taxa (species, or subspecies, etc.) in a community. Given a correlation matrix, $r=(r_{ij})_{v\times v}$, in which the elements are between-taxa correlation coefficients (Jaccard coefficients, or other correlation measures, etc.) that reflect between-taxa dynamic relationships in the community succession, $r_{ij}=r_{ji}$, $-1\leq r_{ij}\leq 1$; and an attribute vector, $a=(a_i)_v$, in which 0 means detrital/neutral taxon, 1 means herbivorous/grazing taxon, 2 means predatory/carnivorous taxon, and 3 means parasitoid/parasite taxon. Some assumptions are made in this method: (1) once a taxon invaded the community, it will establish the community and will not disappear again; (2) all taxa in the pool will finally invaded and established the community; (3) the first taxon, i.e., pioneer taxon is a detrital/neutral or herbivorous/grazing taxon, and each time a randomly chosen taxon invades the community; and (4) the taxon invaded tends to connect (interact with) each of the existing taxa at a probability proportional to its direct correlation with the taxon to be connected, the connected (interactive) taxa probabilistically follow a similar (connection value 1) or inverse (connection value -1) trend in the changes of population size during the community succession; however, an isolated taxon is featured with its distinct and independent population dynamics. The network evolution method for modeling community succession and assembly is described below.

(1) Generate v ordered taxa by producing v random numbers that represent v taxa. If the first taxon is not a detrital/neutral taxon or herbivorous/grazing taxon, repeat the procedure to produce the first taxon, i.e., pioneer taxon. The Java algorithm is:

```
c=0;
for(j=1;j<=v;j++) w[j]=j;
```

```
do {
cs=(int)((v-c)*Math.random()+1);
if ((a[w[cs]]!=1) & (a[w[cs]]!=0) & (c==0)) continue;
x[c+1]=w[cs];
if (cs<(v-c))
for(j=cs+1;j<=v-c;j++) w[j-1]=w[j];
c++; }
while (c<=(v-1));
```

The resultant ordered series of invasion taxa is $x(i)$, $i = 1, 2, \ldots, v$.

(2) Based on ordered taxa $x(i), i = 1, 2, \ldots, v$, let the taxon invade the community one by one. The first taxon invading the community is the pioneer taxon. In the k-th invasion, randomly produce a value p, and

$$\begin{aligned}
&d_{x(k)x(i)} = d_{x(i)x(k)} = 1, \quad \text{if } r_{x(k)x(i)} > 0 \quad \text{and} \quad r_{x(k)x(i)} > p;\\
&d_{x(k)x(i)} = d_{x(i)x(k)} = -1, \quad \text{if } r_{x(k)x(i)} < 0 \quad \text{and} \quad |r_{x(k)x(i)}| > p;\\
&d_{x(k)x(i)} = d_{x(i)x(k)} = 0, \quad \text{if } |r_{x(k)x(i)}| < p;\\
&\qquad i = 1, 2, \ldots k-1.
\end{aligned}$$

where $D = (d_{ij})$ is the adjacency matrix. If $d_{x(k)x(i)} = d_{x(i)x(k)} = 1$, then there is a positive connection (positive interaction) between the taxa $x(k)$ and $x(i)$; if $d_{x(k)x(i)} = d_{x(i)x(k)} = -1$, then there is a negative connection (negative interaction) between the taxa $x(k)$ and $x(i)$. The taxon $x(k)$ is an isolated taxon, if $d_{x(k)x(i)} = d_{x(i)x(k)} = 0$, for $i = 1, 2, \ldots k-1$.

Finally, record all pairs of connected taxa and isolated taxa based on the adjacency matrix.

After the k-th invasion, there are in total k taxa in the community, i.e., $x(i), i = 1, 2, \ldots, k$.

(3) If all taxa have invaded the community, terminate procedure; or else return to (2).

Chapter 14

Mathematical Foundations

Selforganizology covers a large number of mathematical areas, e.g., probability theory, numerical computation, differential equations, etc. Mathematical foundations are thus discussed in this chapter. Some details can be found in Zhang (2010, 2012a).

14.1 Bayes' Rule

Bayes' rule is a foundmental theory in selforganizolgy. Bayes' rule states that the probability that a postulate, A, will be true is positively proportional to the multiplication of the postulate's prior probability and the conditional probability of information, I, being observed given H is true.

Given a discrete sample space, S, and $A_i \in S, i = 1, 2, \ldots$, where $\cup A_i = S$, and $A_i \cap A_j = \phi, i \neq j$. Bayesian rule is:

$$p(A_i/I) = p(I/A_i)p(A_i) \Big/ \sum p(I/A_j)p(A_j)$$

For a continuous sample space, Bayes' rule is expressed as:

$$p(A/I) = p(I/A)p(A) \Big/ \int p(I/A)p(A)dA$$

14.1.1 *Models selection*

Bayesian rule can be used in the selection of models (Mackey, 1992). Suppose there are several models. Given the data or information, I, the posterior probability of model A_i, is thus given by Bayesian rule:

$$p(A_i/I) = p(I/A_i)p(A_i)/p(I)$$

where $p(A_i)$: prior probability of the model $A_i; p(I/A_i)$: the evidence of the model A_i, which can be expressed as:

$$p(I/A_i) = \int p(I/w, A_i)p(w/A_i)dw$$

where $w = (w_1, w_2, \ldots, w_n)$, is a weight vector.

14.1.2 *Bayesian learning*

Without information or samples, I, let $p(w)$ be a prior distribution, of network weights (including various thresholds), and $w = (w_1, w_2, \ldots, w_n)$. Given information or samples, the posterior distribution will be:

$$p(w/I) = p(I/w)p(w) \Big/ \int p(I/w)p(w)dw$$

Without any prior information, the prior distribution, $p(w)$, and conditional distribution, $p(I/w)$, may be represented by exponential distributions (Yan and Zhang, 2000):

$$p(w) = \exp(-cf(w))/h(c)$$
$$p(I/w) = \exp(-ag(I))/q(a)$$

While the posterior distribution of network weights can be achieved by:

$$p(w/I) = \int\int p(w/c, a, I)p(c, a/I)dcda$$

14.1.3 *Bayesian Ying-Yang system*

Bayesian Ying-Yang system was proposed to unify various neural network models and learning algorithms (Xu, 1997). Ying-Yang system is the generalization of Bayesian estimation of weights and Bayesian selection of models.

Perception and association are two of the most important mechanisms in neural network models. They are conducted by supervised or unsupervised learning procedures in the neural networks (Yan and Zhang, 2000), yielding an output (y) from input (x). Perception is targeted to realize pattern recognition and extraction of characteristics, etc. Association is performed by neural networks for classification, function approximation, and control, etc.

The relationship between network input, x, and output, y, can be statistically represented by the joint distribution, $p(x, y)$:

$$p(x, y) = p(x)p(y/x)$$
$$p(x, y) = p(y)p(x/y)$$

There are two models correspondingly, $M_1 = \{M_{y/x}, M_x\}$, and $M_2 = \{M_{x/y}, M_y\}$. M_x and M_y are called Yang model (visible model) and Ying model (invisible model), respectively. The above joint distribution is thus expressed as:

$$p_{M1}(x, y) = p_{Mx}(x)p_{My/x}(y/x)$$
$$p_{M2}(x, y) = p_{My}(y)p_{Mx/y}(x/y)$$

A Yang learning machine and a Ying learning machine are used to realize $p_{M1}(x, y)$ and $p_{M2}(x, y)$, respectively. The learning model is named as a Ying-Yang system. Learning is performed in the following way (Xu, 1997):

(1) obtain an appropriate representation;
(2) design a basic structure of the model;
(3) determine the scale and size of the model; and
(4) learn the parameters in the model.

The learning algorithm tries to achieve

$$p_{Mx}(x)p_{My/x}(y/x) = p_{My}(y)p_{Mx/y}(x/y)$$

Various neural network models, like RBF, multi-tier feed forward network, etc., and learning algorithms, like PCA, Helmholtz machine, EM learning algorithm, etc., are specific realization of this theory (Xu, 1997).

14.2 Linear Regression

14.2.1 *Single-variable linear regression*

A single-variable linear regression, with dependent variable y and independent variable x, is

$$y = a + bx$$

where

$$b = \left(\sum(xy) - \left(\sum x \sum y\right)\Big/ n\right)\Big/ \left(\sum x^2 - \left(\sum x\right)^2\Big/ n\right)$$

$$a = \sum y/n - b\sum x\Big/ n$$

and n: number of samples.

14.2.2 *Multiple-variable linear regression*

A multiple-variable linear regression is (Qi, 2004):

$$y_i = b_{0i} + b_{1i}x_1 + b_{2i}x_2 + \cdots + b_{mi}x_m \qquad i = 1, 2, \ldots, p$$

Suppose we have collected n samples and obtain a matrix, $(x_{ij})_{n\times m}$, for independent variable, and a matrix, $(y_{ij})_{n\times p}$, for the dependent variable. Let $L = (l_{ij}), i, j = 0, 1, \ldots, m; D = (d_{ij}), i = 0, 1, \ldots, m;$ $j = 1, 2, \ldots, p$. Calculate $l_{00} = n$, and

$$l_{0j} = \sum_{k=1}^{n} x_{kj}, \quad l_{i0} = \sum_{k=1}^{n} x_{ki}, \quad l_{ij} = \sum_{k=1}^{n} x_{ki}x_{kj},$$

$$i, j = 1, 2, \ldots, m$$

$$d_{0j} = \sum_{k=1}^{n} y_{kj}, \quad d_{ij} = \sum_{k=1}^{n} x_{ki}y_{kj},$$

$$i = 1, 2, \ldots, m; j = 1, 2, \ldots, p$$

Thus the matrix for correlation coefficients, $b = L^{-1*}D$, where $b = (b_{ij}), i = 0, 1, \ldots, m; j = 1, 2, \ldots, p$.

14.3 Stimulus-Response Functions

The stimulus-response functions include:

(1) Unit impulse function

$$\delta(t) = +\infty, t = 0$$
$$\delta(t) = 0, t \neq 0$$

and

$$\int_{-\infty}^{+\infty} \delta(t)dt = 1$$

(2) Unit slope function

$$r(t) = 0, t < 0$$
$$r(t) = t, t \geq 0$$

(3) Logistic function

$$l(t) = 1/(1 + e^{-t})$$

(4) $y = 1/(a + be^{-t})$
The following is the method to transform the function for obtaining parametrical values through linear regression: $T = e^{-t}, Y = 1/y \rightarrow Y = a + bT$.

(5) $y = at^b + c(a < 0)$
$T = \log t, Y = \log(y - c) \rightarrow Y = \log a + bT$, in which we may take three points $(t_1, y_1), (t_2, y_2), (t_3 = (t_1t_2)^{1/2}, y_3), c = (y_1y_2 - y_3^2)/(y_1 + y_2 - 2y_3)$.

(6) $y = ae^{bt} + c$
$T = t, Y = \log(y - c) \rightarrow Y = \log a + bT$, in which we may take three points $(t_1, y_1), (t_2, y_2), (t_3 = (t_1 + t_2)/2, y_3), c = (y_1y_2 - y_3^2)/(y_1 + y_2 - 2y_3)$.

(7) $y = t/(at + b) + c$
$T = 1/t, Y = 1/(y - c) \rightarrow Y = a + bT$, in which we may take three points $(t_1, y_1), (t_2, y_2), (t_3 = (2t_1t_2)/(t_1 + t_2), y_3), c = (2y_1y_2 - y_3(y_1 + y_2))/(y_1 + y_2 - 2y_3)$.

(8) $y = 1/(at + b)$
$T = t, Y = 1/y \rightarrow Y = a + bT$.

(9) $y = (at + b)/(ct + d)$ $(ad - bc \neq 0)$
Take $(t_0, y_0), T = t, Y = (t - t_0)/(y - y_0) \rightarrow Y = A + BT$.

(10) $y = at^2 + bt + c$
$T_1 = t, T_2 = t^2, Y = y \rightarrow Y = c + bT_1 + aT_2$.

(11) $y = a + b/t + c/t^2$
$T_1 = 1/t, T_2 = 1/t^2, Y = y \rightarrow Y = a + bT_1 + cT_2$.

(12) $y = a\mathrm{e}^{bt} + c\mathrm{e}^{dt}$
Suppose t values form an arithmetic series with common difference h, $u_1 = t + h, u_2 = t + 2h$, their corresponding y are v_1, v_2, respectively. Let $T = v_1/y, Y = v_2/y$, then $Y = (\mathrm{e}^{bh} + \mathrm{e}^{dh})T - \mathrm{e}^{bh}\mathrm{e}^{dh}$, obtain b, dusing linear regression, and let $P = \mathrm{e}^{(b-d)t}, Q = y\mathrm{e}^{-dt} \rightarrow Q = c + aP$.

(13) $y = at\mathrm{e}^{bt}(a > 0)$
$T = t, Y = \ln(y/t) \rightarrow Y = \ln a + bT$.

14.4 Randomization, Bootstrap and Monte Carlo Methods

Randomization, bootstrap and Monte Carlo methods are widely used in computation and intelligence studies. Details on randomization, bootstrap and Monte Carlo techniques can be found in Gentle (2002), Manly (1997) and Zhang (2007a, 2010).

14.4.1 *Random numbers*

14.4.1.1 *General random numbers*

Random numbers are the basis in randomization, bootstrap and Monte Carlo methods.

To produce a series of uniformly distributed random numbers ($0 \sim 1$), Matlab codes are

```
x = rand;
m = rand(5, 6);
```

Randomly permutate integer numbers, $1 \sim n$

$$\mathrm{n} = 10;$$
$$\mathrm{x} = \mathrm{randperm(n)};$$

The random number yielded by Matlab generator is a pseudorandom number. The generator will reset once Matlab is initiated. Thus the sequence of random numbers will be the same. The following codes record the present status (R) of generator, and set the status to R:

$$\mathrm{R} = \mathrm{rand('state')};$$
$$\mathrm{rand('state', R)};$$

Different status of the generator can be set at any time:

$$\mathrm{rand('state', sum(100{*}clock))};$$

Theoretically, Matlab generator may yield 2^{1492} different random numbers, which is enough for general researches.

14.4.1.2 *Probability distributions and random numbers*

The random numbers of normal distribution (norm), χ^2 distribution (chi2), t distribution (t), F distribution (f), β distribution (beta), uniform distribution (unif), and exponential distribution (exp), etc., can be generated in Matlab environment. The commands of probability density, probability distribution, inverse probability distribution, mean and variance, and random number are pdf, cdf, inv, stat, and rnd. Various combinations of corresponding commands will yield different probability distributions and random numbers:

```
x=5;
%Yield a density value of normal distribution with mean 5
%and standard deviation 30
p=normpdf(x,5,30);
%Yield a random number of normal distribution with mean 3
%and standard deviation 25
r=normrnd(3,25);
%Yield a value of probability distribution of t-distribution with
```

```
%degree of freedom 8
s=tcdf(x,8);
%Yield a 0.95 percentile of F-distribution with degrees of freedom
%3 and 9
g=finv(0.95,3,9);
```

The random numbers of a probability distribution can be generated based on uniformly distributed random numbers (Gentle, 2002; Zhang, 2007a, 2010). The often used methods include:

(1) Inverse transformation method. Given the probability distribution, $F(x)$, and its inverse function, F^{-1}, generate an uniformly distributed random number on $[0, 1], X \sim U(0, 1)$, and take $X = F^{-1}(U)$.
(2) Convolution method. Given random variable $Y = X_1+X_2+\cdots+X_n$, where $X_i, i = 1, 2, \ldots, n$, are independent and share the same probability distribution, firstly generate $X_i, i = 1, 2, \ldots, n$, and take $Y = X_1 + X_2 + \ldots + X_n$.
(3) Acceptance-rejection method. Suppose the density function to be found is $f(x)$; firstly get a density function, $g(x)$, and a constant, c, such that $f(x) \leq cg(x)$; then generate a random number, x, of $g(x)$; take $r = cg(x)/f(x)$; finally generate an uniformly distributed random number, u, thus x is the desired random number if $ru < 1$, or else repeat the above procedure.

Matlab codes for generate random numbers of various probability distributions are:

```
%Generate a matrix of uniformly distributed random numbers
rand(3,4)
%Generate a matrix of normally distributed random numbers,
%N(3,25^2)
random('norm',3,25,5,7);
%Generate a matrix of F-distributed random numbers with
%degrees of freedom 10 and 20
random('f',10,20,6,8)
%Generate a matrix of binomially distributed random numbers,
```

```
%B(5,0.3)
random('bino',5,0.3)
```

Latin hypercube is a method to determine inputs (McKay *et al.*, 1979). In Latin hypercube, m factors (i.e., inputs) are all uniformly-distributed random variables, $U(0,1)$. Set the i-th realization of j-th factor as

$$v_j = (p_j(i) + u_j - 1)/n$$

where $p_j(\bullet), j = 1, 2, \ldots, m$, are independent and random selections from $1, 2, \ldots, n; p_j(i)$ is the i-th element of j-th permutation; u_j is a value sampled from $U(0,1)$. n is sample size, and v_j will completely disperse over factor space.

14.4.1.3 *Markov chain and random numbers*

Markov chain can be used to generate random numbers. Using different transition matrices will result in different methods for producing random numbers (Gentle, 2002). One of these methods is Metropolis-Hastings algorithm, i.e., given the probability density function, p_X, of random variable X, and conditional probability density function, $g_{Yt+1|Yt}$, of deviation of Markov chain, then

(1) let $i = 0$, and choose x_i with probability p;
(2) generate y with probability density function $g_{Yt+1|Yt}(y/x_i)$;
(3) calculate Hastings ratio, r

$$r = p_X(y)g_{Yt+1|Yt}(x_i/y)/[p_X(x_i)g_{Yt+1|Yt}(y/x_i)];$$

(4) let $x_i = y$, if $r \geq 1$; or else generate an uniformly distributed random number, v. Let $x_{i+1} = y$, if $v < r$, or else $x_{i+1} = x_i$;
(5) assign $i = i + 1$, and return step (2).

14.4.1.4 *Multivariable random numbers*

Multivariable random numbers can be generated based on single random variables (Gentle, 2002). Suppose $X = (X_1, X_2, \ldots, X_n)$ is a random vector, where $X_i, i = 1, 2, \ldots, n$, are independent and share the same probability distribution with variance 1 and mean 0. Construct a random vector, $Y = AX$, with covariance matrix AA^T, where $|A| \neq 0$. Try to find A, such that $AA^T = \sum$, where $\sum$ is the desired covariance matrix.

14.4.2 *Randomization based data partition*

Data partition includes cross validation, Jackknife, etc. They are used to mine more information from a limited data set (Gentle, 2002; Zhang, 2010).

14.4.2.1 *Cross validation*

The principle of cross validation is to randomly (or systematically) divide the data set, $(x_i, y_i), i = 1, 2, \ldots, n$, into two parts, i.e., training set (T) and validation set (V). Training set is used to determine parameters in fitted model, $y = f(x)$, and validation set is used to validate and test the model. Moreover, training set and validation set can be exchanged to estimate fitting error in x:

$$E(R(Y_0, f(x_0)) = \left(\sum_{i \in V} R(Y_i, f_1(x_i)) \right) + \sum_{i \in T} R(Y_i, f_2(x_i))/n$$

where $f_1(x), f_2(x)$: fitted functions using training set T and V respectively, and $R(Y_0, f(x_0)$: prediction error. Data set can be divided into several intersected subsets (Breiman, 2001), for example, K subsets with similar sizes. Choose a subset as the validation set and the remaining subsets as training set, in order to achieve prediction error. The averaged error for K subsets is the prediction error to be estimated.

14.4.2.2 *Jackknife method*

Data set is systematically divided in jackknife method in order to obtain the estimates, e.g., variance or mean, from data set (Gentle, 2002). Suppose the statistic T of a random sample, $X_1, X_2, \ldots, X_n$, is the estimate of population parameter θ. Divide data set into r subsets, each with m elements (m is always 1 or 2).

Remove subset i from data set and calculate the estimate T_{-i} from the remaining subsets. The estimate of population parameter θ is thus:

$$T' = \sum_{i=1}^{r} T_{-i}/r$$

Moreover, the jackknife $T$is

$$J(T) = T^* \sum_{i=1}^{r} T_i^*/r = rT - (r-1)T'$$

where $T_i^* = rT - (r-1)T_{-i}$. If T_i^* are independent, the variance of T may be estimated from jackknife variance $V(J(T))$:

$$V(J(T)) = \sum_{i=1}^{r} (T_i^* - J(T))^2/(r(r-1))$$

or

$$V(J(T)) = \sum_{i=1}^{r} (T_i^* - T)^2/(r(r-1))$$

Suppose $m = 1$, and the deviation of T may be extended as (Gentle, 2002):

$$D(T) = \sum_{i=1}^{\infty} a_i/n^i$$

$D(T)$ is unbiased if $a_i = 0, i = 1, 2, \ldots; D(T)$ has second-order precision if $a_1 \neq 0$. Jackknife estimate of deviation of T is thus:

$$D(J(T)) = E(J(T)) - \theta = n \sum_{i=1}^{\infty} a_i/n^i - (n-1) \sum_{i=1}^{\infty} a_i/(n-1)^i$$

14.4.3 *Bootstrap method*

The principle of boostrap method is treating observed sample as population and resampling this population. Infer the statistic's distribution according to the conditional distribution of the sample taken from population. In the statistical language (Gentle, 2002), suppose the observed sample $x_i, i = 1, 2, \ldots, n$, and population parameter θ, statistc T is used to estimate θ. First the sampling distribution of T

should be determined in order to achieve the confidence interval of θ. Statistic T is the functional of empirical cumulative distribution function, P_n:

$$T = T(P_n) = \theta(P_n) = \int g(x)dP_n(x)$$

In bootstrap method, the observed sample $x_i, i = 1, 2, \ldots, n$, is resampled to generate a resampling sample, $x_i^*, i = 1, 2, \ldots, n$, and the corresponding statistic is T^*. The variance of T is

$$V(T) = V(T^*) = \sum (T^{*j} - T_{bar}^*)^2/(m-1)$$

where T^{*j} : j-th T^* of T, and T_{bar}^*: average of $T^{*j}, V(T^*)$: variance of m samples, sampled from P_n, with size n.

The confidence interval of θ can be derived based on the relationship between θ and T, e.g., $f(T, \theta)$, and the confidence interval is (Gentle, 2002; Zhang, 2010)

$$P(f_{\alpha/2} \leq f(T, \theta) \leq f_{1-\alpha/2}) = 1 - \alpha$$

$f_{\alpha/2}$ and $f_{1-\alpha/2}$ may be approximated with bootstrap method if the probability of population is not available. $f_{\alpha/2}$ and $f_{1-\alpha/2}$ are determined by percentiles of Monte Carlo samples of $T^* - T_o$, where T_o is the T value of observed sample.

14.4.4 *Monte Carlo Method*

Monte Carlo method is used to test the characteristics of statistical methods, approximate the distribution of statistic with asymptotic approximation, and compute the expectation of function of random variables (Manly, 1997). It is particularly efficient for the hidden models without explicit relationships.

Monte Carlo method has been used in function approximation. For example, we need to estimate $F(x)$ from $f(x)$. Given a set of random inputs x_i, $i = 1, 2, \ldots, n$, and the corresponding outputs are $y_i = f(x_i), i = 1, 2, \ldots, n$. By doing so, the mean and variance of $f(x)$ can be achieved.

Monte Carlo method is also used to estimate the variance and test statistical significance, etc. A procedure for statistical test is as follows (Gentle, 2002; Zhang, 2007a, 2010): generate random sample from observed sample, and compute the characteristic of random sample; repeat this procedure n times (i.e., n randomizations); finally, test null hypothesis, i.e., observed samples are from the same distribution. The statistic, p, is

$$p = r/n$$

or

$$p = (r+1)/(n+1)$$

where r is the number of the characteristic of random sample greater than that of observed sample.

Random sampling of data set and subset generation from data set are additional applications of Monte Carlo method. Moreover, missing data can be generated by Monte Carlo method (Gentle, 2002). A data matrix X is composed of the observed data block and the missing data block. We can generate m missing data and analyze the corresponding m complete data blocks.

14.5 Stochastic Process

Stochastic process and stochastic differential equation are fundamental to the design and analysis of distributable artificial neural networks.

14.5.1 *Stationary stochastic process*

A stationary stochastic process is a stochastic process that meets the condition:

$$p(x(t_i) = c_i | i = 1, 2, \ldots, n) = p(x(t_i + \tau) = c_i | i = 1, 2, \ldots, n),$$
$$\forall t_i, c_i, \tau \in R$$

A stochastic process is reversible if

$$p(x(t_i) = c_i | i = 1, 2, \ldots, n) = p(x(\tau - t_i) = c_i | i = 1, 2, \ldots, n),$$
$$\forall t_i, c_i, \tau \in R$$

14.5.2 *Markov process*

A stochastic process is a Markov process if

$$p(x(t_i) = c_t | x(t_1) = c_1, \ldots,$$
$$x(t_{i-1}) = c_{t-1}) = p(x(t_i) = c_i | x(t_{i-1}) = c_{t-1}),$$
$$\forall t_i, c_i \in R$$

The Markov process with both discrete-time and discrete-state is a Markov chain:

$$p(x(1) = c_1, \ldots, x(t) = c_t) = p(x(t) = c_t | x(t-1)$$
$$= c_{t-1}) p(x(t-1) = c_{t-1} |$$
$$x(t-2) = c_{t-2}) \ldots p(x(2) = c_2 | x(1) = c_1)$$

where $p(x(t) = c_t | x(t-1) = c_{t-1}), t = 2, 3, \ldots$, are transition probabilities of states. Markov chain is stationary if transition probabilities are independent of time. A stationary Markov chain with limited states is a homogeneous Markov chain, if

$$p(x(t) = c_j | x(t-a) = c_i) = p(x(t+a) = c_j | x(t) = c_i)$$

Suppose $\{x(t)\}$ is a stochastic process. Define

$$p(u, \tau | v, t) = p(x(\tau) = u | x(t) = v)$$
$$p(u, \tau) = p(x(\tau) = u)$$
$$\omega(u|v, t) = \lim_{\Delta t \to 0} [p(u, t + \Delta t | v, t) - p(u, t | v, t)] / \Delta t$$

where ω is the velocity of transition probability, and $\omega(u|v) = \omega(u|v, t)$ if ω is time independent.

For a homogeneous Markov chain with continuous-time, suppose $0 \le n \le N$, and define

$$\omega^-(n) = \omega(n-1|n)$$
$$\omega^+(n) = \omega(n+1|n)$$
$$\omega(m|n) = 0, \quad \text{if } m \ne n \pm 1$$

and

$$x = n/N$$
$$\Delta x = \Delta n / N = 1/N = \varepsilon$$
$$p(x, t) = N p(n, t)$$

The Fokker-Planck equation is

$$\partial p(x,t)/\partial t = -\partial[g(x)p(x,t)]/\partial x + (2/\varepsilon)\partial^2[h(x)p(x,t)]/\partial x^2$$

where the drift coefficient $g(x)$, and diffusion coefficient $h(x)$, are defined as

$$g(x) = (\omega^+(n) - \omega^-(n))/N$$
$$h(x) = (\omega^+(n) + \omega^-(n))/N$$

Markov process-based models are not suitable to conduct discriminant decision-making. But the latter can be achieved by artificial neural networks. Therefore, the combination of Markov models and artificial neural networks is advantageous to decision-making. Hideden Markov Model (HMM) may be used to describe the time series (Yan and Zhang, 2000). In a sense, the forward-backward algorithm of HMM is equivalent to BP neural network (Hochberg *et al.*, 1991). A HMM is a Markov chain with n states, $x_i, i = 1, 2, \ldots, n$. Given the state at $t, x(t)$, and possible outputs, $y_i, i = 1, 2, \ldots, m$, a discrete HMM is thus represented by a state transition matrix, $P = (p_{ij})_{n\times n}$, a observation matrix, $B = (b_{ij})_{n\times m}$, and an initial distribution vector, $L = (l_i)_n$. The probabilities of observed output series, $y_{\eta i}, i = 1, 2, \ldots, T$, are calculated from forward-backward algorithm (Rabiner, 1989):

$$\alpha_i = l_i, \quad \alpha_i(t) = \sum_j^i \alpha_j(t-1)p_{ji}b_{i\eta T}$$

$$\beta_i = \delta_{in}, \quad \beta_i(t) = \sum_j^n \beta_j(t+1)p_{ij}b_{j\eta t+1}$$

where $\alpha_i(t)$ is feed forward probability, and $\beta_i(t)$ is backward probability. The probability of entire observed output series is $\sum \alpha_i(t)\beta_i(t)$.

Assume there are n state variables, the probability of initial state is

$$x_0 = (x_{10}, x_{20}, \ldots, x_{n0})$$

and the state transition matrix is $P = (p_{ij}), i, j = 1, 2, \ldots, n$, where $0 \leq p_{ij} \leq 1$, and $\sum_{j=1}^n p_{ij} = 1, i, j = 1, 2, \ldots, n$. The transition process is

$$x_i = P^* x_{i-1} \quad i = 1, 2, \ldots$$

when $i \to \infty, \lim(Px_i) = x_{i+1}$, and x_i is the stable probability of state. The main Java codes of the algorithm are as follows:

```
//m: number of state variables.
String ss;
int i,j,k,m,p;
double ee;
double e[][],z[][],d[][],x[],y[],f[];
e=new double[m+2][m+2];
z=new double[m+2][m+2];
d=new double[m+2][m+2];
x=new double[m+2];
y=new double[m+2];
f=new double[m+2];
//Input the x[1-m]
//Input e[1-m][1-m]
ss="";
for(i=1;i<=m;i++) y[i]=x[i];
for(i=1;i<=m;i++)
for(j=1;j<=m;j++) z[i][j]=e[i][j];
p=1;
for(k=1;k<=m;k++) {
f[k]=0;
for(i=1;i<=m;i++) f[k]+=x[i]*z[k][i]; }
time();
ee=0;
ss+="Step "+p+" ";
for(i=1;i<=m;i++) {
ss+=(int)(f[i]*100)/100.00+" ";
ee+=Math.pow(f[i]-y[i],2);
y[i]=f[i]; }
ss+="\n";
if (ee<1e-08) {
sub(); editt1.appendText(ss); }
```

```
else {
p++;
while (ee>=0) {
for(i=1;i<=m;i++)
for(j=1;j<=m;j++) d[i][j]=z[i][j];
for(i=1;i<=m;i++)
for(j=1;j<=m;j++) {
z[i][j]=0;
for(k=1;k<=m;k++) z[i][j]+=d[i][k]*e[k][j]; }
for(k=1;k<=m;k++) {
f[k]=0;
for(i=1;i<=m;i++) f[k]+=x[i]*z[k][i]; }
ee=0;
ss+="Step "+p+" ";
for(i=1;i<=m;i++) {ss+=(int)(f[i]*100)/100.00+" ";
ee+=Math.pow(f[i]-y[i],2);
y[i]=f[i]; }
ss+="\n";
if (ee<1e-08) {
sub(); editt1.appendText(ss);
break; }
p++; }}
public void sub() {
ss+="Steady state probability:\n";
for(i=1;i<=m;i++) ss+=(int)(f[i]*100)/100.00+" "; }
```

14.6 Interpolation

Interpolation refers to using discrete data for constructing a functional relationship or, to simplify a given functional relationship for computing intermediate values. The functional relationship achieved by interpolation is called interpolation formula. By using interpolation, we can calculate functional values at certain nodes (Li *et al.*, 2001; Burden and Faires, 2001; Zhang, 2007a, 2010).

Interpolation methods are used to generate missing data, smooth data series, construct empirical models, simplify complex models, as well as predict system's dynamics. The interpolation formulae based on algebraic polynomials will be discussed here.

Firstly, introduce the space of continuous functions, $C[a,b]$. It is a space composed of all continuous functions on $[a,b]$. $C[a,b]$ is a normed space, and the norm is defined as a ∞-norm:

$$||f|| = \max_{a \le x \le b} |f(x)|;$$

It is a metric space and, the metric is defined as

$$d(f,p) = ||f-p||, f(x), p(x) \in C[a,b]$$

$C[a,b]$ is also a Banach space.

Suppose there are $n+1$ different nodes, i.e., interpolation nodes, of $f(x) \in C[a,b]$, on $[a,b]$

$$a \le x_0, x_1, \ldots, x_n \le b$$

The functional values at these nodes are $f(x_i), i = 0,1,\ldots,n$. If $L_n(x) \in H_n$ exists, where H_n is the set of polynomials with degree less than or equal to n, such that

$$L_n(x_i) = f(x_i), i = 0,1,\ldots,n$$

then $L_n(x)$ is the interpolation polynomial, i.e., interpolation function, of $f(x)$ on $[a,b]$, based on $(x_i, f(x_i)), i = 0,1,\ldots,n$. $L_n(x)$ is existential and unique.

It is possible to calculate the approximate values of $f(x)$ from $L_n(x)$, if $x \ne x_i, i = 0,1,\ldots,n$. It is called interpolation if $x \in [a,b]$, or else it is called extrapolation.

14.6.1 *Lagrange interpolation*

Lagrange interpolation polymial is

$$L_n(x) = \sum_{i=0}^{n} l_i(x_i) f(x_i)$$

where $l_i(x)$ is basis function and

$$\begin{aligned} l_i(x) = {} & [(x-x_0)(x-x_1)\ldots(x-x_{i-1})(x-x_{i+1})\ldots(x-x_n)]/ \\ & [(x_i-x_0)(x_i-x_1)\ldots(x_i-x_{i-1})(x_i-x_{i+1})\ldots(x_i-x_n)] \\ & i = 0,1,\ldots,n \end{aligned}$$

Lagrange interpolation is a linear interpolation, if $n = 1$; quadratic interpolation, if $n = 2$, and cubic interpolation, if $n = 3$.

The error of interpolation, i.e., the remainder of $L_n(x), R_n(x) = f(x) - L_n(x)$, is given by the following theorem: suppose that $f(x)$ is n-order continuously differentiable on $[a, b]$, and $f^{(n+1)}(x)$ is existential on (a, b), then the remainder term of interpolation polynomial, $L_n(x)$, is

$$R_n(x) = f^{(n+1)}(\xi)\omega_{n+1}(x)/(n+1)!, x \in [a, b]$$

where $\xi = \xi(x) \in (a, b)$, and

$$\omega_{n+1}(x) = (x - x_0)(x - x_1)\ldots(x - x_n)$$

14.6.2 *Newton interpolation*

Newton interpolation is equivalent to Lagrange interpolation, but its calculation is simple.

First, define difference quotient with different order as follows:

First-order difference quotient:

$$f[x_i, x_{i+1}] = (f(x_i) - f(x_{i+1}))/(x_i - x_{i+1}), \quad i = 0, 1, \ldots, n-1$$

Second-order difference quotient:

$$f[x_i, x_{i+1}, x_{i+2}] = (f[x_i, x_{i+1}] - f[x_{i+1}, x_{i+2}])/(x_i - x_{i+2}),\\ i = 0, 1, \ldots, n-2$$

...

n-order difference quotient:

$$f[x_0, x_1, \ldots, x_n] = (f[x_0, x_1, \ldots, x_{n-1}] - f[x_1, x_2, \ldots, x_n])/(x_0 - x_n)$$

Newton interpolation polynomial is thus:

$$N_n(x) = f(x_0) + f[x_0, x_1](x - x_0) + f[x_0, x_1, x_2](x - x_0)(x - x_1)\\ + \ldots + [x_0, x_1, \ldots, x_n](x - x_0)(x - x_1)\ldots(x - x_{n-1})$$

14.6.3 *Hermite interpolation*

If interpolation conditions are as follows:

$$H(x_i) = f(x_i), H'(x_i) = f'(x_i), \quad i = 0, 1, \ldots, n$$

then the following Hermite interpolation polynomial can be used:

$$H(x) = \sum_{i=0}^{n} \alpha_i(x) f(x_i) + \sum_{i=0}^{n} \beta_i(x) f'(x_i)$$

where

$$\alpha_i(x) = [1 - 2(x - x_i) \sum_{\substack{j=0 \\ j \neq i}}^{n} (1/(x_i - x_j))]^*$$

$$\omega_{n+1}^2(x)/\{(x - x_i)^2[\omega'_{n+1}(x_i)]^2\}$$

$$\beta_i(x) = \omega_{n+1}^2(x)/\{(x - x_i)[\omega'_{n+1}(x_i)]^2\}$$

$$\omega'_{n+1}(x_i) = (x_i - x_0)(x_i - x_1) \ldots (x_i - x_{i-1})(x_i - x_{i+1}) \ldots (x_i - x_n)$$

Suppose that $f(x) \in C[a, b]$ has $2n + 2$ order derivative on (a, b), then the remainder term of Hermite interpolation polynomial is

$$R(x) = f(x) - H(x) = f^{(n+2)}(\xi)\omega_{n+1}^2(x)/(2n + 2)!, x \in [a, b]$$

where $\xi = \xi(x) \in (a, b)$.

14.6.4 *Spline interpolation*

The above interpolation methods will become complex and unstable as the order of polynomial increases. For example, the error of interpolation would be considerably large at the end nodes of interpolation interval. Spline interpolation can avoid such issues. Spline interpolation is the most effective interpolation method, in which cubic spline function is the most widely used. Cubic spline interpolation is a piecewise cubic interpolation. Its first- and second-order derivatives are continuous at inner nodes and it is thus relatively smooth.

Suppose there are $n + 1$ different nodes on a given interval $[a, b]$

$$a \leq x_0 < x_1 < \cdots < x_n \leq b$$

If the order of polynomial $S(x)$, on each sub-interval $[x_i, x_{i+1}]$, is less than or equal to m and larger than or equal to 1, and its $m - 1$ order derivative, $S^{(m-1)}(x)$, is continuous at inner nodes, $x_1, \ldots, x_{n-1}$, $S(x)$ is called the m-degree spline function. Known $f(x) \in C[a, b]$, and

$S(x_i) = f(x_i), i = 0, 1, \ldots, n$, then $S(x)$ is called the m-degree spline interpolation polynomial of $f(x)$ on $[a, b]$.

Cubic spline interpolation polynomial

$$S(x) = a_i x^3 + b_i x^2 + c_i x + d_i, x \in [x_i, x_{i+1}], \quad i = 0, 1, \ldots, n-1$$

should meet the following conditions:

(1) there are totally $2n$ conditions for interpolation and continuity;
(2) the first- and second-order derivatives should be continuous at inner nodes.

$$S'(x_i - 0) = S'(x_i + 0), S''(x_i - 0) = S''(x_i + 0), \quad i = 1, 2, \ldots, n-1$$

Moreover, the following three boundary conditions are usually required.

(3) $S'(x_0) = f'(x_0), S'(x_n) = f'(x_n)$
(4) $S''(x_0) = f''(x_0), S''(x_n) = f''(x_n)$
(5) $S'(x_0 + 0) = S'(x_n - 0), S''(x_0 + 0) = S''(x_n - 0)$

Cubic spline interpolation polynomial is expressed as

$$S(x) = M_{i+1}(x - x_i)^3/(6l_i) + M_i(x_{i+1} - x)^3/(6l_i) + (f(x_{i+1})/l_i - M_{i+1}l_i/6)(x - x_i) + (f(x_i)/l_i - M_i l_i/6)(x_{i+1} - x),$$

$$x \in [x_i, x_{i+1}], \quad i = 0, 1, \ldots, n-1$$

where $M_i = S''(x_i), i = 0, 1, \ldots, n; l_i = x_{i+1} - x_i, i = 0, 1, \ldots, n-1$. $M_i, i = 0, 1, \ldots, n$, are obtained by solving three bending moment equation.

Cubic spline function is existential and unique under certain conditions. Moreover, it shows the property of best approximation.

14.7 Function Approximation

We often use the functions with complex structure and tedious calculation. Function approximation can be used to simplify those functions, which will make the calculation simple and rapid without losing accuracy significantly (Li *et al.*, 2001; Burden and Faires, 2001; Zhang and Barrion, 2006; Zhang *et al.*, 2008a; Zhang, 2007a, 2010).

Suppose that $f(x)$ is a function on $[a, b]$. We try to construct a simple function, $p(x)$, to approximate the given function, $f(x)$. It is a problem of function approximation. Function approximation can also be stated as follows: for the given function $f(x) \in C[a,b]$, to find a function $p(x) \in C[a,b]$, making the error between $f(x)$ and $p(x)$ on $[a,b]$ to be the smallest

Often used error metrics include:

(1) uniform approximation: find a function, $p(x)$, so that

$$\lim_{n\to\infty} d(f,p) = 0$$

(2) L_q approximation: find a function, $p(x)$, so that

$$\lim_{n\to\infty} \int_a^b |f(x) - p(x)|^q w(x)dx = 0$$

where $q \geq 1, w(x)$ is the weight function on $[a,b]$. It is the quadratic approximation if $q = 2$. The weight function, $w(x)$, satisfies the following conditions:

(a) $w(x) \geq 0, x \in [a,b]$;
(b) $\int_a^b |x|^n w(x)\mathrm{d}x$ is existential, $n = 0, 1, \ldots$;
(c) for the continuous function, $g(x) \geq 0, g(x) \in C[a,b]$, if

$$\int_a^b g(x)w(x)dx = 0$$

then $g(x) \equiv 0$ on (a,b).

(3) least squares estimation: find a function, $p(x)$, so that

$$\min = \sum_{i=1}^{m} w(x_i)(p(x_i) - f(x_i))^2$$

We discuss the methods using algebraic polynomial $p(x)$ to approximate $f(x)$ in the following.

14.7.1 *Uniform approximation*

14.7.1.1 *Existence of algebraic polynomial of uniform approximation*

For the uniform approximation, Weierstrass theorem can be used here: suppose $f(x) \in C[a,b]$, then for any $\varepsilon > 0$, there is an algebraic polynomial $p(x)$, so that

$$||f(x) - p(x)|| < \varepsilon$$

uniformly holds on $[a, b]$. Weierstrass theorem showed the existence of algebraic polynomial of uniform approximation.

14.7.1.2 *Chebyshev best uniform approximation*

Suppose $1, x, \ldots, x^n$, are a set of linearly independent functions on $[a, b]$, H_n is a set of polynomials with degree less than or equal to $n, H_n = \text{span}\{1, x, \ldots, x^n\}, H_n = C[a, b]$. Any $p_n(x) \in H_n$, can be expressed as

$$p_n(x) = a_0 + a_1 x + \cdots + a_n x^n$$

where $a_i \in R, i = 0, 1, \ldots, n$.

Suppose $f(x) \in C[a, b]$, then $||f(x) - p_n(x)||$ is defined as the deviation between $f(x)$ and $p_n(x)$ on $[a, b]$, and

$$E_n = \min_{p_n(x) \in H_n} ||f(x) - p_n(x)||$$

is the least absolute deviation. Find $p_n^*(x)$, so that

$$||f(x) - p_n^*(x)|| = E_n = \min_{p_n(x) \in H_n} ||f(x) - p_n(x)||$$

This is the Chebyshev best uniform approximation, $p_n^*(x)$ is the Chebyshev polynomial of best uniform approximation.

Chebyshev alternate node pattern: suppose $p(x) \in C[a, b]$, if there is a node pattern $x_i, i = 1, 2, \ldots, n, a \le x_1 < \cdots < x_n \le b$, such that

$$|p(x_i)| = \max_{a \le x \le b} |p(x)|, \quad i = 1, 2, \ldots, n$$

and

$$p(x_i) = -p(x_{i+1}), \quad i = 1, 2, \ldots, n - 1$$

then it is called Chebyshev alternate node pattern of $p(x)$ on $[a, b]$.

The following Chebyshev's theorem shows the existence and uniqueness of Chebyshev polynomial of best uniform approximation: suppose $f(x) \in C[a, b], p(x) \in H_n$, then $p(x)$ is the polynomial of best uniform approximation of $f(x)$, if and only if there is a node pattern of $f(x) - p(x)$ on [a,b], which contains at least $n + 2$ nodes.

By the theorem, we can see that if $f(x) \in C[a, b]$, then there is a unique polynomial of best uniform approximation in H_n, and it is a Lagrange's interpolation polynomial of $f(x)$.

The polynomial of best uniform approximation, $p(x)$, is generally hard to find. However, it can be obtained based on the following theorem: suppose $f(x)$ is $n + 1$-th derivative on $[a, b]$, and $f^{(n+1)}(x)$ does not alter its sign on $[a, b]$, then the end nodes a and b belong to the node pattern of $f(x) - p(x)$, if $p(x) \in H_n$ is the polynomial of best uniform approximation of $f(x)$.

14.7.2 *Quadratic approximation*

If there is a $p_n^*(x)$, such that

$$\int_a^b |f(x) - p_n^*(x)|^2 w(x)dx = \inf_{p_n(x) \in H_n} \int_a^b |f(x) - p_n(x)|^2 w(x)dx$$

then $p_n^*(x)$ is the best quadratic approximation of $f(x)$ on $[a, b]$ in respect to weight function, $w(x)$.

14.7.2.1 *Orthogonal system of functions*

Suppose $f(x), g(x) \in C[a, b]$, and $w(x)$ is the weight function on $[a, b]$, then the inner product of $f(x)$ and $g(x)$ on $[a, b]$ is

$$(f, g) = \int_a^b f(x)g(x)w(x)dx$$

The continuous function space $C[a, b]$ with a definition of inner product is an inner product space. $f(x)$ and $g(x)$ are $w(x)$ weighted orthogonal on $[a, b]$, if $(f, g) = 0$. If functions $\varphi_0(x), \varphi_1(x), \ldots, \varphi_n(x)$, $\ldots$, such that

$$(\varphi_i, \varphi_j) = \int_a^b \varphi_i(x)\varphi_j(x)w(x)dx = 0, \quad i \neq j$$
$$(\varphi_i, \varphi_j) = C_{ij}, \quad i = j$$

$\{\varphi_i\}$ is defined as the orthogonal system of functions on $[a, b]$ with weight function $w(x)$. If $C_{ij} \equiv 1$, it is the orthonormal system of functions. If $\varphi_i(x), i = 0, 1, \ldots$, are algebraic polynomials, then they are called orthogonal polynomials.

Often used orthogonal polynomials are as follows:

(1) Legendre polynomials
Legendre polynomials are defined on $[-1, 1]$, their weight function $w(x) \equiv 1$, and orthogonal system of functions $\{\varphi_i\} = \{1, x, \ldots, x^n, \ldots\}$, then Legendre polynomials are

$$\begin{aligned} \varphi_0(x) &= 1, \varphi_1(x) = x, \\ \varphi_{n+1}(x) &= ((2n+1)x\varphi_n(x) - n\varphi_{n-1}(x))/(n+1), x \in [-1, 1], \\ n &= 1, 2, \ldots \end{aligned}$$

(2) Chebyshev polynomials
Chebyshev polynomials are defined on $[-1, 1]$, the weight function $w(x) = 1/(1-x^2)^{1/2}$, and orthogonal system of functions $\{\varphi_i\} = \{1, x, \ldots, x^n, \ldots\}$, then Chebyshev polynomials are

$$\begin{aligned} \varphi_0(x) &= 1, \varphi_1(x) = x, \\ \varphi_{n+1}(x) &= 2x\varphi_n(x) - \varphi_{n-1}(x), \quad x \in [-1, 1], \\ n &= 1, 2, \ldots \end{aligned}$$

(3) Laguerre polynomials
Laguerre polynomials are defined on $[0, \infty)$, the weight function $w(x) = \mathrm{e}^{-x}$, and orthogonal system of functions $\{\varphi_i\} = \{1, x, \ldots, x^n, \ldots\}$, then Laguerre polynomials are

$$\begin{aligned} \varphi_0(x) &= 1, \varphi_1(x) = 1 - x, \\ \varphi_{n+1}(x) &= (2n+1-x)\varphi_n(x) - n^2\varphi_{n-1}(x), x \in [0, \infty) \\ n &= 1, 2, \ldots \end{aligned}$$

(4) Hermite polynomials
Hermite polynomials are defined on $(-\infty, \infty)$, the weight function $w(x) = \mathrm{e}^{-x*x}$, and orthogonal system of functions $\{\varphi_i\} = \{1, x, \ldots, x^n, \ldots\}$, then Hermite polynomials are

$$\begin{aligned} \varphi_0(x) &= 1, \varphi_1(x) = 2x, \\ \varphi_{n+1}(x) &= 2x\varphi_n(x) - 2n\varphi_{n-1}(x), x \in (-\infty, \infty) \\ n &= 1, 2, \ldots \end{aligned}$$

(5) Trigonometric functions
Trigonometric functions are defined on $[0, 2\pi]$, and weight function $w(x) \equiv 1$. They have the following form:

$$\begin{aligned}
&\varphi_0(x) = 1,\\
&\varphi_1(x) = \cos(x), \varphi_2(x) = \sin(x),\\
&\varphi_3(x) = \cos(2x), \varphi_4(x) = \sin(2x),\\
&\varphi_5(x) = \cos(3x), \varphi_6(x) = \sin(3x), \ldots, x \in [0, 2\pi]
\end{aligned}$$

14.7.2.2 *Linearly independent system of functions*

If $\varphi_k(x) \in C[a, b], k = 0, 1, \ldots, n-1$, such that

$$a_0\varphi_0(x) + a_1\varphi_1(x) + \cdots + a_{n-1}\varphi_{n-1}(x) = 0,$$

if and only if $a_0 = a_1 = \cdots = a_{n-1} = 0$, we call $\varphi_k(x), k = 0, 1, \ldots, n-1$, are linearly independent. $\{\varphi_i\}$ is a linearly independent system of functions, if a finite number of arbitrary $\varphi_k(x)$ in the system of functions $\{\varphi_i\}$ are linearly independent. Subspace expanded from linearly independent system of functions, $\varphi_k(x), k = 0, 1, \ldots, n-1$,

$$\begin{aligned}
S = \operatorname{span}\{\varphi_0, \varphi_1, \ldots, \varphi_{n-1}\} = \{&a_0\varphi_0(x) + a_1\varphi_1(x) + \cdots\\
&+ a_{n-1}\varphi_{n-1}(x) | a_0, a_1, \ldots, a_{n-1} \in R\}
\end{aligned}$$

is the subspace of $C[a, b]$.

14.7.2.3 *Best quadratic approximation*

The following theorem shows the existence and uniqueness of best quadratic approximation, L_2: suppose $f(x) \in C[a, b]$, then $f(x)$ has one and only one best quadratic approximation.

Suppose $\varphi_k(x), k = 0, 1, \ldots, n$, are linearly independent, solve the following equations:

$$\begin{gathered}
(\varphi_i, \varphi_0)a_0 + (\varphi_i, \varphi_1)a_1 + \cdots + (\varphi_i, \varphi_n)a_n = (\varphi_i, f)\\
i = 0, 1, \ldots, n
\end{gathered}$$

The best quadratic approximation of $f(x)$ is thus

$$a_0\varphi_0(x) + a_1\varphi_1(x) + \cdots + a_n\varphi_n(x)$$

A simple way is taking $\varphi_i(x) = x^i, w(x) \equiv 1$.

Orthogonal system of functions, in particular orthogonal polynomials, is usually used as the linearly independent system of functions. For this situation, we obtain the solution of above equations:

$$a_i = (\varphi_i, f)/(\varphi_i, \varphi_i)$$

that is

$$a_i = \int_a^b \varphi_i(x) f(x) w(x) dx \Big/ \int_a^b (\varphi_i(x))^2 w(x) dx$$
$$i = 0, 1, \ldots, n$$

and a_i are said to be generalized Fourier coefficients of $f(x)$ in respect to the orthogonal system of functions, $\{\varphi_i\}$. The expansion of $f(x)$:

$$\sum_{i=0}^{\infty} a_i \varphi_i(x)$$

is a generalized Fourier series. As a result, any function, $f(x) \in C[a, b]$, can be expanded as a generalized Fourier series.

14.7.3 *Least squares approximation*

Find the relationship between independent and dependent variables from experimental data $(x_i, f(x_i)), i = 0, 1, \ldots, m$, so that the error:

$$\sum_{i=0}^{m} w(x_i)(p(x_i) - f(x_i))^2$$

is minimal. It is the least squares approximation, $p(x)$. Here, $w(x)$ is the weight function representing the weights of various data nodes.

Construct orthogonal functions, $\varphi_k(x), k = 0, 1, \ldots, n, n \leq m$, the least squares approximation is

$$\sum_{j=0}^{n} a_j \varphi_j(x)$$

Define inner product as follows:

$$(\varphi_i, \varphi_j) = \sum_{k=0}^{m} w(x_k) \varphi_i(x_k) \varphi_j(x),$$

$$(\varphi_i, f) = \sum_{k=0}^{m} w(x_k) \varphi_i(x_k) f(x_k),$$

then

$$a_i = (\varphi_i, f)/(\varphi_i, \varphi_i)$$

As an example, suppose the population dynamics of an animal is $f(t) \in C[a, b]$, conduct best quadratic approximation of $f(t)$ based on trigonometric system of functions. The expansion of $f(t)$ is

$$p(t) = a_0/2 + \sum_{i=1}^{\infty}(a_i \cos it + b_i \sin it)$$

where

$$a_i = (1/\pi)\int_0^{2\pi} f(t)\cos it\ dt, \quad i = 0, 1, 2, \ldots$$
$$b_i = (1/\pi)\int_0^{2\pi} f(t)\sin it\ dt, \quad i = 1, 2, \ldots$$

This is just Fourier series. Fourier series will uniformly converge to $f(t)$, if $f'(t)$ is piecewise continuous on $[0, 2\pi]$. Thus its partial sum can be taken as the best quadratic approximation of $f(t)$. Periodicity of population dynamics can also be analyzed according to Fourier series.

14.8 Optimization Methods

14.8.1 *Steepest descent method*

The steepest descent method is a basic unconstrained optimization technique. Given an unconstrained optimization problem:

$$\min f(x)$$

where $x = (x_1, x_2, \ldots, x_n)^T$, and $f(x)$ is the non-linear function of x. Calculate the $t^i \in R$, such that

$$f(x^i + t^i p^i)_t = \min f(x^i + tp^i)$$

The direction of search, p^i, is determined by:

$$p^i = -\nabla f(x^i) = -(\partial f/\partial x_1, \partial f/\partial x_2, \ldots, \partial f/\partial x_n)|x = x^i$$

Finally the next point, $x^{i+1} = x^i + t^i p^i$, is achieved. The following is the Java class for steepest descent method:

```
import java.io.*;
class steepestDescent {
public static int n;
public static double la,ff;
public static double x[];
public static double f[];
public static double v[];
public static void main(String args[]) {
double s,pr;
int nu=0,i,max;
n=2;max=10;
pr=1e-06;
x=new double[n+2];
f=new double[n+2];
v=new double[n+2];
point1: while (nu<max) {
derivative();
s=0;
for(i=1;i<=n;i++) s+=Math.pow(f[i],2);
if (s<=pr) break point1;
derivative();
for(i=1;i<=n;i++) v[i]=x[i];
steplength();
for(i=1;i<=n;i++) x[i]=v[i]-la*f[i];
nu++; }
objectivefun();
for(i=1;i<=n;i++) System.out.println("x("+i+")="+x[i]+" ");
System.out.println("f(x)="+ff);
System.out.println("PR="+pr);
System.out.println("Number="+nu); }

public static double objectivefun() {
ff=-4-3*x[1]-2*x[2]-5*x[1]*x[2]+5*Math.pow(x[1],2)+2*Math.pow(x[2],2);
return ff; }
```

```
public static double[] derivative() {
f[1]=-3-5*x[2]+10*x[1];
f[2]=-2-5*x[1]+4*x[2];
return f; }

public static double steplength() {
double la1,la2;
la1=Math.pow(f[1],2)+Math.pow(f[2],2);
la2=f[1]*(-20+125*x[1]-70*x[2])+f[2]*(7-70*x[1]+41*x[2]);
la=la1/la2;
return la;}
}
```

14.8.2 *Conjugate gradient method*

Suppose the objective function, $f(x)$, is approximately a quadratic function in the neighborhood of the extreme point, x^*:

$$f(x) \approx a + b^T x + x^T Ax/2$$

Calculate

$$\begin{aligned} p_0 &= -b - Ax_0 \\ g_i &= b + Ax_i \\ \beta_{i-1} &= ||g_i||^2/||g_{i-1}||^2 \\ p_i &= -g_i + \beta_{i-1}p_{i-1} \end{aligned}$$

such that

$$f(x^i + t^i p^i) = \min_t f(x^i + tp^i)$$

the next point, $x^{i+1} = x^i + t^i p^i$, is thus achieved. The iteration terminates if $||g_i|| \leq \varepsilon$.

14.8.3 *Newton's method*

Suppose the objective function, $f(x)$, can be expanded as a quadratic Taylor polynomial near in the neighborhood of the point, x^i:

$$f(x) \approx f(x^i) + \nabla f(x^i)^T \Delta x + \Delta x^T A^i \Delta x/2$$

where $\Delta x = x - x^i$. Calculate $t^i \in R$, such that

$$f(x^i + t^i p^i) = \min_t f(x^i + tp^i)$$

The direction of search, p_i, is determined by

$$p^i = -(A^i)^{-1} \nabla f(x^i)$$

The next point, $x^{i+1} = x^i + t^i p^i$, is therefore achieved.

14.8.4 *DFP method*

Given an unconstrained optimization problem:

$$\min f(x)$$

where $x = (x_1, x_2, \ldots, x_n)^T$, and $f(x)$ is the non-linear function of x. Assume initial values $x_0 = (x_{10}, x_{20}, \ldots, x_{n0})$ and iterative error $e > 0$. $H_0 = I$, and I is the unit matrix. Search along the direction $-H_0 f'(x_0)$, in order to determine λ_0:

$$\min f(x_0 - \lambda H_0 f'(x_0)) = f(x_0 - \lambda_0 H_0 f'(x_0))$$

then solution $x_1 = x_0 - \lambda_0 H_0 f'(x_0)$. Repeat the process, we obtain x_k. Let

$$g'_{k-1} = f'(x_k) - f'(x_{k-1}), dx_{k-1} = x_k - x_{k-1}$$

Calculate the scaling matrix:

$$\begin{aligned} H_k = H_{k-1} &+ dx^*_{k-1} dx_{k-1} / (g'_{k-1} dx_{k-1}) \\ &- H_{k-1} g'_{k-1} g'_{k-1} H_{k-1} / (g'_{k-1} H_{k-1} g'_{k-1}) \end{aligned}$$

and search along the direction $-H_k f'(x_k)$ to determine λ_k

$$\min f(x_k - \lambda H_k f'(x_k)) = f(x_k - \lambda_k H_k f'(x_k))$$

then the solution is $x_{k+1} = x_k - \lambda_k H_k f'(x_k)$. If $||dx_k|| < e, x_{k+1}$ is the optimization solution; or else, let $k = k + 1$, repeat the process.

Java codes of DFP method are as follows:

```
import java.io.*;
class dfp {
public static int n,i,j;
public static double f,lm;
```

```
public static double x[];
public static double y[];
public static double z[];
public static double b[];
public static double c[];
public static double d[];
public static double p[];
public static double e[];
public static double s[];
public static double h[][];
public static double g[];
public static double a[];
public static void main(String args[]) throws Exception{
double ss=0,t,r,pr,v;
int max,it,q=100,qp=10,l;
n=2;max=10;
pr=1e-05;
x=new double[n+2];
y=new double[n+2];
z=new double[n+2];
b=new double[n+2];
c=new double[n+2];
d=new double[n+2];
p=new double[1000];
e=new double[n+2];
s=new double[n+2];
h=new double[n+2][n+2];
g=new double[n+2];
a=new double[n+2];
it=0;
for(i=1;i<=n;i++) {for(j=1;j<=n;j++) h[i][j]=0;h[i][i]=1;}
derivative();
point1: while (it>=0) {
for(i=1;i<=n;i++) s[i]=g[i];
for(i=1;i<=n;i++) z[i]=x[i];
for(l=1;l<=q;l++) {lm=l*1.0/qp;
```

```
for(i=1;i<=n;i++) {e[i]=0;
for(j=1;j<=n;j++) e[i]+=h[i][j]*g[j];
y[i]=x[i]-lm*e[i];
for(i=1;i<=n;i++) x[i]=y[i];
p[l]=objectivefun();
for(i=1;i<=n;i++) x[i]=z[i]; }
v=1e+10;
for(l=1;l<=q;l++) if (p[l]<v) {v=p[l];lm=l;}
lm/=qp;
for(i=1;i<=n;i++) {e[i]=0;
for(j=1;j<=n;j++) e[i]+=h[i][j]*g[j];
y[i]=x[i]-lm*e[i]; }
it++;
ss=0;
for(i=1;i<=n;i++) ss+=Math.pow(y[i]-x[i],2);
ss=Math.sqrt(ss);
if ((ss<pr)|(it>=max)) break point1;
for(i=1;i<=n;i++) {a[i]=y[i]-x[i];x[i]=y[i];}
derivative();
for(i=;i<=n;i++) b[i]=g[i]-s[i];
t=0;
r=0;
for(i=1;i<=n;i++) {c[i]=0;d[i]=0;
for(j=1;j<=n;j++) {c[i]-=h[i][j]*b[j];d[i]+=h[j][i]*b[j];}
t+=a[i]*b[i];
r+=d[i]*b[i];}
for(i=1;i<=n;i++) for(j=1;j<=n;j++)
h[i][j]+=a[i]*a[j]/t+c[i]*d[i]/r; }
for(i=1;i<=n;i++) {x[i]=y[i];System.out.println("x("+i+")="+x[i]+" ");}
System.out.println("f(x)="+objectivefun());
System.out.println("E="+ss);
System.out.println("Number="+it);
}
public static double objectivefun() {
f=Math.pow(x[1]-2,2)+Math.pow(x[2]-3,2)+x[1]*x[2]-x[1]-13;
return f; }
```

```
public static double[] derivative() {
g[1]=x[2]+2*(x[1]-2)-1;
g[2]=x[1]+2*(x[2]-3);
return g; }
}
```

14.8.5 *Dynamic programming*

The foundation of dynamic programming is Bellman's principle (Bellman, 1957), i.e., the optimal decision sequence of a multi-phase decision process has such a property, that is, regardless of how the initial phase, initial state, and initial decision are, take the phase and state generated in the first decision as the initial condition, the subsequent decisions must constitute the optimal sequences with respect to the corresponding problem (Norton, 1972; Li *et al.*, 1982).

The discrete and deterministic dynamic programming is described as follows (Zhang, 2012a).

First, divide the procedure into n phases, $k = 1, 2, \ldots, n$, and determine the state x_k of phase k, x_k is an initial state of phase k.

Second, determine the decision variables of each phase. Suppose $u_k(x_k)$ is the decision variable of phase k when state is x_k. $u_k(x_k) \in D_k(x_k)$, where $D_k(x_k)$ is the admissible decision set of phase k. The decision function series from phase k to the end point, i.e., the substrategies are:

$$P_{kn} = \{u_k(x_k), u_{k+1}(x_{k+1}), \ldots, u_n(x_n)\}$$

Third, determine the rules for state transition. Given the state variable x_k of phase k, use the decision variable u_k, then the state x_{k+1} of phase $k+1$ can be determined, i.e., $x_{k+1} = T_k(x_k, u_k)$.

Fouth, define the goal function. The goal function V_{kn} is used to evaluate the goodness of the procedure:

$$V_{kn} = V_{kn}(x_k, u_k, x_{k+1}, \ldots, x_{n+1}), k = 1, 2, \ldots n,$$

where the optimal value of V_{kn} is the optimal goal function $f_k(x_k)$. Calculation formula of the goal function is

$$V_{kn} = v_k(x_k, u_k) + V_{k+1\ n}(x_{k+1}, \ldots, x_{n+1})$$

where $v_k(x_k, u_k)$ is the goal value of phase k. Goal function is the function of initial condition and strategies, so its calculation formula can be written as:

$$V_{kn}(x_k, P_{kn}) = v_k(x_k, u_k) + V_{k+1n}(x_{k+1}, P_{k+1\ n})$$

where $P_{kn} = \{u_k(x_k), P_{k+1\ n}(x_{k+1})\}$.

Finally, make reverse sequence optimization:

$$\text{opt}(P_{kn})V_{kn}(x_k, P_{kn}) = \text{opt}(u_k)\{v_k(x_k, u_k) + \text{opt}(P_{k+1\ n})V_{k+1\ n}$$
$$k = n, n-1, \ldots, 1$$
$$f_1(x_1) = \text{opt}(P_{1n})V_{1n}(x_1, P_{1n})$$

or

$$f_k(x_k) = \text{opt}(u_k \in D_k(x_k))\{v_k(x_k, u_k) + f_{k+1}(x_{k+1})\}$$
$$k = n, n-1, \ldots, 1$$
$$f_{n+1}(x_{n+1}) = 0$$

where opt is minimization (min) or maximization (max). Starting from $k = n$, calculate toward ahead until $f_1(x_1)$ is obtained. Then optimal strategy and the optimal value of goal function will be obtained.

The Java codes, dynProg, for the discrete and deterministic dynamic programming can be found in Zhang (2012a).

14.8.6 *Shortest path problem*

The objective of shortest path problem is to find a path with minimal weight sum (shortest path) in a weighted graph. The shortest path refers to the one between two given vertices, or the one between a given vertex to every vertices. The weights in weighted graph are distance, cost, flux, etc.

To find between-vertex shortest path of an undirected graph, Dijkstra algorithm can be used (Dijkstra, 1959). First, define the weight matrix $D = (d_{ij})$ of an undirected graph X, where d_{ij} is the weight of the edge e_{ij}. $d_{ij} = 0$, if $i = j; d_{ij} > 0$, if there exists an edge e_{ij}, and $d_{ij} = \infty$, if there is not an edge e_{ij}. Suppose the two vertices are A (starting vertex) and B (terminal vertex), then Dijkstra algorithm is as follows:

(1) Label v_A as $v_A = 0$, and the other vertex v_i as $v_i = \infty, i \neq A$.
(2) Label the unlabeled vertex v_j as

$$v_j = (j_{\text{old}}, i_{\text{old}} + d_{ij})$$

(3) Find the minimum of labels and take the label as the fixed label of the vertex; return to (2), until B is labeled. The shortest path and its length are thus achieved.

The Java codes, Dijkstra, for Dijkstra algorithm can be found in Zhang (2012a).

Other optimization algorithms on graphs and networks can also be found in Zhang (2012a).

14.9 Functional Analysis

14.9.1 *Functional representation*

The functional relationship, $s = f(r)$, where $s = y(t), r = x(t)$, is called an operator. The functional relationship, $y(t) = f(t, x(\tau)|\tau \leq t)$, is generally called a functional (Yan and Zhang, 2000; Zhang, 2010). Usually a linear functional can be represented by:

$$f(t, x(\tau)|\tau \leq t) = \int g(t, \tau)x(\tau)d\tau$$

A common non-linear functional is the n-order regular homogeneous functional:

$$f(t, x(\tau)|\tau \leq t) = \int\int \dots \int g(t, \tau_1, \tau_2, \dots, \tau_n)x(\tau_1)x(\tau_2)\dots x(\tau_n)d\tau_1 d\tau_2 \dots d\tau_n$$

A functional, $f(t, x(\tau)|\tau \leq t)$, can be expanded as a Volterra series:

$$\begin{aligned} f(t, x(\tau)|\tau \leq t) &= \sum \int \dots \int g(t, \tau_1, \tau_2, \dots, \tau_n)x(\tau_1)x(\tau_1)\dots x(\tau_n)d\tau_1 d\tau_2 \dots d\tau_n \\ &= g_0(t) + \int g(t, \tau)x(\tau)d\tau \end{aligned}$$

$$+\int\int g(t,\tau_1,\tau_2)x(\tau_1)x(\tau_2)d\tau_1 d\tau_2$$

$$+\ldots+\int\ldots\int g(t,\tau_1,\tau_2,\ldots,\tau_n)x(\tau_1)x(\tau_1)\ldots$$

$$x(\tau_n)d\tau_1 d\tau_2\ldots d\tau_n+\cdots$$

which is equivalent to the three-tier feed forward artificial neural network (Yan and Zhang, 2000).

14.9.2 *Functional analysis*

Some methods of functional analysis may be used in the analysis of mathematical models. Some principles and methods of functional analysis are discussed in the following (Chen, 1987; Rudin, 1991; Liu, 2000; Men and Feng, 2005; Zhang, 2007a, 2010).

Deflation Principle 1: suppose (X, d) is a complete metric space, a mapping T: $X \to X$, such that:

$$d(Tx, Ty) \leq \theta d(x, y), \quad \forall\, x, y \in X$$

where $\theta \in (0, 1)$, then there is only one fixed point $x' \in X$, such that $Tx' = x'$.

Deflation Principle 2: suppose (X, d) is a complete metric space, and T: $X \to X$, is a mapping, if there is a natural number n_0, such that

$$d(T^{n0}x, T^{n0}y) \leq \theta d(x, y), \quad \forall\, x, y \in X$$

where $\theta \in [0, 1]$, then there is only one fixed point $x' \in X$, such that $Tx' = x'$.

According to Deflation Principle 1, take $x_0 \in X$, and conduct iterative calculation, $x_{n+1} = Tx_n$, if $\{x_n\}$ is sequential convergent, then the limit of $\{x_n\}$ is the fixed point x', moreover, the error to approximate x' with x_n is

$$d(x_n, x') \leq \theta^n d(x_0, Tx_0)/(1-\theta)$$

The nearer x_0 approaches Tx_0, the smaller the error.

As an example, consider a differential equation:

$$\mathrm{d}x/\mathrm{d}t = f(x,t), x|_{t=t0} = x_0$$

where $f(x,t)$ is continuous on R^2, and satisfies with the condition:

$$|f(x_1,t) - f(x_2,t)| \leq K|x_2 - x_1|$$

in respect to x. Let us try to show that the above problem has a unique solution in the neighborhood of t_0.

Take $\delta > 0$, such that $K\delta < 1$. Define an operator on $C[t_0 - \delta, t_0 + \delta]$

$$Tx(t) = \int_{t0}^{t} f(x(\tau),\tau)d\tau + x_0$$

then T is the mapping of R^2 to itself, and

$$\begin{aligned} d(Tx_1, Tx_2) &= \max_{|t-t0|\leq\delta} \left| \int_{t0}^{t} (f(x_1(\tau),\tau) - f(x_2(\tau),\tau))d\tau \right| \\ &\leq \max_{|t-t0|\leq\delta} \left| \int_{t0}^{t} K|x_2(\tau) - x_1(\tau)|d\tau \right| \\ &\leq K\delta \max_{|t-t0|\leq\delta} |x_2(\tau) - x_1(\tau)| \\ &= K\delta d(x_1, x_2) \end{aligned}$$

The space $C[t_0 - \delta, t_0 + \delta]$ is complete, and $0 \leq K\delta < 1$. The existence and uniqueness of solution is thus shown by Deflation Principle.

Suppose X and Z are normed spaces, and $S(T)$ is the linear subspace of X. If the mapping T: $S(T) \to Z$, satisfies the following conditions:

$$T(x+y) = Tx + Ty, \quad T(\alpha x) = \alpha Tx$$
$$\forall x, y \in S(T), \quad \alpha \in K$$

then T is said to be linear operator from inside X to inside Z. $S(T)$ is the domain of definition of T, then T is said to be linear operator on X into Z, if $S(T) = X$. f is called linear functional if it is a linear operator on normed space X into number field K. For the linear operator T on normed space X into normed space $Z, \exists M \in K$, such that $||Tx|| \leq M||x||, \forall x \in X, T$ is called bounded linear operator.

T is a linear operator on normed space X into normed space Z, then T is continuous if and only if T is bounded.

Hahn-Banach Theorem on Real Space: suppose M is the linear subspace of linear real space $X, g : X \to R$, and

$$g(x+y) \leq g(x) + g(y), g(\alpha x) = \alpha g(x)$$
$$\forall x, y \in X, \quad \alpha \geq 0$$

Moreover, f is a linear functional on M and

$$f(x) \leq g(x), \quad x \in M$$

then there is a linear functional $p(x)$ on X, such that:

$$p(x) = f(x), \quad x \in M$$
$$-g(-x) \leq p(x) \leq g(x), \quad x \in X$$

Riesz Theorem 1: suppose f is the bounded linear functional on $C[a,b]$, then there is a function of bounded variation, $v(t)$, on $[a,b]$, such that:

$$f(x) = \int_a^b x(t)dv(t), \quad x \in C[a,b]$$
$$||f|| = V(v)$$

where $V(v)$ is the total variation of $v(t)$ on $[a,b]$. Moreover, based on any function of bounded variation, $v(t)$ on $[a,b]$, we may define a bounded linear functional on $C[a,b]$ through above expression.

Riesz Theorem 2: suppose H is a Hilbert space, f is an arbitrary bounded linear functional on H, then there is only one $y_f \in H$, such that:

$$f(x) = (x, y_f), \quad \forall x \in H$$
$$||f|| = ||y_f||$$

Hilbert-Schmidt Theorem: suppose T is a self-conjugate compact operator on Hilbert space H, then there is an orthonormal system $\{e_n\}$, which is composed of eigenvectors corresponding to eigenvalues $\{\lambda_n\}, \lambda_n \neq 0$, such that:

$$x = \sum \alpha_n e_n + x_0, \quad Tx_0 = 0,$$
$$Tx = \sum \lambda_n \alpha_n e_n$$

and if $\{e_n\}$ is infinite, then

$$\lim_{n\to\infty} \lambda_n = 0$$

The decomposition of eigenspace is an important topic in signal processing.

14.10 Algebraic Topology

Homology method (Lin, 1998) has been used to solve local minimum problem.

Using homotopy method to find the zero point of nonlinear function $f(x)$, we need to get a related and simpler function $g(x)$.

First try to obtain zero point of $g(x)$, and gradually transit to get zero point of $f(x)$. In homotopy method, construct a homotopy function:

$$H(t, x) = (1 - t)g(x) + tf(x)$$

where t is the parameter variable. In the training t will gradually change from 0 to 1, and

$$H(0, x) = g(x), \quad \text{if } t = 0; \text{ zero point is easy to obtain}$$

$$H(1, x) = f(x), \quad \text{if } t = 1; \text{ zero point is that we need}$$

Trace the trajectory $x_0(t)$ of zero point of $H(t, x)$ when t changes from 0 to 1, and thus make solution transit from $x_0(0)$ to $x_0(1)$ to obtain zero point.

14.11 Differential Equations and Motion Stability

Given the differential equation of a dynamical (motion) system:

$$dx/dt = f(x, t)$$

where

$$x(t) = (x_1(t), x_2(t), \ldots, x_n(t))^T$$

$$f(x, t) = (f_1(x, t), f_2(x, t), \ldots, f_n(x, t))^T$$

The system can be categorized by the form of $f(x, t)$: $f(x, t) = f(x, t)$, non-linear and non-stationary system; $f(x, t) = f(x)$, non-linear stationary system; $f(x, t) = Ax$, linear stationary system; $f(x, t) = A(t)x$, non-stationary linear system (Zhang, 2010).

We can use Runge-Kutta method to obtain numerical solution of the differential equation. Runge-Kutta method is calculated as follows:

$$
\begin{aligned}
K_1 &= hf(x_k, t_k) \\
K_2 &= hf(x_k + K_1/2, t_k + h/2) \\
K_3 &= hf(x_k + K_2/2, t_k + h/2) \\
K_4 &= hf(x_k + K_3, t_k + h) \\
h &= t_{k+1} - t_k
\end{aligned}
$$

where h is the step length of iteration. The numerical solution is $x_{k+1} = x_k + (K_1 + 2K_2 + 2K_3 + K_4)/6, k = 0, 1, 2, \ldots$

The following is the Java class for numerical solution of the differential equation using Runge-Kutta algorithm.

```
class ordinaryDifferentialEquation {
public static int i,n;
public static double ti,tm,h;
public static double x[];
public static double x1[];
public static double f[];
public static double k1[];
public static double k2[];
public static double k3[];
public static double k4[];
public static void main(String args[]) {
n=6;
ti=1;tm=15;h=1;
x=new double[n+2];
x1=new double[n+2];
f=new double[n+2];
k1=new double[n+2];
k2=new double[n+2];
k3=new double[n+2];
k4=new double[n+2];
while (ti<=tm)
{for(i=1;i<=n;i++) x1[i]=x[i];
```

```
farray();
for(i=1;i<=n;i++) k1[i]=h*f[i];
ti=ti+h/2;
for(i=1;i<=n;i++) x[i]=x1[i]+k1[i]/2;
farray();
for(i=1;i<=n;i++) k2[i]=h*f[i];
for(i=1;i<=n;i++) x[i]=x1[i]+k2[i]/2;
farray();
for(i=1;i<=n;i++) k3[i]=h*f[i];
ti=ti+h/2;
for(i=1;i<=n;i++) x[i]=x1[i]+k3[i];
farray();
for(i=1;i<=n;i++) k4[i]=h*f[i];
for(i=1;i<=n;i++) x[i]=x1[i]+(k1[i]+2*k2[i]+2*k3[i]+k4[i])/6;
System.out.println();
System.out.println("t="+ti);
for(i=1;i<=n;i++) System.out.println("x"+i+"="+x[i]); }
}

public static double[] farray() {
f[1]=-0.1057444*Math.sin(3.1+0.0172142*(ti-1));
f[2]=x[1]+0.04*x[3]-(0.0098+0.00027*Math.exp(0.012*ti)
   +0.01176*Math.sin(2*ti-0.7))*x[2];
f[3]=(0.0084+0.01176*Math.sin(2*ti-0.7))*x[2]
   -(0.0405+0.011*Math.sin(ti+2))*x[3];
f[4]=0.00027*x[2]*Math.exp(0.012*ti)
   -(0.0018+0.0018*Math.sin(2*ti-1.56))*x[4];
f[5]=18-x[5];
f[6]=-18+x[5]+(0.0018+0.0018*Math.sin(2*ti-1.56))*x[4]+0.0014*x[2]+(0.00055
   +0.011*Math.sin(ti+2))*x[3];
return f;
}}
```

14.11.1 *Motion stability and discrimination*

Suppose $x \in \Omega \subset R^n, t \in I = (t_1, t_2)$, the domain of definition of $f(x,t)$ is $\Omega \times I$ and $f(x,t)$ is continuous on $\Omega \times I$. If there is a constant, K, such that

$$|f(x,t) - f(y,t)| \leq K|x-y|, \quad \forall\, x, y \in \Omega, t \in I$$

then $f(x,t)$ is called satisfying with Lipschitz condition on $\Omega \times I$. $f(x,t)$ satisfies with Lipschitz condition, if $\partial f_i(x,t)/\partial x_i, i = 1, 2, \ldots, n$, are finite. Given a system, $\mathrm{d}x/\mathrm{d}t = f(x,t)$, if $f(x,t)$ is continuous and satisfies with Lipschitz condition on $\Omega \times I$, then there is a constant $c > 0$, and an unique solution $x = x(t)$ on $[t_0 - c, t_0 + c]$ for any $(x_0, t_0) \in \Omega \times I$, where $x(t)$ is continuous and $x(t_0) = x_0$.

Suppose $z(t), y(t)$, and $x(t)$ are the given motion, perturbation, and observed motion, respectively, i.e., $z(t) = x(t) - y(t)$. It is obvious that

$$\mathrm{d}y/\mathrm{d}t = f(y(t) + z(t), t) - f(z(t), t) = g(y, t)$$

Given the perturbation equation of a given system, $\mathrm{d}z/\mathrm{d}t = f(z,t)$, is as the following:

$$\mathrm{d}y/\mathrm{d}t = g(y,t), g(0,t) = 0$$

Inside a B-neighborhood of $y(t) = 0$, i.e.,

$$\{(y_1(t), y_2(t), \ldots, y_n(t)) | y_i(t) < B,\ i = 1, 2, \ldots, n\}$$

given a real value $0 < \varepsilon < B$; if there is a real value, $\delta = \delta(\varepsilon, t_0)$, such that the perturbed motion, $y_i(t)$, satisfies with

$$|y_i(t)| < \varepsilon, \quad i = 1, 2, \ldots, n; \quad \forall\, t \geq t_0$$

when $|y_i(0)| \leq \delta, i = 1, 2, \ldots, n$; the given motion, $z(t)$, is called to be stable. If the given motion, $z(t)$, is stable and

$$\lim_{t \to \infty} y_i(t) = 0, \quad i = 1, 2, \ldots, n$$

it is asymptotic stable. If there is a real value $\varepsilon > 0$, such that a value, δ, that satisfies with the stability condition, is not existent, the given motion is unstable.

If $x(t) = x_e = (x_{1e}, x_{2e}, \ldots, x_{ne})^T$, or $f(x,t) = 0$, the system is called to be in a equilibrium state, and x_e is the equilibrium state. For any real value $\varepsilon > 0$, if there is a value, δ, such that

$$|x_i(t) - x_{ie})| < \varepsilon, \quad i = 1, 2, \ldots, n; \ \forall\, t \geq t_0$$

when $|x_i(0) - x_{ie})| \leq \delta, i = 1, 2, \ldots, n; x_e$ is stable. If x_e is stable and

$$\lim_{t \to \infty} x_i(t) = x_{ie}, \quad i = 1, 2, \ldots, n;$$

the state, x_e is asymptotic stable. If there is a real value $\varepsilon > 0$, such that a value, δ, that satisfies with the stability condition, is not existent, the state x_e is unstable (Yan and Zhang, 2000).

If the non-linear function, $f(x,t)$, is considerably smooth, the system can be linearized around the equilibrium state, x_e:

$$d\Delta x/dt = A(t)\Delta x(t)$$

by taking

$$x(t) = x_e + \Delta x(t)$$
$$f(x,t) = x_e + A(t)\Delta x(t)$$

where $A(t) = \partial f(x,t)/\partial x|_{x=xe}$. Suppose $A(t) = A$, and $|A^{-1}| \neq 0$: if the eigenvalues of A are negative real values, x_e is a stable node; if the eigenvalues are conjugate complex numbers with negative real parts, x_e is a stable focal point.

Given a real function $V(x)$ defined in a neighborhood, Ω, of the origin. $V(0) = 0$. V is single valued on Ω, and is continuously derivative in respect to $x_i, i = 1, 2, \ldots, n$. $V(x)$ is positive definite (negative definite), if for any $x \in \Omega$, in exception of $x = 0, V(x) > 0$ $(V(x) < 0); V(x)$ is half positive definite (half negative definite), if for any $x \in \Omega, V(x) \geq 0$ $(V(x) \leq 0); V(x)$ is sign changing, if $V(x)$ will be positive, zero, or negative values for different $x \in \Omega$.

Based on the definition of $V(x)$, the Liapunov theorem, used to discriminate the stability of non-linear systems, is as follows:

(1) the system is stable on the origin, if there is a positive (negative) definite function $V(x)$ on the neighborhood, Ω, of the origin, and the derivative of $V(x)$ is half negative (positive) definite;
(2) the system is asymptotic stable on the origin, if there is a positive (negative) definite function $V(x)$ on the neighborhood, Ω, of the origin, and the derivative of $V(x)$ is negative (positive) definite;
(3) the system is unstable on the origin, if there is a function $V(x)$ on the neighborhood, Ω, of the origin, and the derivative of $V(x)$ is positive (negative) definite, but $V(x)$ itself is not half negative (positive) definite.

For the linear stationary system, the Java method to determine systematic stability, systemStability, is as the following:

```
public void systemStability(int m, double a[][]) {
double bb=0,ep,yy,ee,ff,p1,q1,q,r,ss,vv=0,ww,xx,cc1,ll1;
int n2=m*2,n3=n2+1,n4=m*n2,ll,aa,cc,it,dd,gg;
double s[],p[],w[],t[],v[],f[];
p=new double[2*(int)Math.pow(m,2)+m+2];
w=new double[m+3];
t=new double[m+3];
f=new double[2*m+2];
v=new double[m+2];
s=new double[2*(int)Math.pow(m,2)+m+2];
for(i=1;i<=m;i++)
for(j=1;j<=m;j++)
{s[i+(j-1)*n2]=a[i][j];
p[i+(j-1)*n2]=a[i][j]; }
w[m+1]=1;
ll=1;
while (ll<=m) {
bb=0;
for(i=1;i<=n4;i+=n3) bb-=p[i];
bb=bb/ll;
w[m+1-ll]=bb;
for(i=1;i<=n4;i+=n3) p[i]+=bb;
if (ll==m) break;
for(i=1;i<=m;i++) {
aa=1;
for(j=i;j<=n4;j+=n2) {
bb=0;
cc=j-i;
for(k=i;k<=n4;k+=n2) {
cc+=1;
bb+=p[k]*s[cc]; }
t[aa]=bb; aa+=1; }
aa=1;
for(j=i;j<=n4;j+=n2) {
```

```
p[j]=t[aa];aa+=1; }}
ll+=1; }
ep=1e-05;
aa=50;
p1=0;
q1=0;
for(i=1;i<=m+1;i++) t[i]=w[m-i+2];
for(i=1;i<=m+1;i++) {
w[i]=t[i];
t[i]=0;}
it=0;
yy=w[1];
for(i=1;i<=m;i++) w[i]=w[i+1]/yy;
nm5: while (m>=0) {
if (m<=1) {
it++;
dd=2*it;
f[dd]=0;
f[dd-1]=-w[1];
m=it;
eigen();
break nm5; }
if (m<=2) {
ee=w[1];
ff=w[2];
yy=-0.5*ee;
it+=2;
ww=ff-Math.pow(yy,2);
xx=Math.sqrt(Math.abs(ww));
dd=2*it;
if (ww>0) {
f[dd]=-xx;
f[dd-1]=yy;
f[dd-2]=xx;
f[dd-3]=yy;
m-=2;
```

```
if (m>0) {
for(i=1;i<=m;i++) w[i]=t[i];
continue nm5; }
if (m<=0) m=it; eigen(); break nm5; }}
if (ww<=0) {f[dd]=0; f[dd-1]=yy-xx; f[dd-2]=0; f[dd-3]=yy+xx; m-=2;
if (m>0) {
for(i=1;i<=m;i++) w[i]=t[i];
continue nm5; }
if (m<=0) {m=it;eigen(); break nm5; }}
}
if (m>2) {ee=p1;ff=q1;gg=1;
t[1]=w[1]-ee;
t[2]=w[2]-ee*t[1]-ff;
dd=m-1;
v[1]=t[1]-ee;
v[2]=t[2]-ee*v[1]-ff;
if (dd==2) dd=3;
for(j=3;j<=dd;j++) {
t[j]=w[j]-ee*t[j-1]-ff*t[j-2];
v[j]=t[j]-ee*v[j-1]-ff*v[j-2]; }
dd=m-1;
ll1=v[dd]-t[dd];
cc1=-ll1;
if (m!=3) cc1*=v[m-3];
cc1=cc1+Math.pow(v[m-2],2);
if (cc1<=1e-08) {m=it; break nm5; }
nm4: while (cc1!=0) {
t[m]=w[m]-ee*t[m-1]-ff*t[m-2];
q=-t[m];
if (m!=3) q*=v[m-3];
q=(q+t[m-1]*v[m-2])/cc1;
r=(t[m]*v[m-2]-ll1*t[m-1])/cc1;
ee+=q;
ff+=r;
ss=Math.abs(q)+Math.abs(r);
if (gg==1) vv=ss;
```

```
if ((gg==5) & (ss>vv)) {m=it; break nm5;}
if (ss>ep) {if (gg<aa) {gg++;
t[1]=w[1]-ee;
t[2]=w[2]-ee*t[1]-ff;
dd=m-1;
v[1]=t[1]-ee;
v[2]=t[2]-ee*v[1]-ff;
if (dd==2) dd=3;
for(j=3;j<=dd;j++) {
t[j]=w[j]-ee*t[j-1]-ff*t[j-2];
v[j]=t[j]-ee*v[j-1]-ff*v[j-2]; }
dd=m-1;
||1=v[dd]-t[dd];
cc1=-||1;
if (m!=3) cc1*=v[m-3];
cc1=cc1+Math.pow(v[m-2],2);
if (cc1<=1e-08) {m=it; break nm5;}
continue nm4;}
if (gg>=aa) {m=it;break nm5; }}
yy=-0.5*ee;
it+=2;
ww=ff-Math.pow(yy,2);
xx=Math.sqrt(Math.abs(ww));
dd=2*it;
if (ww>0) {f[dd]=-xx; f[dd-1]=yy; f[dd-2]=xx; f[dd-3]=yy; m-=2;
if (m>0) {
for(i=1;i<=m;i++) w[i]=t[i];
continue nm5; }
if (m<=0) {m=it; eigen(); break nm5;}}
if (ww<=0) {f[dd]=0; f[dd-1]=yy-xx; f[dd-2]=0; f[dd-3]=yy+xx; m-=2;
if (m>0) {
for(i=1;i<=m;i++) w[i]=t[i];
continue nm5; }
if (m<=0) {m=it; eigen(); break nm5;}}
}}}
}
```

```
public void eigen() {
int sd,zw=0;
output.editt1.appendText("Eigenvalues of System Matrix\n\n");
output.editt1.appendText("Real part"+" "+"Image part\n");
i=1;
while (i<=2*m)
{output.editt1.appendText(String.valueOf((long)(f[i]*1000000)/
1000000.00)+" "+String.valueOf((long)(f[i+1]*10000)/10000.00)+"\n");i+=2;}
sd=0;
for(i=1;i<=2*m;i+=2) {
if (f[i]<0) sd+=1;
else if (f[i]>0) {output.editt1.appendText("System is
unstable\n");zw=1;break;} }
if ((sd<m) & (zw==0)) output.editt1.appendText("System is stable\n");
else if ((sd>=m) & (zw==0)) output.editt1.appendText("System is
asymptotic stable\n"); }
```

14.11.2 *Stochastic differential equation*

When a continuous dynamical system is perturbed by an external random variable, a chaotic term, $E(t)$, must be incorporated in its differential equation (Langevin equation):

$$dx/dt = f(x,t) + V(t)$$

The chaotic term, $E(t)$, is a discrete function of time. The differential equation has a solution as the following:

$$x(t) = x(t_0) + \int_{t0}^{t} f(x,\tau)d\tau + v(t) - v(t_0)$$

It is equivalent to:

$$dx(t) = f(x,t)dt + dv(t)$$

The above equation is a stochastic differential equation.

A coefficient, $w(t)$, may be included in chaotic term to generate a new equation:

$$dx(t) = f(x,t)dt + w(t)dv(t)$$

and corresponding solution is

$$x(t) = x(t_0) + \int_{t0}^{t} f(x,\tau)d\tau + \int_{t0}^{t} w(t)dv(t)$$

where the later integral term is defined by Wiener integral. If $w(t)$ is also dependent on x, the following Ito integral and Stratonovich integral should be used:

Ito integral

$$\int_{t0}^{t} w(x(t))dv(t) = \lim_{ti\to 0} \sum_i w(x(t_{i-1}))(v(t_i) - v(t_{i-1}))$$

Stratonovich integral:

$$\int_{t0}^{t} w(x(t))dv(t) = \lim_{\tau i\to 0} \sum_i w(x(\tau_i))(v(t_i) - v(t_{i-1}))$$

where $\tau_i = (t_i - t_{i-1})/2$. The Ito and Stratonovich stochastic differential equation and their solution expressions are represented by

$$dx_i = f_i(x(t),t)dt + \sum_j w_{ij}(x(t))dv_j(t)$$

$$x_i(t) = x_i(s) + \int_s^t f_i(x,\tau)d\tau + \sum_j \int_s^t w_{ij}(x(\tau))dv_j(\tau)$$

$$i = 1, 2, \ldots, n$$

14.11.3 *Stability of feedback networks*

Feedback neural networks are non-linear dynamic systems and therefore the stability of system is an important topic (Yan and Zhang, 2000). The stability of neural networks can be analyzed in two ways, i.e., treating a neural network as a deterministic system and using a group of non-linear differential equations to describe it, or treating it as a stochastic system and using a group of non-linear stochastic differential equations to describe it.

A feedback neural network may be equivalent to a model of continuous time-continuous state, discrete time-discrete state, discrete time-continuous state, and continuous time-discrete state.

Given a completely connected feedback neural network that contains n neurons, in which any neuron is connected to the remaining $n-1$ neurons, the weight between neuron i to neuron j is w_{ij}, The output of i-th neuron at time t is $x_i(t) = 1$, or $-1, i = 1, 2, \ldots, n$. The behavior of feedback neural network can be considered to be a process of state transition, i.e., a dynamic process (Luo, 2004), which is represented by

$$dx/dt = x(t) + f(wx)$$

The stability of the system can be discriminated by using Liapunov theorem.

14.12 Diffusion Equation

Diffusion equation is a partial differential equation that describes density dynamics in a material undergoing diffusion. It is also used to describe processes exhibiting diffusive-like behaviour. Diffusion equation is represented by the following expression (Qi *et al.*, 2004):

$$\partial u/\partial t = a^2 \partial^2 u/\partial x^2 \quad x > 0, t > 0$$
$$u(x, 0) = 0, u(0, t) = f(t)$$

where $u = u(x, t)$ is the concentration of material at position x and time t, a is the diffusion coefficient. The solution of diffusion equation is

$$u(x,t) = x/(2^* a^* (\pi)^{1/2})^* \int_0^t f(\tau)/(t-\tau)^{1.5} \times e^{-x^* x/(4^* a^* a^* (t-\tau))} d\tau$$

using Chebyshev integral algorithm, let

$$s = \int_0^t f(\tau) d\tau$$

then we have

$$s = t/2^* \int_{-1}^{1} f(t^* \tau/2 + t/2) d\tau$$

and the integral term is

$$2/n^* \sum_{k=1}^{n-1} f(t^* T_k/2 + t/2)$$

where, $n = 5, T_0 = -0.8324975, T_1 = -0.3745414, T_2 = 0, T_3 = 0.3745414, T_4 = 0.8324975$.

The Java method, diffusionEquation, is as the following:

```
public void diffusionEquation (int tim, double dif, double xmin, double xstep, double xmaxx, double tt[], double fff[]) {
//dif: diffusion coefficient. //The diffusion scope on direction x is (0, oo).
//xmin: start location for diffusion; xmaxx: the farthest location for diffusion;
//xstep: the interval between two adjacent locations on direction x; tim: time duration.
//tt[]: time t.
//fff[]: the population size on location x=0 at time t, i.e., f(t).
int i,k,kw;
double xxx,uu,sum1,su1,tao,ftao[],q[];
q=new double[10];
ftao=new double[tim+1];
output.editt1.appendText("------------------------------------------\n");
output.editt1.appendText("     "+"x      t                          u(x,t)"+"\n");
output.editt1.appendText("------------------------------------------\n");
q[1]=-0.8324975;
q[2]=-0.3745414;
q[3]=0;
q[4]=0.3745414;
q[5]=0.8324975;
for(k=1;k<=tim;k++) {
xxx=xmin;
for(kw=1;kw<=5;kw++)
{tao=tt[k]/2*q[kw]+tt[k]/2;
ftao[kw]=interp(tim,tao,tt,fff); }
i=1;
do {
sum1=xxx/(2*dif*Math.sqrt(3.1415926));
su1=0;
for(kw=1;kw<=5;kw++)
{tao=tt[k]/2*q[kw]+tt[k]/2;
```

```
su1=su1+ftao[kw]/Math.exp(1.5*Math.log(tt[k]-tao))*Math.exp(-Math.exp
(2*Math.log(xxx))/(4*Math.exp(2*Math.log(dif))*(tt[k]-tao))); }
    su1=su1*tt[k]/5;
    uu=su1*sum1;
    i++;
    output.editt1.appendText("     "+String.valueOf((int)(xxx*10000)
/10000.00)+" "+String.valueOf((int)(k*10000)/10000.00)+"     "+String.valueOf
((int)(uu*10000)/10000.00)+"\n");
    xxx+=xstep; }
    while (xxx<=xmaxx);
    output.editt1.appendText("----------------------------------------\n");
    }}

    public double interp (int tim, double tao, double tt[], double fff[]) {
    double thre=0;
    for(int kk=1;kk<=tim;kk++) {
    if (Math.abs(tao-tt[kk])<=1e-08) thre=fff[kk];
    else if ((tao>=tt[kk-1]) && (tao<tt[kk]))
thre=(fff[kk]-fff[kk-1])/(tt[kk]-tt[kk-1])*(tao-tt[kk-1])+fff[kk-1];
    }
    return thre; }
```

14.13 Entropy of Systems

The probability model of a dynamic system can be developed according to the principle of maximum entropy and the principle of minimum relative information (Zhang, 2010, 2012a).

14.13.1 *Principle of maximum entropy*

If the prior distribution of micro-state is not given, the principle of maximum entropy will be expressed as the following optimum problem:

$$\max H = -\sum_{i=1}^{n} p_i \log p_i$$

$$\sum_{i=1}^{n} p_i f(x_i) = f(x_i)$$
$$\sum_{i=1}^{n} p_i = 1$$

where $x(t) = (x_1(t), x_2(t), \ldots, x_n(t))^T$; $f(x) = (f(x_1), f(x_2), \ldots, f(x_n))^T \geq 0$, is the state function, which is an analogue of energy function; $f(x)$ is the mean of $f(x)$, and p_i is the occurrence probability of micro-state x_i. Using Lagrange multiplier method, the maximum entropy can be achieved as the following:

$$H_{\max} = \mu + \mu f(x)$$

where $\mu \geq 0$, is the Lagrange parameter ($1/\mu$ is the analogue of temperature in thermodynamics), and

$$L(\mu) = \sum_{i=1}^{n} \exp(-\mu f(x_i))$$
$$\mu = \log L(\mu)$$
$$p_i = \exp(-\mu f(x_i))/L(\mu), \quad i = 1, 2, \ldots, n$$

14.13.2 *Principle of minimum relative information*

If the prior distribution of micro-state, $(p_1^0, p_2^0, \ldots, p_n^0)$, is given, the principle of minimum relative information will be represented by

$$\min I = \sum_{i=1}^{n} p_i \log(p_i/p_i^0)$$
$$\sum_{i=1}^{n} p_i f(x_i) = f(x_i)$$
$$\sum_{i=1}^{n} p_i = 1$$

The principle of minimum relative information minimizes the difference between $\{p_i\}$ and $\{p_i^0\}$. This principle is the generalization of principle of maximum entropy. The former is applicable to both continuous and discrete systems. For a continuous system, the principle of minimum relative information is represented by

$$\min I = \int p(x)\log(p(x)/p^0(x))dx$$
$$\int p(x)f(x)dx = f(x)$$
$$\int p(x)dx = 1$$

The solution of this problem is

$$L(\mu) = \int p^0(x)\exp(-\mu f(x))dx$$
$$p(x) = p^0(x)\exp(-\mu f(x))/L(\mu)$$
$$I_{\min} = -\mu f(x) - \log L(\mu)$$

14.13.3 *Principle of minimum mean energy*

The principle of minimum mean energy describes the convergence degree of system to its limit state, given a certain degree of disorder. The principle of minimum mean energy is represented by

$$\min f(x) = \sum_{i=1}^{n} p_i f(x_i)$$
$$-\sum_{i=1}^{n} p_i \log p_i = E$$
$$\sum_{i=1}^{n} p_i = 1$$

The solution is

$$L(\mu) = \sum_{i=1}^{n} \exp(-\mu f(x_i))$$
$$p_i = \exp(-\mu f(x_i))/L(\mu), \quad i = 1, 2, \ldots, n$$

14.13.4 *Probability distribution with maximum entropy*

The entropy of exponential distribution:

$$p(x) = \lambda\exp(-\lambda x), \quad x > 0, \quad \lambda > 0$$
$$p(x) = 0, \quad x \leq 0$$

is $H = \log \lambda - 1$. Among all of the density functions with the same mean, the exponential distribution has the maximum entropy.

The entropy of the normal distribution is

$$H = -\int_{-\infty}^{\infty} p_0(x) \log p(x) dx$$

where $p_0(x)$ is an arbitrary density function having the same mean and variance with the normal distribution, $p(x)$. Among all of the density functions with the same mean and variance, the normal distribution has the maximum entropy.

14.14 Multi-attribute Decision Making

Suppose there are n plans and m indices, and the weight of every indice is $w_1, w_2, \ldots, w_m$, respectively (Zhang, 2007a; Zhang and Gu, 1996). First, given the data matrix $(x_{ij})_{n\times m}$, and thereafter construct the normalized matrix (y_{ij}). If an index is benefit-type (i.e., greater values are preferred), then let

$$y_{ij} = (x_{ij} - x_j^{min})/(x_j^{max} - x_j^{min})$$

or else (i.e., cost-type index, which means that small values are preferred), then let

$$y_{ij} = (x_j^{max} - x_{ij})/(x_j^{max} - x_j^{min})$$

where $x_j^{max} = \max x_{ij}, x_j^{min} = \min x_{ij}, i = 1, 2, \ldots, n; j = 1, 2, \ldots, m$.

By the normalization process, we have $0 \le y_{ij} \le 1, i = 1, 2, \ldots, n$; $j = 1, 2, \ldots, m$.

For Residual Value Weighting Algorithm (REVAWEA), calculate between-plan differences in utility:

$$z_{ij} = \sum_{y_{ik} > x_{ik}} w_i(y_{ik} - y_{jk}) \quad i, j = 1, 2, \ldots, n; \;\; i \neq j$$

and the utility for each plan:

$$d_k = 1/(n-1) \sum_{\substack{j=1 \\ j \neq k}}^{n} z_{kj} \quad k = 1, 2, \ldots, n$$

For Simple Additive Weighting Algorithm (SAWA), calculate between-plan differences in utility:

$$d_k = \sum_{i=1}^{m} w_i \times y_{ki} \quad k = 1, 2, \ldots, n$$

Rank d_k at the decreasing order, then the pros and cons of order of plans are achieved.

The following are some of the Java codes for the methods:

```
public void decisionMaking (int m, int n, int sele, int chh[], double w[],
    double y[][]) {
//sele=1 means SAWA; sele=2 means REVAWEA.
//m plans and n indices.
//chh[]: benefit- or cost-type indices; 1: benefit-type index; 0: cost-type
    index.
//w[]: weights for indices.
//y[][]: dataset.
int i,j,k,choc;
double v[],ww[];
v=new double[m+1];
ww=new double[m+1];
subthree(n,chh,y);
choc=0;
if (sele==1) {
output.editt1.appendText("---------------------------------------------\n");
output.editt1.appendText("SAWA Result\n"); }
for(j=1;j<=m;j++){
v[j]=0;
for(i=1;i<=n;i++) v[j]+=w[i]y[i][j]; }
choc++;
subfour(m,choc,v); }
if (sele==2) {
output.editt1.appendText("---------------------------------------------\n");
output.editt1.appendText("REVAWEA Result\n");
for(i=1;i<=m;i++) {
for(j=1;j<=m;j++) {
ww[j]=0;
```

```
for(k=1;k<=n;k++) if (y[k][i]>=y[k][j]) ww[j]+=w[k]*(y[k][i]-y[k][j]); }
v[i]=0;
for(k=1;k<=m;k++) v[i]+=ww[k]/(m-1); }
choc++;
subfour(m,choc,v); }}

public double[][] subthree(int n, int chh[], double y[][]) {
int i,j,choc;
double l[],kkk[];
l=new double[n+1];
kkk=new double[n+1];
for(i=1;i<=n;i++) {
l[i]=1e+150;kkk[i]=-1e+150; }
for(j=1;j<=n;j++)
for(i=1;i<=m;i++) {
if (y[j][i]>=kkk[j]) kkk[j]=y[j][i];
if (y[j][i]<=l[j]) l[j]=y[j][i]; }
for(i=1;i<=n;i++)
for(j=1;j<=m;j++)
if (chh[i]!=0) y[i][j]=(y[i][j]-l[i])/(kkk[i]-l[i]);
else y[i][j]=(kkk[i]-y[i][j])/(kkk[i]-l[i]);
output.editt1.appendText("Normalized Matrix\n");
for(i=1;i<=m;i++) {
for(j=1;j<=n;j++)
output.editt1.appendText(String.valueOf((int)(y[j][i]*10000)/10000.00)
   +"  ");
output.editt1.appendText("\n"); }
return y; }

public void subfour(int m, int choc, double v[]) {
int i,j,kk,ll;
double u;
int pp[][];
pp=new int[m+1][m+1];
for(i=1;i<=m;i++) pp[choc][i]=i;
for(i=1;i<=m-1;i++) {
kk=i;
```

```
for(j=i;j<=m-1;j++) if (v[j+1]>v[kk]) kk=j+1;
ll=pp[choc][i];
pp[choc][i]=pp[choc][kk];
pp[choc][kk]=ll;
u=v[i];
v[i]=v[kk];
v[kk]=u; }
output.editt1.appendText("Rank Plan Utility\n");
for(i=1;i<=m;i++)
output.editt1.appendText(String.valueOf(i)+"  "+String.valueOf(Math.round
(pp[choc][i]))+" "+String.valueOf((int)(v[i]*10000)/10000.00)+"\n");
}
```

References

Albert R, Barabási AL. 2002. Statistical mechanics of complex networks. *Reviews of Modern Physics*, 74: 47–97

Aleinikov AG, Smarsh DA. 2010. Law of conservation of intensivity. *Proceedings of the Academy of Strategic Management*, 9(1): 3

Aleinikov AG, Gera R. Mathematical predictions of organizology, the new science of organization. *Proceedings of the Fall 2006 International Conference*, Reno, Nevada, USA, 2006, 11–12

Aleinikov AG. 2004. Organizology, new science of organization: A breakthrough in scientific explanation of the world. *Proceedings of the Allied Academies Spring International Conference*, 1–2

Aleinikov AG. 2005. Organizology: The main unit of measurement for the new science. *Proceedings of the Academy of Creativity and Innovation*, 2(1): 15

Alexandridis A, Sarimveis H. 2005. Nonlinear adaptive model predictive control based on self-correcting neural network models. *AIChE Journal*, 51(9): 2495–2506

Alizadeh R. 2011. A dynamic cellular automaton model for evacuation process with obstacles. *Safety Science*, 49: 315–323

Amari SI. 1983. Field theory of self-organizing neural nets. *IEEE Transactions on Systems, Man and Cybernetics*, 13: 741–748

Amari SI. 1990. Mathematical foundation of Neurocomputing. *Proceedings of the IEEE on Neural Networks*, 78: 1443–1463

Amari S, Murata N, Muller K, Finke M, Yang HH. 1997. Asymptotic statistical theory of overtraining and cross-validation. *IEEE Transactions on Neural Networks*, 85: 985–996

Amburgey TL, Rao H. 1996. Organizational ecology: past, present, and future directions. *Academy of Management Journal*, 39(5): 1265–1286

Amoroso S, Patt YN. 1972. Decision procedures for surjectivity and injectivity of parallel maps for tessellation structures. *Journal of Computer and System Sciences*, 6: 448–464

Andriani P, McKelvey B. 2009. Perspective — from gaussian to paretian thinking: causes and implications of power laws in organizations, *Organization Science*, 20(6): 1053–1071

Angeline PJ. 1998. Using selection to improve particle swarm optimization. *Proceedings of IEEE International Conference on Neural Networks*, 84–89

Anthony M, Biggs N. *Computational Learning Theory.* Cambridge University Press, Boston, USA, 1992

Arnold DV, Beyer HG. 2006. Optimum tracking with evolution strategies. *Evolutionary Computation*, 14(3): 291–308

Ashby WR. 1947. Principles of the self-organizing dynamic system. *Journal of General Psychology*, 37(2): 125–128

Audia P, Bielefeld W, Dowell M. 2006. Organizational foundings in community context: instruments manufacturers and their interrelationship with other organization. *Administrative Science Quarterly*, 51(3): 381–419

Bailey R, Bras B, Allen JK. 2004. Applying ecological input-output flow analysis to material flows in industrial system: Part 1: tracing flows. *Journal of Industrial Ecology*, 8(1/2): 45–68

Bakke WE. Concept of the social organization. In: *Modern Organization Theory*(Haire M, ed). Chapman and Hall, USA, 1959

Bakshy E, Wilensky U. NetLogo Team Assembly model. Center for Connected Learning and Computer-Based Modeling, Northwestern University, Evanston, IL, USA, 2007

Ball P. 2014. One rule of life: Are we poised on the border of order? http://www.newscientist.com/article/mg22229660.700-one-rule-of-life-are-we-poised-on-the-border-of-order.html?cmpid=RSS|NSNS|2012-GLOBAL|online-news. Accessed 30 April 2014

Ballestores F Jr, Qiu ZY. An integrated parcel-based land use change model using cellular automata and decision tree. *Proceedings of the International Academy of Ecology and Environmental Sciences*, 2(2): 53–69

Barabasi AL. 2002. *The New Science of Networks.* Perseus Publishing, Cambridge, Massachusetts, USA, 2012, 79–92

Barabasi AL. 2009. Scale-free networks: a decade and beyond. *Science*, 325: 412–413

Barabasi AL, Albert R. 1999. Emergence of scaling in random networks. *Science*, 286(5439): 509–512

Barzilai K. 2010. Organizational theory. http://www.cwru.edu/med/epidbio/mphp439/Organizational_Theory.htm. Retrieved August 20, 2010

Bastolla U, Lassig M, Manrubia SC, Valleriani A. 2001. Diversity patterns from ecological models at dynamical equilibrium. *Journal of Theoretical Biology*, 212: 11–34

Baum JAC, Amburgey TL. Organizational ecology. In: *The Blackwell Companion to Organizations* (Baum JAC, ed). Wiley, USA, 2005

Beattie PD, Bishop JM. 1998. Self-Localisation in the 'Senario' autonomous wheelchair. *Journal of Intelligent and Robotic Systems*, 22: 255–267

Bednar JA, Choe Y, Paula JD, *et al.* 2004. Modeling cortical maps with topographica. *Neurocomputing*, 58: 1129–1135

Begon M, Harper JL, Townsend CR. *Ecology: Individuals, Populations, Communities.* Blackwell Science Press, Boston, USA, 1994

Bell SS, McCoy ED, Mushinsky HR. *Habitat Structure: The Physical Arrangements of Objects in Space.* Chapman and Hall, New York, USA, 1991

Bellman RE. 1957. *Dynamic Programming.* Princeston University Press, Princeston, USA

Berry T, Linoff J. *Data Mining Techniques.* JohnWiley and Sons, New York, USA, 1997

Beyer HG. 2001. On the performance of (1, λ)-evolution strategies for the ridge function class. *IEEE Transactions on Evolutionary Computation*, 5(3): 218–235

Bian ZQ, Zhang XG. *Pattern Recognition (2nd edition).* Tsinghua University Press, Beijing, China, 2000

Bishop JM. *Anarchic Techniques for Pattem Classification.* UK University of Reading, UK, 1989

Bissette A. 2013. Life on Earth Was Not a Fluke. *The Conversation*, 19452

Bjumenefld LA. *Problems of Biological Physics.* Springer-Verlag, Berlin, Germany, 1985

Bonabeau E. 2002. Agent-based modeling: methods and techniques for simulating human systems. *Proceedings of the National Academy of Sciences*, 99(Suppl.3): 7280–7287

Bonabeau E, Dorigo M, Theraulaz G. *Swarm Intelligence: From Natural to Artificial Systems.* Oxford University Press, UK, 1999

Boone C, van Witteloostuijn A, Carroll GR. 2002. Resource distributions and market partitioning: Dutch daily newspapers, 1968 to 1994. *American Sociological Review*, 67(3): 408–431

Bormann FH, Likens GE. *Pattern and Process in a Forested Ecosystem.* Springer Verlag, New York, USA, 1979

Bossuyt B, Honnay O, Hermy M. 2005. Evidence for community assembly constraints during succession in dune slack plant communities. *Plant Ecology*, 178: 201–209

Bradon RN. *Adaptation and Environment.* Princeton University Press, Chicago, USA, 1990

Bradshaw J. *An Introduction to Software Agents.* AAAI Press, CA, USA, 1997

Breiman L. 2001. Statistical modeling: the two cultures (with discussion). *Statistical Science*, 16: 199–231

Brogliato B, Lozano R, Maschke B, *et al. Dissipative Systems Analysis and Control: Theory and Applications (2nd Ed).* Springer Verlag, London, UK, 2007

Bullnheimer B, Hartl RF, Strauss C. 1999. A new rank based version of the ant system: A computational study. *Central European Journal for Operations Research and Economics*, 7(1): 25–28

Bullnheimer B, Hartl RF, Strauss C. 1999. An improved ant system algorithm for the vehicle routing problem. *Annals of Operations Research*, 89: 319–328

Burden R, Faires JD. *Numerical Analysis (Seventh Edition).* Thomson Learning, Inc., New York, USA, 2001

Burrell G, Morgan G. *Sociological Paradigms and Organizational Analysis.* Ashgate Publishing Limited, USA, 1979

Camazine S. *Self-Organization in Biological Systems.* Princeton University Press, USA, 2003

Caplat P, Anand M, Bauch C. 2008. Symmetric competition causes population oscillations in an individual-based model of forest dynamics. *Ecological Modeling*, 211: 491–500

Carpenter GA, Grossberg S. 1987. A massive parallel architecture for a self-organing neural pattern recognition machine. *Computer Vision, Graphics, and Image Processing*, 37: 54–115

Carpenter GA, Grossberg S. 1987. ART2: self-organization of stable category recognition codes for analog input patterns. *Applied Optics*, 26(23): 4019–4930

Carpenter GA, Grossberg S. 1990. ART3: hierachical search using chemical transmitters in self-organizing pattern recognition architectures. *Neural Networks*, 3(23): 129–152

Carpenter GA, Grossberg S, Reynolds J. 1991. ARTMAP: supervised real-time learning and classification of nonstationary data by a self-organizing neural network. *Neural Networks*, 4(5): 169–181

Carroll GR. 1984. Organizational ecology. *Annual Reviews of Sociology*, 10: 71–98

Case TJ. 1990. Invasion resistance arises in strongly interacting species-rich model competition communities. *Proceedings of the National Academy of Sciences of USA*, 87: 9610–9614

Castellano G, Fanelli AM, Pelillo M. 1997. An iterative pruning algorithm for feedforward neural networks. *IEEE Transactions on Neural Networks*, 8(3): 519–531

Casti J. *Would-Be Worlds: How Simulation Is Changing the World of Science.* Wiley, New York, USA, 1997

Cavender-Bares J, Kozak KH, Fine PVA, Kembel SW. 2009. The merging of community ecology and phylogenetic biology. *Ecology Letters*, 12: 693–715

Chaitin G. *Meta Math! The Quest for Omega.* Vintage Press, New York, USA, 2006

Charnell MA. 2008. An individual-based model of a tritrophic ecology. *Ecological Modeling*, 218: 95–206

Chan JX. *Lectures on Foundations of Algebraic Topology. Higher Education* Press, Beijing, China, 1987

Chang HS. 1991. Indirect gradient analysis, quantitative classification and environmental interpretation of plant communities in Ngari, Xizang (Tibet). *Acta Phytoecologica et Geobotanica Sinica*, 15(2): 101–113

Chase JM. 2003. Community assembly: when should history matter? *Oecologia*, 136: 489–498

Chase JM. 2005. Towards a really unified theory for meta communities. *Functional Ecology*, 19: 182–186

Chase JM, Leibold MA. *Ecological Niches: Linking Classical and Contemporary Approaches.* University of Chicago Press, Chicago, Illinois, USA, 2003

Chen J, Chen Q. 2006. Ant colony algorithm for multiple sequence alignment. *Journal of Computer Applications*, 26(S1): 124–128

Chen J, Ma XH. 2013. Research of ant colony algorithm based on Travelling Salesman Problem. *Modern Computer (Professional Edition)*, 3: 8–11

Chen JX. *Course Notes on Algebraic Topology.* Higher Education Press, Beijing, China, 1987

Chen WQ, Deng BW. 2001. On relationship and difference between the phenotype and molecular evolution of living things. *Journal of Hanzhong Teachers College (Natural Science)*, 19(1): 82–85

Chen Y, Hu XC, Liu BB, *et al.* 2008. Modeling and simulation of economic system based on UML and Swarm. *Computer Engineering and Design*, 29(15): 4040–4042, 4060

Chen GZ, Wang JQ, Li CJ, *et al.* 2009. An improved artificial fish swarm algorithm and its applications. *Systems Engineering*, 27(12): 105–110

Chen YH, Lu XK, Chen YF. 2014. Temporal mortality-colonization dynamic can influence the coexistence and persistence patterns of cooperators and defectors in an evolutionary game model. *Computational Ecology and Software*, 4(1): 12–21

Cherkassky V, Lari-Najafi H. 1991. Constrained topological mapping for nonparametric regression analysis. *Neural Networks*, 4: 27–40

Chia R. Organizational theory as a postmodern science. In: *The Oxford Handbook of Organizational Theory* (Tsoukas H, Knudsen C, eds). Oxford University Press, UK, 2003

Chipperfield A, Fleming P, Pohlheim H, *et al. Genetic Algorithm Toolbox User's Guide.* MathWorks, USA, 2014

Chisholm RA, Burgman MA. 2004. The unified neutral theory of biodiversity and biogeography: commment. *Ecology*, 85(11): 3172–3174

Clegg SR, Hardy C. 1999. *Studying Organization.* SAGE Publication Ltd, UK

Clements FE. 1916. *Plant Succession: An Analysis of the Development of Vegetation.* Carnegie Institute of Washington Publication, 242: 1–512

Clerc M. Discrete particle swarm optimization, illustrated by the traveling salesman problem. In: *New Optimization Techniques in Engineering.* Springer, Netherlands, 2004, 219–239

Clerc M. *Particle Swarm Optimization.* ISTE (International Scientific and Technical Encyclopedia), 2006

Clyde DE, Corado MS, Wu X, *et al.* 2003. A self-organizing system of repressor gradients establishes segmental complexity in Drosophila. *Nature*, 426(6968): 849–853

Cohen JE, Beaver RA, Cousins SH, *et al.* 1993. Improving food webs. Ecology, 74: 252–258

Cohen JE, Newman CM. 1985. A stochastic theory of community food webs I. models and aggregated data. *Proceeding of the Royal Society of London Series B (Biological Sciences)*, 224(1237): 421–448

Coleman AC, Beierle JM, Stuart MC, *et al.* 2011. Light-induced disassembly of self-assembled vesicle-capped nanotubes observed in real time. *Nature Nanotechnology*, 6(9): 547–552

Colorni A, Dorigo M, Maniezzo V. Distributed optimization by ant colonies. In: *Proceedings of the First European Conference on Artificial Life.* Paris, France, 1991, 134–142

Colorni A, Dorigo M, Maniezzo V. 1992. An investigation of some properties of an ant algorithm. In: *Proceedings of the Parallel Problem Solving from Nature*. Brussels, Belgium, 509–520

Colorni A, Dorigo M, Maniezzo V. 1994. Ant system for job-shop scheduling. *Statistics and Computer Science*, 34: 39–53

Colomi A, Maniezzo V. Distributed optimization by ant colonies. In: *The First European Conference on Artificial Life*. Elsevier, France, 1991, 134–142

Corning PA. 2014. Systems theory and the role of synergy in the evolution of living systems. *Systems Research and Behavioral Science*, 31(2): 181–196

Connell JH, Slatyer RO. 1977. Mechanism of succession in natural communities and their role in community stability and organization. *American Naturalist*, 111: 1119–1144

Coombes S. *The Geometry and Pigmentation of Seashells*. Department of Mathematical Sciences, University of Nottingham, Nottingham, UK, 2009

Cormen T, Leiserson C, Rivest Z. Greedy algorithms. In: *Introduction to Algorithms*. 329, MIT Press, USA, 1990

Cornwell WK, Ackerly DD. 2009. Community assembly and shifts in plant trait distributions across an environmental gradient in coastal California. *Ecological Monographs*, 79(1): 109–126

Couzin LD. 2006. Behavioral ecology: social organization in fission-fusion societies. *Current Biology*, 16(5): 170

Cunliffe AL. *Organizational Theory*. SAGE Publications Inc., USA, 2008

Daniel S, Spenko M, Parness A, *et al.* 2007. Directional adhesion for climbing: theoretical and practical considerations. *Journal of Adhesion Science and Technology*, 21(12–13): 1317–1341

Davies J. *Life Unfolding: How the human body creates itself*. Oxford University Press, UK, 2014

Daft RL, Armstrong A. 2007. *Organizational Theory and Design*. Graphic World Inc., USA

Darwin C. *The Origin of Species (6th edition)*. John Murry, London, UK, 1911

Davis GF, Marquis C. 2005. Prospects for organizational theory in the early twenty-first century: institutional fields and mechanisms. *Organization Science*, 16(4): 332–343

den Besten M, Stutzle T, Dorigo M. 2000. Ant colony optimization for the total weighted tardiness problem. *Lecture Notes in Computer Science*, 1917: 611–620

Denning PJ. 2003. Great principles of computing. *Communications of the ACM* 46: 15–20

Denning PJ. 2007. Computing is a natural science. *Communications of the ACM* 50: 15–18

Denning PJ. 2010. The great principle of computing. *American Scientist*, 98(5): 369–372

Denning PJ, Freeman P. 2009. Computing's paradigm. *Communications of the ACM*, 52: 28–30

Diamond JM. Assembly of species communities. In: *Ecology and Evolution of Communities* (Cody ML, Diamond JM, eds). Belknap Press, Harvard University Press, Cambridge, USA, 1975, 342-444

Dijkstra EW. 1959. A note on two problems in connection with graphs. *Numerche Math*, 1: 269–271

Dimopoulos I, Chronopoulos J, Chronopoulou-Sereli A, Lek S. 1999. Neural network models to study relationships between lead concentration in grasses and permanent urban descriptors in Athens city (Greece). *Ecological Modelling*, 120: 157–165

Dobrev SD, van Witteloostuijn A, Baum JAC. 2006. Introduction: ecology versus strategy or strategy and ecology? *Advances in Strategic Management*, 23: 1–26

Docherty JP. Organizational theory. In: *Textbook of Administrative Psychiatry: New Concepts for a Changing Behavioral Health System* (Talbott JA, Hales RE, eds). American Psychiatric Publishing Inc., USA, 2001

Doctor S, Venayagamoorthy GK, Gudise VG. 2004. Optimal PSO for collective robotic search applications. *Proceedings of the 2004 Congress on Evolutionary Computation*, 1 and 2: 1390–1395

Donaldson L. Organizational theory as a positive science. In: *The Oxford Handbook of Organizational Theory* (Tsoukas H, Knudsen C, eds). Oxford University Press, UK, 2003

Dong GA. 1999. On interpretion principles of adaptationism. *Journal of Harbin University*, 2: 16–21

Dorigo M. Optimization, learning and natural algorithms. PhD Thesis, Department of Electronics, Politecnico di Milano, Italy, 1992

Dorigo M, Gambardella LM. 1997. Ant colony system: a cooperative learning approach to the traveling salesman problem. *IEEE Transactions on Evolutionary Computation*, 1(1): 53–66

Dorigo M, Maniezzo M, Colorni A. 1996. The ant system: optimization by a colony of cooperating agents. *IEEE Transactions on Systems, Man, and Cybernetics-Part B*, 26(1): 29–41

Dorigo M, Caro GD, Gambardella LM. 1999. Ant algorithms for discrete optimization. *Artificial Life*, 5: 137–172

Dreo J, Siarry P. A new ant colony algorithm using the heterarchical concept aimed at optimization of multminima continuous functions. *The Third Intemational Workshop on Ant Algorithms.* Brussels, Belgium, 2002

Duan HB. *Principles and Applications of Ant Colony Algorithms.* Science Press, Beijing, China, 2005

Duan HB, Wang DB, Zhu JQ, *et al.* 2004. Development on ant colony algorithm theory and its application. *Control and Decision*, 19(12): 1321–1326, 1340

Duan HB, Wang DB, Yu XF. 2007. Ant colony algorithm: Survey and prospect. *Engineering Science*, 2: 98–102

Duncan AB, *et al.* 2011. Adaptation of *Paramecium caudatum* to variable conditions of temperature stress. *Research in Microbiology*, 162(9): 939–944

Durand R. 2008. Logics of organizational theory: audiences, codes, and ecologies. *Contemporary Sociology* , 37(6): 605–606

EACH Unit. 1997. *Evolution of Altruistic and Cooperative Habits: Learning About Complexity in Evolution.* http://ccl.northwestern.edu/cm/EACH/ for more information.

Eberhart RC, Shi YH. *Computational Intelligence: Concepts to Implementations.* Morgan Kaufmann, MA, USA, 2007

Eberhart RC, Shi Y. Comparing inertia weights and constriction factors in particle swarm optimization. In: *Proceedings of the Congress on Evolutionary Computation. IEEE*, USA, 2000, 84–88

Eberhart RC, Shi YH. 2001. Particle swarm optimization: Developments, applications and resources. *Proceedings of the 2001 Congress on Evolutionary Computation*, 1 and 2: 81–86

Egler FE. 1954. Vegetation science concepts: Initial floristic composition, a factor in old-field vegetation development. *Vegetation*, 4: 412–417

Ehrlieh PR, Raven PH. 1964. Butterflies and Plants: a study in coevolution. *Evolution*, 18: 586–608

Fang KT. 1979. Several cluster methods for ordered samples. *Journal of Applied Mathematics*, 5: 94–101

Fath BD. 2004. Network analysis applied to large scale cyber-ecosystems. *Ecological Modelling*, 171(4): 329–337

Fath BD, Scharler UM, Ulanowicz RE, *et al.* 2007. Ecological network analysis: network construction. *Ecological Modelling*, 208(1): 49–55

Fecit. *Analysis and Design of Neural Networks in MATLAB 6.5.* Electronics Industry Press, Beijing, China, 2003

Feng YJ, Zhang WJ. 2004. Network software of Genetic BP algorithm and its application in biodiversity research. *Application Research of Computers*, Suppl: 387–389

Fenichela EP, Castillo-Chavezb C, Ceddiac MG. 2011. Adaptive human behavior in epidemiological models. *Proceedings of the National Academy of Sciences of USA*, 108(15): 6306–6311

Ferrarini A. 2011. A fitter use of Monte Carlo simulations in regression models. *Computational Ecology and Software*, 1(4): 240–243

FIPA (Foundation for Intelligent Physical Agents). 2005.

FIPA Home Page. http://www.fipa.org/

Foerster HV. 1960. On self-organising systems and their environments. *Self Organising Systems*, 31–50

Foerster HV. 1996. Self-organization, the progenitor of conversation and interaction theories. *Systems Research and Behavioral Science*, 13(3): 349–362

Fogel LJ, Owens AJ. Walsh M.J. *Artificial Intelligence through Simulated Evolution.* John Wiley & Sons, New York, USA, 1966

Forman RTT. *Land Mosaics: The Ecology of Landscapes and Regions.* Cambridge University Press, UK, 1995

Frean M. 1990. The upstart algorithm: a method for constructing and training feed forward neural networks. *Neural Computation*, 2(2): 198–209

Freeman JH, Audia PG. 2006. Community ecology and the sociology of organizations. *Annual Review of Sociology*, 32: 145–169

Fukami T. 2004. Assembly history interacts with ecosystem size to influence species diversity. *Ecology*, 85(12): 3234–3242

Fukami T. 2010. Community assembly dynamics in space. In: *Community Ecology: Processes, Models, and Applications*, 45–54

Futuyma DJ, Slatkin M. *Coevolution.* Sinauer Associates Inc, Massachusetts, USA, 1983, 1–13

Garbarino S, Caviglia G, Brignone M, Massollo M, *et al.* 2013. Estimate of FDG excretion by means of compartmental analysis and ant colony optimization of nuclear medicine data. *Computational and Mathematical Methods in Medicine*, 2013: 793142

Gardner M. 1970. The fantastic combinations of John Conway's new solitaire game "Life". *Scientific American*, 223: 120–123

Ge XJ, Hsu TW, Hung KH, *et al.* 2012. Inferring multiple refugia and phylogeographical patterns in *Pinus massoniana* based on nucleotide sequence variation and DNA fingerprinting. *PLoS ONE*, 7(8): e43717

Gentle JE. *Elements of Computational Statistics.* Springer Science+Business Media, Inc., Netherlands, 2002

Gevrey I, Dimopoulos I, Lek S. 2003. Review and comparison of methods to study the contribution of variables in artificial neural network models. *Ecological Modelling*, 160: 249–264

Gevrey M, Dimopoulos I, Lek S. 2006. Two-way interaction of input variables in the sensitivity analysis of neural network models. *Ecological Modelling*, 195: 43–50

Giacomini HC, De Marco P Jr, Petrere M Jr. 2009. Exploring community assembly through an individual-based model for trophic interactions. *Ecological Modeling*, 220: 23–29

Gibney E. 2014. Termite-inspired robots build castles. *Nature*, doi:10.1038/nature.2014.14713. Available at http://www.nature.com/news/termite-inspired-robots-build-castles-1.14713

Gillison AN, Brewer KRW. 1985. The use of gradient directed transects or gradsects in natural resource surveys. *Journal of Environmental Management*, 20: 103–127

Glansdorff P, Prigogine I. *Thermodynamic Theory of Structure, Stability and Fluctuations.* Wiley-Interscience, New York, USA, 1971

Goh CS, Bogan AA, Joachimiak M, *et al.* 2000. Co-evolution of proteins with their interactions partners. *Journal of Molecular Biology*, 299: 283–293

Goldberg DE. 1989a. Genetic algorithms and Walsh functions: Part I A gentle introduction. *Complex Systems*, 3: 129–152

Goldberg DE. *Genetic Algorithms in Search, Optimization, and Machine Learning.* Addison-Wisely, Reading, MA, USA, 1989b

Gong GQ, Huang ZL, Huang JX, *et al.* 2011. How individual species structure the community in Dinghushan 20 ha forest plot? *Ecology and Environmental Sciences*, 20(6-7): 991–995

González RER, Coutinho S, dos Santos RMZ, de Figueirêdo PH. 2013. Dynamics of the HIV infection under antiretroviral therapy: A cellular automata approach. *Physica A*, 392(19): 4701–4716

Goslee SC, Peters DPC, Beck KB. 2006. Spatial prediction of invasion success across heterogeneous landscapes using an individual-based model. *Biological Invasions*, 8:193–200

Graham R, Wunderlin A. 1987. *Lasers and Synergetics.* Springer-Verlag, New York, USA

Gravel D, Canham CD, Beaudet M, *et al.* 2006. Reconciling niche and neutrality: the continuum hypothesis. *Ecology Letters*, 9: 399–409

Griebeler EM. 2011. Are individual based models a suitable approach to estimate population vulnerability? — a case study. *Computational Ecology and Software*, 1(1): 14–24

Grime JP. 1977. Evidence for the existence of three primary strategies in plants and its relevance to ecological and evolutionary theory. *American Naturalist*, 111: 1169–1174

Grime JP. *The C-S-R Model of Primary Plant Strategies-Origins, Implications and Tests.* ChanPman & Hall, London, UK, 1988

Grimma V, Berger U, Bastiansen F. 2006. A standard protocol for describing individual-based and agent-based models. *Ecological Modelling*, 198: 115–126

Grosenick L, Clement TS, Fernald RD. 2007. Fish can infer social rank by observation alone. *Nature*, 445: 429–432

Gu HZ, Takahashi H. 1996. Towards more practical average bounds on supervised learning. *IEEE Transactions on Neural Networks*, 7(44): 953–968

Guimera R, Uzzi B, Spiro J, Amaral L. 2005. Team assembly mechanisms determine collaboration network structure and team performance. *Science*, 308(5722): 697–702

Guzy MR, Smith CL, Bolte JP, Hulse DW, Gregory SV. 2008. Policy research using agent-based modeling to assess future impacts of urban expansion into farmlands and forests. *Ecology and Society*, 13(1): 37

Hackwoods S, Beni G. Self-organization of sensors for swarm intelligence. In: *Proceedings of the 1992 IEEE International Conference on Robotics and Automation.* IEEE, Nice, France, 1992, 819–829

Hagan MT, Demuth HB, Beale MH. *Neural Network Design.* PWS Publishing Company, Boston, USA, 1996

Haken H. *Synergetics: An Introduction.* Springer, Berlin, Germany, 1978

Haken H. *Synergetics, an Introduction: Nonequilibrium Phase Transitions and Self-Organization in Physics, Chemistry, and Biology (3rd rev. enl. ed).* Springer-Verlag, New York, USA, 1983

Haken H. 2004. Future trends in synergetics. *Solid State Phenomena,* 97–98: 3–10

Han KH, Kim JH. Genetic quantum algorithm and its application to combinatorial optimization problem. *Proceedings of the Congress on Evolutionary Computation.* LaJolla, USA, 2000, 112–117

Han KH, Park KH, Lee CH. 2001. Parallel quantum inspired genetic algorithm for combinatorial optimization problem. *IEEE Trans on Evolutionary Computation,* 5(1): 1422–1429

Hannan MT. 2005. Ecologies of organizations: diversity and identity. *Journal of Economic Perspectives*, 19(1): 51–70

Hannan MT. *Logics of Organizational Theory: Audiences, Codes, and Ecologies.* Princeton University Press, USA, 2007

Hannan MT, Freeman J. 1977. The population ecology of organizations. *American Journal of Sociology*, 82(5): 929–964

Hannan MT, Freeman J. *Organizational Ecology.* Harvard University Press, USA, 1993

Hannah ST, Lester PB. A multilevel approach to building and leading learning organizations. *The Leadership Quarterly*, 20: 34–48

Hansen AJ, de Castro F, Naiman RJ. 1988. Ecotones: when and why? *Biology International* (Special Issue), 17: 9–46

Harmer GE, Abbott D. 1999a. Parrondo's paradox. *Statistical Science*, 14(2): 206–213

Harmer GP, Abbott D. 1999b. Losing strategies can win by Parrondo's paradox. *Nature*, 402(6764): 864–870

Harada A, Kobayashi R, Takashima Y, *et al.* 2011. Macroscopic self-assembly through molecular recognition. *Nature Chemistry,* 3(1): 34–37

Harte J. 2004. The value of null theories in ecology. *Ecology,* 85: 1792–1794

Hatch MJ, Yanow D. Organizational theory as an interpretive science. In: *The Oxford Handbook of Organizational Theory* (Tsoukas H, Knudsen C, eds). Oxford University Press, UK, 2003

Haykin S. *Neural Nwtworks: A Comprehensive Foundation.* Macmillan College Publishing Company, New Jersey, USA, 1994

Hazen R. *Genesis: The Scientific Quest for Life's Origins.* Joseph Henry Press, Washington DC, USA, 2007

Hemmatian H, Fereidoon A, Sadollah A, Bahreininejad A. 2013. Multi-objective optimization of laminate stacking sequence for minimizing weight and cost using ant colony optimization. *Advances in Engineering Software,* 57: 8–11

Hess B, Mikhailov A. 1994. Self-organization in living cells. *Science,* 264(5156): 223–224

Heyligen F. The science of self-organisation and adaptivity. In: *The Encyclopedia of Life Support Systems.* UNESCO Publishing-Eolss Publisher, Oxford, UK, 2002

Hinton GE, Sejnowski TJ, Ackley DH. *Boltzmann Machines: Constraint Satisfaction Networks that Learn. Carnegie-Mellon University Computer Science Technical Report: CMU-CS-84-119,* Carnegie-Mellon University, Pittsburgh, USA, 1984

Hochberg MM *et al.* HMM/NN training techniques for connected alphadigit speech recognition. *Proc. IEEE ICASSP'91,* Toronto, Canada, 1991, 109–112

Holland JH. *Adaptation in Natural and Artificial Systems. An Introductory Analysis with Applications to Biology, Control and Artificial Intelligence.* University of Michigan Press, Ann Arbor, MI, USA, 1975

Holland MM, Risser PG, Naiman RJ. *Ecotones: the Role of Landscape Boundaries In The Management and Restoration of Changing Environments.* Chapman and Hall, New York, USA, 1991

Hopfield JJ. 1982. Neural networks and physical systems with emergent collective computational properties. *PNAS,* 79: 2554–2558

Hopfield JJ. 1984. Neurons with graded response have collective computational properties like those of two-state neurons. *PNAS,* 81: 3088–3092

Hopfield JJ, Tank DW. 1985. Neural computation of decisions in optimization problems. *Biological Cybernetics,* 52: 141–154

Hou GK, Deng WA. *An Introduction to Artificial Intelligence.* Sun Yat-sen University Press, Guangzhou, China, 1993

Hraber PT, Milne BT. 1997. Community assembly in a model ecosystem. *Ecological Modelling,* 103: 267–285

Hu XP, Din QL, Li YX. 2008. A review on ant colony system. *Journal of Industrial Engineering/Engineering Management,* 22(2): 74–79

Hu ZG, Li J. 2008. The progress of swarm intelligence algorithms. *Control Theroy and Applications,* 27(2): 13–15

Hubbell SP. *The Unified Neutral Theory of Biodiversity and Biogeography.* Princeton University Press, Princeton, NJ, USA, 2001

Huang GQ, Liu JF, Yao YX. 2012. Global convergence proof of artificial fish swarm algorithm. *Computer Engineering,* 38(2): 204–206

Ilachinski A. *Cellular Automata: A Discrete Universe.* World Scientific, Singapore, 2001

Imran M, Hashim R, Abd Khalid NE. 2013. An overview of particle swarm optimization variants. *Procedia Engineer,* 53: 491–496

Ingber L. 1993. Simulated annealing: practice versus theory. *Mathematical Computer and Modelling,* 18: 29–57

Ingber L, Rosen B. 1992. Generic algorithms and very fast simulated reannealing: a comparison. *Ibid,* 16: 87–100

Iourinski D, Starks SA, Kreinovich V. Swarm intelligence: Theoretical proof that empirical techniques are optimal. In: *Proceedings of the 2002 World Automation Congress WAC 2002.* Orlando, FL, USA, 2002, 107–112

Iwamatsu M. 2002. Generalized evolutionary programming with Levy-type mutation. *Computer Physics Communications,* 147(1–2): 729–732

Izaguirre JA, Chaturvedi R, Huang C, *et al.* 2004. CompuCell, a multimodel framework for simulation of morphogenesis. *Bioinformatics,* 20(7):1129–1137

Janson S, Knuth DE, Luczak T, Pittel B. 1993. The birth of the giant component. *Random Structures and Algorithms,* 4(3): 233–358

Jaynes ET. 1957. Information theory and statistical mechanics. *Physics Review,* 106: 620–630

Jazon DH. 1980. When is it coevolution. *Evolution,* 34: 611–612

Jennings NR. 2000. On agent-based software engineering. *Artificial Intelligence,* 117: 277–296

Ji MJ, Tang HW, Guo J. 2004. A single-point mutation evolutionary programming. *In-formation Processing Letters,* 90(6): 293–299

Jia ZJ. Polymorphism of HLA-DRB1 gene in southern Chinese populations. PhD Thesis. 46-47, Sun Yat-sen University, Guangzhou, China, 2001

Jiang H. 2011. Hyper-heuristic algorithms: A trans-discipliniary problem-solving model. *Letters of Chinese Society of Computers,* 7(3): 63–70

Jiao Y, Torquato S. 2011. Emergent behaviors from a cellular automaton model for invasive tumor growth in heterogeneous microenvironments. *PLOS Computational Biology,* 7(12): e1002314

Jin BH, Zhang L, He ZY. 2011. A method of contingency logistics path planning based on max-min ant system. *Journal of Zhongyuan University of Technology,* 22(2): 14–17

Jones C, Munro R. 2005. Organizational theory, 1985–2005. *Sociological Review,* 53(1): 1–15

Jong WR, Lai PJ. 2013. Applying ant colony system algorithm in the navigation process for plastic injection mould manufacturing scheduling optimisation. *International Journal of Production Research,* 52(9): 2530–2549

Jukes TH. 2000. The neutral theory of molecular evolution. *Genetics,* 154: 955–958

Kamps J. 2009. Logics of organizational theory: Audiences, codes, and ecologies. *Administrative Science Quarterly,* 54(2): 350–353

Karsenti E. 2008. Self-organization in cell biology: A brief history. *Natural Reviews Molecular Cell Biology,* 9(3): 255–262

Katsuhiko A, Hill, JP, Lee, MV, *et al.* 2008. Challenges and breakthroughs in recent research on self-assembly. *Science and Technology of Advanced Materials,* 9: 014109

Kemp SJ, Zaradic P, Hansen F. 2007. An approach for determining relative input parameter importance and significance in artificial neural networks. *Ecological Modelling,* 204: 326–334

Kennedy J, Eberhart, R. Particle Swarm Optimization. In: *Proceedings of IEEE International Conference on Neural Networks IV.* Perth, Australia, 1995, 1942–1948

Kerner BS, Klenov SL, Schreckenberg M. 2011. Simple cellular automaton model for traffic breakdown, highway capacity, and synchronized flow. *Physical Review E,* 84(4 pt 2): 046110

Kennedy J. 2000. Stereotyping: Improving particle swarm performance with cluster analysis. *Proceedings of the 2000 Congress on Evolutionary Computation, 1 and 2:* 1507–1512

Kerr JT, Kharouba HM, Currie DJ. 2007. The macroecological contribution to global change solutions. *Science,* 316: 1581–1584

Khalili A, Tinati MA, Rastegarnia A. 2010. Analysis of incremental RLS adaptative networks with noisy links. *Electronics Express,* 8(9): 623–628

Kim YK, Cook S, Tuladhar, SM, *et al.* 2006. A strong regioregularity effect in self-organizing conjugated polymer films and high-efficiency polythiophene: fullerene solar cells. *Nature Materials,* 5(3): 197–203

Kimura M. 1968. Evolutionary rate at the molecular level. *Nature,* 217: 624–626

Kimura M. *The Neutral Theory of Molecular Evolution.* Cambridge, UK, 1983

King BG, Felin T, Whetten DA. 2010. Finding the organization in organizational theory: A meta-theory of the organization as a social actor. *Organization Science,* 21(1): 290–305

Kohonen T. 1982. Self-organizing formation of topologically correct feature maps. *Biological Cybernatics,* 43: 59–69

Kohonen T. *Self-organization and Associative Memory (3rd edition).* Springer-Verlag, New York, USA, 1988

Kohonen T. 1990. The self-organizing map. *Proceedings of the IEEE,* 78(9): 1464–1480

Kohonen T. *Self-Organizing Maps.* Springer-Verlag, Heidelberg, Germany, 1995

Kondrashov FA. 2012. Gene duplication as a mechanism of genomic adaptation to a changing environment. *Proceedings of the Royal Society B,* 279: 1749, 5048–5057

Koza JR. *Genetic Programming: On the Programming of Computers by Means of Natural Selection.* MIT Press, Cambridge, USA, 1992

Krebs CJ. *Ecological Methodology.* Happer Collins Publishers Inc, New York, USA, 1989

Krink T, Løvbjerg M. The Life Cycle Model: Combining particle swarm optimisation, genetic algorithms and hill climbers. In: *Proceedings of Parallel Problem Solving from Nature VII (PPSN).* IEEE, USA, 2002, 621–630

Kronfeldner M. 2007. Is cultural evolution Lamarckian? *Biology and Philosophy,* 22: 493–512

Ku XTX. 2011. *Graph Search Algorithm.* http://mindlee.com/2011/10/28/graph-search/

Kuhn T, Jackson MH. 2008. Accomplishing knowledge: A framework for investigating knowing in organizations. *Communication Quarterly,* 21(4): 454–485

Kulik BW, Baker T. 2008. Putting the organization back into computational organizational theory: A complex Perrowian model of organizational action. *Computational and Mathematical Organizational Theory,* 14(2): 84–119

Kutschera U. 2009. *Charles Darwin's Origin of Species.* Naturwissenschaften, 96: 1247–1263

Ladislav H. 2010. An outline of soil nematode succession on abandoned fields in South Bohemia. *Applied Soil Ecology,* 46(3): 355–371

Lando DY, Teif VB. 2000. Long-range interactions between ligands bound to a DNA molecule give rise to adsorption with the character of phase transition of the first kind. *Journal of Biomolecular Structure and Dynamics,* 17(5): 903–911

Lebrija-Trejos E, Meave JA, Poorter L, *et al.* 2010. Pathways, mechanisms and predictability of vegetation change during tropical dry forest succession. *Perspectives in Plant Ecology, Evolution and Systematics,* 12(4): 267–275

Lee WT, Tenorio MF. 1991. On an asymptotically optimal adaptive classifier design criterion. *IEEE Transactions on Pattern Analysis and Machine Intelligence,* 15(3): 312–318

Lehn JM. 1988. Perspectives in supramolecular chemistry-from molecular recognition towards molecular information processing and self-organization. *Angewandte Chemie International Edition in English,* 27(11): 89–121

Lehn JM. 1990. Supramolecular chemistry-scope and perspectives: Molecules, supermolecules, and molecular devices (Nobel Lecture). *Angewandte Chemie International Edition in English,* 29(11): 1304–1319

Leibold MA and McPeek MA. 2006. Coexistence of the niche and neutral perspectives in community ecology. *Ecology,* 87(6): 1399–1410

Lek S, Belaud A, Baran P, Dimopoulos I, Delacoste M. 1996. Role of some environmental variables in trout abundance models using neural networks. *Aquatic Living Resources,* 9: 23–29

Lewin A, Volberda H. The future of organization studies: Beyond the selection — adaptation debate. In: The *Oxford Handbook of Organizational Theory* (Tsoukas H, Knudsen C, eds). Oxford University Press, UK, 2003, 568–595

Lewens T. 2010. Natural selection then and now. *Biological Reviews,* 85(4): 829–835

Li D. 2009. A study on Lamarckian theory of evolution. *Journal of Chongqing University of Science and Technology* (*Social Sciences Edition*), 12: 36–37

Li D, *et al. Operational Research.* Tsinghua University Press, Beijing, China, 1982

Li HL, Chen H, Jin SY. 2007. Resear ch of modeling method for Agent-based complex systems distributed simulation. *Computer Engineering and Applications,* 43(8): 209–213

Li QY, Wang NC, Yi DY. *Numerical Analysis* (*Fourth Edition*). Tsinghua University Press, Springer, Beijing, China, 2001

Li R, Chen DT, Liang GW, *et al.* 2014. Species diversity and aggregation patterns of plant species in a grass community. *Selforganizology,* 1(2): 78–88

Li XL. A New Intelligent Optimization Method-Artificial Fish School Algorithm. PhD Thesis. Zhejiang University, Hangzhou, China, 2003

Li XL, Lu F, Tian GH, Qian JX. 2004. Applications of artificial fish school algorithm in combinatorial optimization problems. *Chinese Journal of Shandong University* (*Engineering Science*), 34(5): 65–67

Li Y, Ma SF. 2006. Modeling of agent-based simulation system. *Journal of Systems Engineering,* 21(3): 225–231

Li Y, Wang XY, Jin W, *et al.* 2010. Adaptive evolution of digestive RNASE1 genes in leaf-eating monkeys revisited: New insights from ten additional colobines. *Molecular Biology and Evolution,* 27(1): 121–131

Li XR. *A New Intelligence Optimization Algorithm-Artificial Fish Swarm Algorithm.* Zhejiang University Press, China, 2003

Lidicker WZ Jr. 2007. Levels of organizations in biology: on the nature and nomenclature of ecology's fourth level. *Biological Review,* 83: 71–78

Lin JK. *Foundations of Topology.* Science Press, Beijing, China, 1998

Lin JZ, Ye DY. 2013. Optimization and improvement based on ant colony clustering algorithm. *Computer Systems and Applications,* 22(12): 93–99

Ling SH, Iu HHC, Leung FHF, *et al.* 2008. Improved hybrid particle swarm optimized wavelet neural network for modeling the development of fluid dispensing for electronic packaging. *IEEE Transactions on Industrial Electronics,* 5: 3447–3460

Lippuner D, Moschytz GS. 2004. The Kalman filter in the context of adaptive filter theory. *International Journal of Circuit Theory and Applications,* 32(4): 223–235

Liu BC. *Functional Analysis.* Science Press, Beijing, China, 2000

Liu CH. 2007. Summary of research on applications of adaptive control. *Modular Machine Tool and Automatic Manufacturing Technique,* 1: 1–4

Liu LZ, Zhou YQ. 2009. Hybrid global optimization algorithm based on artificial fish swarm and cultural algorithm. *Application Research of Computers,* 12: 4446–4448

Liu RJ, Wang LJ, Shi Y. 2013. Multi-colony ant algorithm applied to the rectangular pieces layout optimization. *Journal of Jiangnan University (Natural Science Edition),* 12(3): 288–291

Liu X, Wang DM, Wang XH. 2004. An overview on the development of adaptive control. *Oil-Gas field Surface Engineering,* 23(1): 31

Liu YD, Gu FH. 2013. Research advances of differential evolution algorithm. *Ke Ji Guang Chang,* 3: 20–23

Liu YZ, Xuan HY. 2004. Application of the ant algorithm to vehicle routing problem. *Information and Control,* 33(2): 249–252

Lockwood JL, Powell RD, Nott MP, Pimmental SL. 1997. Assembling ecological communities in space and time. *Oikos,* 80: 549–553

Luhmann N. *Soziale Systeme-Grundriß einer allgemeinen Theorie. suhrkamp taschenbuch Wissenschaft,* Germany, 1991

Luo LM, Zhang WJ. 2014. A review on biological adaptation: with applications in engineering science. *Selforganizology,* 1(1): 23–30

Luo SW. *Theoretical Principles of Large-scale Artificial Neural Networks.* Tsinghua University Press, Northern Jiaotong University Press, China, 2004

Ma G. 1995. Identification of some concepts in animal evolution. *Bulletin of Biology,* 30(5): 18–19

Ma L, Jiang F. 1999. Ant algorithm for the degree-constrained minimum spanning tree. *Journal of Systems Engineering,* 14(3): 211–214

Ma M, Li B, Chen WK. 2006. Convergent adaptation of desert plants to their arid habitats. *ACTA ECOLOGICA SINICA,* 26(11): 3861–3869

Ma XM, Liu N. 2013. An improved artificial fish-swarm algorithm based optimization method for city taxi scheduling system. *Journal of Computational Information Systems,* 9(16): 6351–6359

Macal CM, North MJ. Tutorial on agent-based modeling and simulation. In: *Proceedings of the 37th Winter Simulation Conference* (Kuhl ME, Steiger NM, Armstrong FB, Joines JA, eds). Orlando, FL, USA, 2005, 2–15

Machol RE, Gray P. 1964. Recent developments in information and decision processes. *Econometrica,* 32(4): 710

Mackey DJC. 1992. Bayesian interpolation. *Neural Computation,* 4: 415–447

Maherali H, Klironomos JN. 2007. Influence of phylogeny on fungal community assembly and ecosystem functioning. *Science,* 316(5832): 1746–1748

Mahony S, Benosa PV, Smith TJ, *et al.* 2006. Self-organizing neural networks to support the discovery of DNA-binding motifs. *Neural Networks,* 19: 950–962

Malerba F. 2006. Innovation and the evolution of industries. *Journal of Evolutionary Economics,* 16: 3–23

Manly BFJ. *Randomization, Bootstrap and Monte Carlo Methods in Biology (2nd Edition).* Chapman & Hall, London, UK, 1997

Marchand M, Golea M, Rujoriginalán P. 1990. A convergence theorem for sequential learning in two-layer perceptrons. *Europhysics Letters,* 11(65): 487–492

Margalef R. *Perspectives in Ecological Theory.* University of Chicago Press, Chicago, USA, 1968

Marshak RJ. 2004. Morphing: the leading edge of organizational change in the twenty-first century. *Organization Development Journal,* 22(3): 8

Masatoshi N, Yoshiyuki S, Masafumi N. 2010. The neutral theory of molecular evolution in the genomic era. *The Annual Review of Genomics and Human Genetics,* 11: 265–289

Mathworks. *MATLAB 6.5: Neural Network Toolbox.* Mathworks, Natick, USA, 2002

Mathworks. *MATLAB 2007: Neural Network Toolbox.* Mathworks, Natick, USA, 2007

May RM. *Stability and Complexity in Model Ecosystems.* Princeton University Press, USA, 1973

McKay MD, Conover WJ, Beckman RJ. 1979. A comparison of three methods for selecting values of input variables in the analysis of output from a computer code. *Technometrics,* 21: 239–245

McKelvey B, Lichtenstein BB. Leadership in Four Stages of Emergence. In: *Complex Systems Leadership Theory: New Perspectives from Complexity Science on Social and Organizational Effectiveness* (Hazy JK, Goldstein J, Lichtenstein BB, eds). ISCE Publishing Company, Boston, MA, USA, 2007, 93–107

Mellouli S, Mineau G, *et al.* Laying the foundations for an agent modelling methodology for fault tolerant multi-agent systems. In: *Fourth International Workshop Engineering Societies in the Agents World.* Imperial College London, UK, 2003

Men SP, Feng JH. *Applied Functional Analysis.* Science Press, Beijing, China, 2005

Meyer KD. *Foundation of Stochastic Diffusion Search.* University of Reading, UK, 2004

Meyer KD, Nasuto SJ, Bishop JM. *Stochastic Diffusion Search: Partial Function Evaluation in Swarm Intelligence Dynamic Optimisation. Studies in Computational Intelligence Series.* Springer-Verlag, Cambridge, USA, 2006, 185–207

Mezard M, Nadal JP. Learning in feedforward layered network: the tiling algorithm. *Journal of Physics A,* 22: 2191–2204

Miled AB. 2010. Individual Based model for grouper populations. *Acta Biotheor.*, 58: 247–264

Min YJ, Akbulut M, Kristiansen K, *et al.* 2008. The role of interparticle and external forces in nanoparticle assembly. *Nature Materials,* 7(7): 527–538

Mirjalili S, Mirjalili SM, Lewis A. 2014. Grey Wolf Optimizer. *Advances in Engineering Software,* 69: 46–61

Misteli T. 2001. The concept of self-organization in cellular architecture. *Journal of Cell Biology,* 155(2): 181–185

Morgan G. *Images of Organization.* Sage Publications Inc., UK, 2006

Morton D, Law R, Pimmental SL, Drake JA. 1996. On models for assembling ecological communities. *Oikos,* 75: 493–499

Motegi F, Zonies S, Hao Y, *et al.* 2011. Microtubules induce self-organization of polarized PAR domains in *Caenorhabditis elegans* zygotes. *Nature Cell Biology,* 13(11): 1361–1367

Myatt DR, Bishop JM, Nasuto SJ. 2004. Minimum stable convergence criteria for stochastic diffusion search. *Electronics Letters,* 40(2): 112–113

Nagendra SMS, Khare M. 2006. Artificial neural network approach for modelling nitrogen dioxide dispersion from vehicular exhaust emissions. *Ecological Modelling,* 190: 99–115

Narayanan A, Moore M. Quantum inspired genetic algorithms. Proceedings of *IEEE International Conference on Evolutionary Computation.* Nagoya, Japan, 1996, 61–66

Nasuto SJ, Bishop JM, Lauria S. Time complexity analysis of the stochastic diffusion search. *IFAC Symposium on Neural Computation.* Vienna, Austria, 1998, 260–266

Nedorezov LV. 2012. Continuous-discrete model of population dynamics with time lag in a reaction of intra-population self-regulative mechanisms. *Network Biology,* 2(4): 139–147

NetLogo. 2004-2005. http://ccl.northwestern.edu/netlogo

Newman MEJ. 2006. *Power laws, Pareto distributions and Zipf's law.* http://arxiv.org/pdf/cond-mat/0412004.pdf

Nicolis G, Prigogine I. *Self-Organization in Non-Equilibrium Systems: From Dissipative Structures to Order Through Fluctuations.* Wiley Interscience, New York, USA, 1977

Niknam T, Amiri B. 2010. An efficient hybrid approach based on PSO, ACO and k-means for cluster analysis. *Applied Soft Computing,* 10(1): 183–197

Norton M. *Modern Control Engineering.* Pergamon Press, UK, 1972

Oakley MT, Richardson EG, Carr H, Johnston RL. 2013. Protein structure optimization with a "lamarckian" ant colony algorithm. IEEE/ACM *Transactions on Computational Biolology and Bioinformatics,* 10(6): 1548–1552

Olden JD. 2000. An artificial neural network approach for studying phytoplankton succession. *Hydrobiology.* 436: 131–143

Olden JD, Jackson DA. 2002. Illuminating the "black box": A randomization approach for understanding variable contributions in artificial neural networks. *Ecological Modelling,* 154: 135–150

Olden JD, Joy MK, Death RG. 2004. An accurate comparison of methods for quantifying variable importance in artificial neural networks using simulated data. *Ecological Modelling,* 178: 389–397

Özesmi SL, Özesmi U. 1999. An artificial neural network approach to spatial habitat modelling with interspecific interaction. *Ecological Modelling,* 116: 15–31

Özesmi SL, Tan CO, Özesmi U. 2006. Methodological issues in building, training, and testing artificial neural networks in ecological applications. *Ecological Modelling,* 195: 83–93

Pal NR, Bezdek JC, Tsao ECK. 1993. Generalized clustering networks and Kohonen's self-organizing scheme. *IEEE Transcations on Neural Networks,* 4: 549–557

Pascual A, BaHrcena M, Merelo JJ, *et al.* 2000. Mapping and fuzzy classification of macromolecular images using self-organizing neural networks. *Ultramicroscopy,* 84: 85–99

Pazos F, Valencia A. 2001. Similarity of phylogenetic trees as indicators of protein-protein interaction. *Protein Engineering,* 14: 609–614

Peak D. 2004. Evidence for complex, collective dynamics and emergent, distributed computation in plants. *Proceedings of the National Academy of Sciences of USA,* 101(4): 918–922

Penczak T, Glowacki L, Kruk A, Galicka W. 2012. Implementation of a self-organizing map for investigation of impoundment impact on fish assemblages in a large, lowland river: Long-term study. *Ecological Modelling,* 227: 64–71

Perterson C, Anderson JR. 1987. A mean field theory learning algorithm for NN. *Complex Systems,* 1(5): 995–1019

Pfeffer J, Salancik GR. *The External Control of Organizations: A Resource Dependence Perspective.* Harper and Row, USA, 2003

Phillips DC. *The Social Scientist's Bestiary: A Guide to Fabled Threats to, and Defenses of, Naturalistic Social Science.* Pergammon Press, Oxford, UK, 1992

Pimm SL, Lawton JH. 1978. On feeding on more than one trophic level. *Nature,* 275: 542–544

Pimmental SL. 1979. Complexity and stability: another look at MacArthur's original hypothesis. *Oikos,* 35: 139–149

Pimmental SL. 1980. Food web design and the effect of species deletion. *Oikos,* 35: 139–149

Piyaratne Mkdk, Zhao HY, Hu ZQ, *et al.* 2014. Catastrophic behavior of aphid population dynamics: An analysis of swallowtail model. *Computational Ecology and Software,* 4(3): 135–146

Pokroy B, Kang SH, Mahadevan L, *et al.* 2009. Self-organization of mesoscale bristle into ordered, hierarchical helical assemilies. *Science,* 323: 237–240

Porter TB. 2006. Coevolution as a research framework for organizations and the natural environment. *Organization and Environment,* 19(4): 1–26

Pourtakdoust SH, Nobahari H. An extension of ant colony system to continuous optimization problems. *The* 4th *International Workshop Ant Colony Optimization and Swarm Intelligence,* Brussels, Belgium, 2004

Priscila VS, Capriles Z, Fonseca LG, Barbosa HJC, *et al.* 2007. Rank-based ant colony algorithms for truss weight minimization with discrete variables. *Communications in Numerical Methods in Engineering,* 23(6): 553–575

Puranam P. 2012. A future for the science of organization design. *Journal of Organization Design,* 1(1): 18–19

Purves DW, Pacala SW. 2005. Ecological drift in niche-structured communities: neutral pattern does not imply neutral process. *Biotic Interactions in the Tropics,* 107–138

Qi YH. Habitat Heterogeneity and Pest Diffusion in Different Habitats: Models and Algorithms. MSc Thesis. Sun Yat-sen University, China, 2003

Qi YH. 2004. A network software of multivariable regression model used in statistical analysis of information. *Information Science,* 22(1): 104–106, 114

Qi YH, Zhang WJ. 2002. A percolation model for pest perturbation in diverse habitat and net work computing software. *Modern Computer,* 133: 16–19

Qi YH, Zhang ZG, Zhang WJ. 2004. Simulation models and network software for describing pest diffusion in homogeneous habitat. *Computer Applications Research,* 24(4): 131–134

Qian XS, Song J. *Engineering Cybernetics.* Science Press, Beijing, China, 1983

Qu LD, He DX, Huang Y. 2011. Novel heuristic artificial fish swarm algorithm. *Computer Engineering,* 17: 140–142

Rabiner LR. A tutorial on HMM and selected applications in speech recognition. *Proc. IEEE,* 1989, 77: 257–286

Radwan J, Babik W. 2012. The genomics of adaptation. *Proceedings of the Royal Society B,* 279: 5024–5028

Railsback SF. 2001. Concepts from complex adaptive systems as a framework for individual-based modelling. *Ecological Modelling,* 139: 47–62

Rank ON. 2008. Formal structures and informal networks: structural analysis in organizations. *Scandinavian Journal of Management,* 24: 145–161

Rashman L, Withers E, Hartley J. 2009. Organizational learning and knowledge in public service organizations: A systematic review of the literature. *International Journal of Management Reviews,* 11(4): 463–494

Rechenberg I. Cybernetic solution path of an experimental problem. *Royal Aircraft Establishment,* Farnborough p. Library Translation 1122, UK, 1965

Rechenberg I. Evolutionsstrategie: Optimierung technischer Systeme nach der biologischen Evolution. Dr-Ing Thesis, Technical University of Berlin, Germany, of Berlin, Germany, 1971

Rechenberg I. *Evolutionsstrategie: Optimierung technischer Systeme nach Prinzipien der biologischen Evolution.* Fommann-Holzboog, Stuttgart, Germany, 1973

Recknagel F, French M, Harkonen P, Yabunaka K. 1997. Artificial neural network approach for modelling and prediction of algal blooms. *Ecological Modelling,* 96: 11–28

Reed R. 1993. Prunning algorithm-a survey. *IEEE Transactions on Neural Networks,* 4(5): 740–747

Reed R. 1993. Prunning algorithm-a survey. *IEEE Transactions on Neural Networks,* 4(5): 740–747

Ren WJ, Chen JL, Han D, *et al.* A review on ant colony algorithms. In: *Proceedings of the 2007 Annual Conference of Control and Decision of China.* Beijing, China, 2007

Repast. 2004. http://repast.sourceforge.net/

Resnik D. 1997. Adaptationism: Hypothesis or heuristic. *Biology and Philosophy,* 12: 39–50

Resnick M. *Turtles, Termites and Traffic Jams: Explorations in Massively Parallel Microworlds.* MIT Press, MA, Cambridge, USA, 1994

Reuter H, Jopp F, Munkemuller, T, *et al.* 2010. Ecological hierarchies and self-organization — pattern analysis, modeling and process integration across scales. *Basic and Applied Ecology,* 11: 572–581

Reyjol Y, Lim P, Belaud A, Lek S. 2001. Modelling of microhabitat used by fish in natural and regulated flows in the river Garonne (France). *Ecological Modelling,* 146: 131–142

Reynolds CW. 1987. Flocks, herds and schools: A distributed behavioral model. *Computer Graphics,* 21(4): 25–34

Rifkin J. *The Third Industrial Revolution: How Lateral Power Is Transforming Energy, The Economy, and The World.* Palgrave Macmillan, USA, 2013

Roberts J. 2007. Knowledge in the organization of contemporary business and economy. *Journal of Management Studies,* 44(4): 656–668

Rosvall M, Bergstrom CT. 2011. Multilevel compression of random walks on networks reveals hierarchical organization in large integrated systems. *Plos One,* 6(4): e18209

Roughgarden J. *Theory of Population Genetics and Evolutionary Ecology: An Introduction.* Macmllna, New York, USA, 1979

Rudin W. *Functional Analysis (Second Edition).* McGraw-Hill, Columbus, USA, 1991

Sanei A, 2004. French M. Towards a performance theory of robust adaptive control. *International Journal of Adaptive Control and Signal Processing,* 18: 403–421

Sante I, Gacia AM, Miranda D, *et al.* 2010. Cellular automata models for the simulation of real-world urban processes: A review and analysis. *Landscape and Urban Planning,* 96: 108–122

Santos FM, Eisenhardt KM. 2005. Organizational boundaries and theories of organization. *Organization Science,* 16(5): 491–508

Saunders PT. *An Introduction to Catastrophe Theory.* Cambridge University Press, UK, 1980

Savage M, Askenazi M. 1999. *Arborscapes: a Swarm-based Multi-agent Ecological Disturbance Model.* http://citeseerx.ist.psu.edu/viewdoc/summary?doi=10.1.1.50.1188

Scardi M. 1996. Artificial neural networks as empirical models for estimating phytoplankton production. *Marine Ecology Progress Series,* 139: 289–299

Scardi M, Harding Jr LW. 1999. Developing an empirical model of phytoplankton primary production: A neural network case study. *Ecological. Modelling,* 120: 213–223

Schoenharl TW. An Agent Based Modeling Approach for the Exploration of Self-organizing Neural Networks. MS Thesis, University of Notre Dame, USA, 2005

Schoenly KG, Zhang WJ. IRRI Biodiversity Software Series. V. RARE, SPPDISS, and SPPANK: Programs for detecting between-sample difference in community structure. *IRRI Technical Bulletin No. 5.* International Rice Research Institute, Manila, Philippines, 1999

Schwefel HP. Kybernetische Evolution als Strategie der exprimentellen Forschung in der Strömungstechnik. Master's thesis, Technical University of Berlin, Germany, 1965

Schwefel HP. Evolutionsstrategie und numerische Optimierung. Dissertation, TU Berlin, Germany, 1975

Schwefel HP. *Numerical Optimization of Computer Models.* John Wiley & Sons, New York, USA, 1981

Sebelius F, Eriksson L, Holmberg H, *et al.* 2005. Classification of motor commands using a modified self-organising feature map. *Medical Engineering and Physics,* 27: 403–413

Setiono R. 1997. A penalty-function approach for pruning feedforward NN. *Neural Computation,* 9(1): 185: 204

Sheu JS, Woo TK, Wen JH. 2012. A novel convergence accelerator for the LMS adaptive filter. *Circuits, Systems and Signal Processing,* 31: 283–300

Shi YH, Eberhart RC. 1998. A modified particle swarm optimizer. In: *Proceedings of IEEE International Conference on Evolutionary Computation.* IEEE, Piscataway, NJ, USA, 1998, 69–73

Shi YH, Eberhart RC. 2001. Fuzzy adaptive particle swarm optimization. *Proceedings of the 2001 Congress on Evolutionary Computation, 1 and 2:* 101–106

Shields LM. 1951. The evolution mechanism in leaves of certain xeric grasses. *Phytomorphology,* 1(3): 225–241

Simon HA. 1962. The architecture of complexity. *Proceedings of the American Philosophical Society,* 106: 467–482

Simonson TS, McClain DA, Jorde LB, Prchal JT. 2012. Genetic determinants of Tibetan high-altitude adaptation. *Human Genetics,* 131: 527–533

Singh JV, Lumsden CJ. 1990. Theory and research in organizational ecology. *Annual Review of Sociology,* 16: 161–195

Smith JM. 1979. Optimization in evolution. In: *Conceptual Issues in Evolutionary Biology* (Sober E, ed) (2nd edition; 1994). MIT Press, MA, USA

Solnon C. 2002. Ants can solve constraint satisfaction problems. IEEE *Transactions on Evolutionary Computation,* 6(4): 347–357

Solow AR. 1993. A simple test for change in community structure. *Journal of Animal Ecology,* 62: 191–193

Song MY, Hwang HJ, Kwark IS *et al.* 2007. Self-organizing mapping of benthic macroinvertebrate communities implemented to community assessment and water quality evaluation. *Ecological Modelling,* 203: 18–25

Sontag ED. VC dimension of neural networks. In: *Neural Networks and Machine Learning* (Eds Bishop CM). 69-95, NATO ASI Series, Springer, Netherlands, 1998

Spanier EH. *Algebraic Topology.* Springer-Verlag, New York, USA, 1966

Spearman C. 1904. The proof and measurement of association between two things. *American Journal of Psychology,* 15: 72–101

Sreeja NK, Sankar A. An ant colony optimization based approach of feature selection for efficient classification of very small datasets by mining patterns. In: *Self-organization: Theories and Methods* (WenJun Zhang, ed). Nova Science Publishers, New York, USA, 2013, 13–30

StartLogo. 2004. http://www.media.mit.edu/starlogo

Stefanovic I, Prokić S, Vukosavljević D. 2011. The response to the changing landscape of tomorrow: Reconfigurable organizations. *African Journal of Business Management,* 5(35): 13344–13351

Stinchcombe JR. 2002. Can tolerance traits impose selection on herbivores. *Evolutionay Ecology,* 16: 595–602

Stonedahl F, Wilensky U. *NetLogo Diffusion on a Directed Network model.* Center for Connected Learning and Computer-Based Modeling, Northwestern University, Evanston, IL, USA, 2008a

Stonedahl F, Wilensky U. *NetLogo Particle Swarm Optimization Model.* Center for Connected Learning and Computer-Based Modeling, Northwestern University, Evanston, IL, USA, 2008b

Stonedahl F, Wilensky U. *NetLogo Simple Genetic Algorithm Model.* Center for Connected Learning and Computer-Based Modeling, Northwestern University, Evanston, IL, USA, 2008c

Storn R, Price K. 1997. Differential evolution — A simple and efficient heuristic for global optimization over continuous spaces. *Journal of Global Optimization,* 11(4): 341–359

Strand E, Huse G, Giske J. 2002. Artificial evolution of life history and behavior. *The American Naturalist,* 159: 624–644

Strauss SY, Rudgers JA, Lau JA, *et al.* 2002. Direct and ecological costs of resistance to herbivory. *Tree,* 17: 278–285

Stützle T, Hoos HH. 2000. MAX-MIN ant system. *Future Generation Computer Systems,* 16(8): 889–914

Sun DC, Lin FY. *An Introduction to Systems Engineering.* Tsinghua University Press, Beijing, China, 2004

SWARM. 2004. http://wiki.swarm.org/

Swingler K. *Applying Neural Networks: A Practical Guide.* Academic Press, London, UK, 1996

Tang JF, Ma YY, Guan J, Yan CJ. 2013. A Max–Min ant system for the split delivery weighted vehicle routing problem. *Expert Systems with Applications,* 40(16): 6661–6673

Thom R. 1969. Topological models in biology. *Topology,* 8: 313–335

Thom R. *Structural Stability and Morphogenesis.* Benjamin, New York, USA, 1975

Thom R. 1977. Structural stability, catastrophe theory and applied mathematics. *SIAM Review,* 19(2): 189–201

Tilman D. 1985. The resource-ratio hypothesis of plant secession. *American Naturalist,* 125: 827–852

Tomassini M, Sipper M, Perrenoud M. 2000. On the generation of high-quality random numbers by two-dimensional cellular automata. *IEEE Transactions on Computers,* 40(10): 1146–1151

Thompson J, Michael T. *Instabilities and Catastrophes in Science and Engineering.* Wiley, New York, USA, 1982

Tompkins JR. *Organizational Theory and Public Management.* Clark Baxter Press, USA, 2005

Tomoyoshi K, Tsutomu M. 2002. High speed computation of three dimensional cellular automaton with FPGA. *FPL,* 2438: 1126–1130

Topping CJ, Hansen TS, Jensen TS, *et al.* 2003. ALMaSS, an agent-based model for animals in temperate European landscapes. *Ecological Modeling,* 167: 65–82

Toral R. 2001. Cooperative Parrondo's games. *Fluctuation and Noise Letters,* 1: 7–12

Toral R. 2002. Capital redistribution brings wealth by Parrondo's paradox. *Fluctuation and Noise Letters,* 2: 305–311

Tu WJ. Protein-protein interactions of lung cancer related proteins. MSc Thesis, Sun Yat-sen University, Guangzhou, China, 2006

Üsdiken B, Leblebici H. Organizational theory. In: *Handbook of Industrial, Work and Organizational Psychology.* Sage Publication Press, UK, 2001

Valiant LG. 1984. A Theory of the Learnable. *Communications of the ACM,* 27(11): 1134–1142

Vapnik V. *The Nature of Statistical Learning Theory.* Springer-Verlag, New York, USA, 1995

Vapnik V. 1999. An overview of statistical learning theory. *IEEE Transactions on Neural Networks,* 10: 988–999

Vlachos A. 2013. Ant colony system algorithm solving a thermal generator maintenance scheduling problem. *Journal of Intelligent & Fuzzy Systems,* 24: 713–723

Venkatech S. Computation and Learning in the Context of Neural Network Capacity. In: *Neural Networks for Perception Vol 2* (Ed Wechsler H). Academic Press, USA, 1992, 173–207

Von der Malsburg. Network self-organization. In: *An Introduction to Neural and Eletric Network* (Zornetger SF *et al.*, eds). Academic Press, USA, 1990

Walsh JP, Meyer, AD, Schoonhoven CB. 2006. A future for organizational theory: living in and living with changing organizations. *Organization Science,* 17(5): 657–671

Wang LG, Hong Y, Zhao FQ, *et al.* 2009a. Improved artificial fish swarm algorithm. *Computer Engineering,* 19: 192–194

Wang LG, Hong Y, Zhao FQ, *et al.* 2009b. Simplified artificial fish swarm algorithm. *Mini-micro Systems,* 8: 1663–1667

Wang LG, Shi QH. 2010. Parameters analysis of artificial fish swarm algorithm. *Computer Engineering,* 36(24): 169–171

Wang LG, Shi QH, Hong Y. 2010. Hybrid optimization algorithm of PSO and AFSA. *Computer Engineering,* 5: 176–178

Wang R, Zhang DL, Wei W, *et al.*, 2014. Interspecific associations of weed species around rice fields in Pearl River Delta, China: A regional survey. *Selforganizology,* 1(3-4): 139–201

Wang XP, Cao LM. *Generic Algorithms: Theory, Applications and Software Implementation.* Xian Jiaotong University Press, Xian, China, 2002

Wang Y. *Discrete Control System.* Science Press, Beijing, China, 1987

Wang YP. 2013. Adaptive ant colony algorithm for the VRP solution of logistics distribution. *Research Journal of Applied Sciences, Engineering and Technology,* 6(5): 807–811

Wang Z, Chen Y, Chen Y. 2009. Functional grouping and establishment of distribution patterns of invasive plants in China using self-organizing maps and indicator species analysis. *Archives of Biological Science,* 61: 71–78

Wang ZJ, Wang NC, Feng F, Tian WF. 2007. Research on evolved behavior of cellular automata. *Application Research of Computers,* 24(8): 38–41

Wang ZY, Nin F, Huang SG, *et al.* 2014. Ant colony optimization based routing and resources assignment in mixed-line-rate IP over OTN networks. *Acta Photonica Sinica,* 43(Suppl.): 01060051-01060054

Warner M. 2007. Kafka, Weber and organizational theory. *Human Relations,* 60(7): 1019–1038

Warren PH, Law R, Weatherby AJ. 2003. Mapping the assembly of protist communities in microcosms. *Ecology,* 84(4): 1001–1011

Warren CP, Pascual M, Lafferty KD, *et al.* 2010. The inverse niche model for food webs with parasites. *Theoretical Ecology,* 3(4): 285–294

Watts DJ. *Six Degrees: The Science of a Connected Age.* W.W. Norton & Company, New York, USA, 2003, 43–47

Watts MJ, Worner SP. 2009. Estimating the risk of insect species invasion: Kohonen self-organising maps versus k-means clustering. *Ecological Modelling,* 220: 821–829

Webb CO, Ackerly DD, McPeek MA, *et al.* 2002. Phylogenies and community ecology. *Annual Review of Ecology and Systematics,* 33: 475–505

Wei N, Huang XY, Liu SD. 2011. Research on principle and improved strategy of quantum evolutionary algorithm. *Computer Engineering,* 37(20): 223–226, 229

Weiher E, Keddy PA. *Ecological Assembly Rules: Perspectives, Advances, Retreats.* Cambridge University Press, USA, 1999

Weinberg S. *Dreams of a Final Theory: The Scientist's Search for the Ultimate Laws of Nature.* Knopf Doubleday Publishing Group, USA, 2011

Whittaker RH. 1979. Classifying species according to their demographic strategy. *American Naturalist,* 113: 185–200

Wikipedia. 2014. *Self-organization.* http://en.wikipedia.org/wiki/Self- organization. Accessed Mar 12, 2014

Wilensky U. *NetLogo Sand Model.* Center for Connected Learning and Computer-Based Modeling, Northwestern University, Evanston, IL, USA, 1996

Wilensky U. *NetLogo Ants Model.* Center for Connected Learning and Computer-Based Modeling, Northwestern University, Evanston, IL, USA, 1997a

Wilensky U. *NetLogo Termites Model.* Center for Connected Learning and Computer-Based Modeling, Northwestern University, Evanston, IL, USA, 1997b

Wilensky U. *NetLogo Cooperation Model.* Center for Connected Learning and Computer-Based Modeling, Northwestern University, Evanston, IL, USA, 1997c

Wilensky U. *NetLogo DLA Model.* Center for Connected Learning and Computer-Based Modeling, Northwestern University, Evanston, IL, USA, 1997d

Wilensky U. *NetLogo Tumor Model.* Center for Connected Learning and Computer-Based Modeling, Northwestern University, Evanston, IL, USA, 1998

Wilensky U. *NetLogo Crystallization Basic Model.* Center for Connected Learning and Computer-Based Modeling, Northwestern University, Evanston, IL, USA, 2002

Wilensky U. *NetLogo Giant Component Model.* Center for Connected Learning and Computer-Based Modeling, Northwestern University, Evanston, IL, USA, 2005

Wilensky U. *NetLogo Hex Cell Aggregation Model.* Center for Connected Learning and Computer-Based Modeling, Northwestern University, Evanston, IL, Evanston, IL, USA, 2007

Wilensky U, Reisman K. 1999. Connected science: learning biology through constructing and testing computational theories — an embodied modeling approach. *International Journal of Complex Systems,* 234: 1–12

Williams GC. *Natural Selection: Domains, Levels and Challenges.* Oxford University Press, Oxford, UK, 1992

Williams RJ. 2010. Simple MaxEnt models explain food web degree distributions. *Theoretical Ecology,* 3: 45–52

Williams RJ, Martinez ND. 2000. Simple rules yield complex food webs. *Nature,* 404: 180–183

Willmott H. Organizational Theory as a critical science? In: *The Oxford Handbook of Organizational Theory* (Tsoukas H, Knudsen C, eds). Oxford University Press, UK, 2003

Wilson JB, Lee WG. 1994. Niche overlap of congeners: A test using plant altitudinal distribution. *Oikos,* 69: 469–475

Wilson JB, Roxburgh SH. 1994. A demonstration of guild-based assembly rules for a plant community, and determination of intrinsic guilds. *Oikos,* 69: 267–276

Wilson EO. *Sociobiology: The New Synthesis. Belknap Press,* Cambridge, MA, USA, 1975

Wilson MA, Bhalla US, Uhley JD, *et al.* 1988. Genesis: A system for simulating neural networks. *NIPS,* 485–492

Witten Jr TA, Sander LM. 1981. Diffusion-limited aggregation, a kinetic critical phenomenon. *Physical Review Letters,* 47: 1400

Wolfram S. *A New Kind of Science.* Wolfram Media, Champaign, IL, USA, 2002

Wolin S. *Politics and Vision.* Allen and Unwin, UK, 1961

Woodcock AER, Davis M. *Catastrophe Theory.* E. P. Dutton, New York, USA, 1978

Xiao JM, Zheng XM, *et al.* A modified artificial fish-swarm algorithm. *IEEE Proceedings of WCICA.* Piscataway, USA, 2006, 3456–3460

Xiao X, Wang P, Chou KC. 2011. Cellular automata and its applications in protein bioinformatics. *Current Protein and Peptide Science,* 12(6): 508–519

Xie XF, Zhang WJ, Yang ZL. 2003. Overview of particle swarm optimization. *Control and Decision,* 18(2): 129–134

Xu L. Bayesian ying-yang system and theory as a unified statistical learning approach (1): for unsupervised and semi-unsupervised learning. In: *Brain-like Computing and Intelligent Information Systems* (Amari S and Kasabov N eds). Springer-Verlag, Germany, 1997, 241–274

Yan PF, Zhang CS. *Artificial Neural Networks and Computation of Simulated Evolution.* Tsinghua University Press, Beijing, China, 2000

Yang W, Li QQ. 2004. Survey on particle swarm optimization algorithm. *Engineering Science,* 6(5): 87–94

Yang XS. *Nature-Inspired Optimization Algorithms.* Elsevier, Netherlands, 2014

Yang XS, Deb S, Loomes M, *et al.* 2013. A framework for self-tuning optimization algorithm, *Neural Computing and Applications,* 23(7-8): 2051–2057

Yao X, Liu Y, Lin GM. 1999. Evolutionary programming made faster. *IEEE Transactions on Evolutionary Computation,* 3(2): 82–102

YashRoy RC. 1987. 13-C NMR studies of lipid fatty acyl chains of chloroplast membranes. *Indian Journal of Biochemistry and Biophysics,* 24(6): 177–178

Young TP, Chase JM, Huddleston RT. 2001. Community succession and assembly comparing, contrasting and combining paradigms in the context of ecological restoration. *Ecological Restoration,* 19(1): 5–18

Yu XJ, Li JH. 2012. Is the natural selection omnipotent? Biological and philosophical controversies on adaptationism in evolutionary. *Studies in Dialectics of Nature,* 28(6): 25–30

Zeeman EC. 1976. Catastrophe theory. *Scientific American,* 234: 65–83

Zeeman EC. *Catastrophe Theory-Selected Papers 1972–1977.* Addison-Wesley, MA, USA, 1977

Zhang J, Liu YJ. 2006. Biological adaptation and biological evolution. *Journal of Shangqiu Teachers College,* 22(5): 146–147

Zhang MF, Shao C, Gan Y, *et al.* 2006. Hybrid artificial fish swarm optimization algorithm based on mutation operator and simulated annealing. *Acta Electronica Sinica,* 34(8): 1381–1385

Zhang Q, Kang LS, Li DN. 2008. A summary for swarm intelligence algorithm and its application. *Journal of Huanggang Normal University,* 28(6): 44–48

Zhang Q, Li CH, Liu Y, *et al.* Fast multi-swarm optimization with Cauchy mutation and crossover operation. In: *Lecture Notes in Computer Science.* Springer, Berlin, Germany, 2007, 344–352

Zhang W, Liu YT, Clerc M. 2006. Adaptive particle swarm for reactive power optimization in power systems. *Engineering Intelligent Systems for Electrical Engineering and Communications,* 14: 161–167

Zhang WJ. 1993. Two-dimensional order clustering and its application in the regionalization of agriculture: Taking Yanan Prefecture in Shaanxi Province as an example. *Bulletin of Soil and Water Conservation,* 13(1): 34–41

Zhang WJ. *Methodology on Ecology Research.* Sun-Yat-sen University, Guangzhou, China, 2007a

Zhang WJ. 2007b. Pattern classification and recognition of invertebrate functional groups using selforganizing neural networks. *Environmental Monitoring and Assessment,* 130: 415–422

Zhang WJ. 2007c. Computer inference of network of ecological interactions from sampling data. *Environmental Monitoring and Assessment,* 124: 253–261

Zhang WJ. *Computational Ecology: Artificial Neural Networks and Their Applications.* World Scientific, Singapore, 2010

Zhang WJ. 2011a. A Java algorithm for non-parametric statistic comparison of network structure. *Network Biology,* 1(2): 130–133

Zhang WJ. 2011b. A Java program for non-parametric statistic comparison of community structure. *Computational Ecology and Software,* 1(3): 183–185

Zhang WJ. 2011c. A Java program to test homogeneity of samples and examine sampling completeness. *Network Biology,* 1(2): 127–129

Zhang WJ. 2011d. Constructing ecological interaction networks by correlation analysis: hints from community sampling. *Network Biology,* 1(2): 81–98

Zhang WJ. *Computational Ecology: Graphs, Networks and Agent-based Modeling.* World Scientific, Singapore, 2012a

Zhang WJ. 2012b. How to construct the statistic network? An association network of herbaceous plants constructed from field sampling. *Network Biology,* 2(2): 57–68

Zhang WJ. 2012c. Modeling community succession and assembly: A novel method for network evolution. *Network Biology,* 2(2): 69–78

Zhang WJ. 2013a. Selforganizology: A science that deals with self-organization. *Network Biology,* 3(1): 1–14

Zhang WJ. An overview on theories and methods of self-organization. In: *Self-organization: Theories and Methods.* Nova Science Publishers, New York, USA, 2013b

Zhang WJ. *Self-organization: Theories and Methods.* Nova Science Publishers, New York, USA, 2013c

Zhang WJ. 2014. Selforganizology: A more detailed description. *Selforganizology,* 1(1): 31–46

Zhang WJ. 2014a. A framework for agent-based modeling of community assembly and succession. *Selforganizology,* 1(1): 16–22

Zhang WJ. 2014b. Interspecific associations and community structure: A local survery and analysis on a grass community. *Selforganizology,* 1(2): 89–129

Zhang WJ. 2014c. Selforganizology: A more detailed description. *Selforganizology,* 1(1): 31–46

Zhang WJ. 2014d. Research advances in theories and methods of community assembly and succession. *Environmental Skeptics and Critics,* 3(3): 52–60

Zhang WJ. 2015. Coevolution: A synergy in biology and ecology. *Selforganizology,* 2(2): 35–38

Zhang WJ. 2015a. A hierarchical method for finding interactions: Jointly using linear correlation and rank correlation analysis. *Network Biology,* 5(4) (Accept)

Zhang WJ. 2015b. Calculation and statistic test of partial correlation of general correlation measures. *Selforganizology,* 2(4) (Accept)

Zhang WJ, Qi YH, Zhang ZG. 2015a. A cellular automaton for population diffusion in the homogeneous rectangular area. *Selforganizology,* 2(1): 13–17

Zhang WJ, Bai CJ, Liu GD. 2007. Neural network modeling of ecosystems: A case study on cabbage growth system. *Ecological Modelling,* 201: 317–325

Zhang WJ, Barrion AT. 2006. Function approximation and documentation of sampling data using artificial neural networks. *Environmental Monitoring and Assessment,* 122: 185–201

Zhang WJ, Gu DX. 1996. A kind of algorithm for stochastic decision-making in insect pest management. *Bulletin of Science and Technology.* 12(5): 288–293

Zhang WJ, Gu DX. 2001. A non-linear partial differential equation to describe spatial and temporal changes of insect population. *Ecologic Science,* 20(4): 1–7

Zhang WJ, Jiang LQ, Chen WJ. 2014a. Effect of parasitism on food webs: Topological analysis and goodness test of cascade model. *Network Biology,* 4(4): 170–178

Zhang WJ, Li QH. 2007. Development of topological functions in neural networks and their application in SOM learning to biodiversity. The *Proceedings of the China Association for Science and Technology,* 4(2): 583–586

Zhang WJ, Li X. 2015. Linear correlation analysis in finding interactions: Half of predicted interactions are undeterministic and one-third of candidate direct interactions are missed. *Selforganizology,* 2(3): 39–45

Zhang WJ, Liu GH, Dai HQ. 2008a. Simulation of food intake dynamics of holometabolous insect using functional link artificial neural network. *Stochastic Environmental Research and Risk Assessment,* 22: 123–133

Zhang WJ, Pang Y, Qi YH, Chen QJ, 1997. Simulation model for epizootic disease of *Spodoptera litura* F. baculovirus. *Acta Scientiarum Naturalium Universitatis Sunyatseni,* 36(1): 54–59

Zhang WJ, Qi YH. 2005. Pattern classification of HLA-DRB1 alleles and human races using Self-Organizing Competitive Neural Network. *Modern Computer,* 4: 10–13, 30

Zhang WJ, Qi YH, Zhang ZG. 2004. An algorithm and network software used to detect homogeneity of ecological habitat and its randomization statistic test. *Computer Applications and Software,* 21(11): 66–69

Zhang WJ, Qi YH, Zhang ZG. A Cellular Automata Method for Species Migration Process in Heterogeneous Environment. In: *Self-organization: Theories and Methods* (WenJun Zhang, ed). Nova Science Publishers, New York, USA, 2013, 195–200

Zhang WJ, Qi YH, Zhang ZG. 2014b. Two-dimensional ordered cluster analysis of component groups in self-organization. *Selforganizology,* 1(2): 62–77

Zhang WJ, Schoenly KG. IRRI Biodiversity Software Series. II. COLLECT1 and COLLECT2: Programs for Calculating Statistics of Collectors' Curves. *IRRI Technical Bulletin* No. 4. 1–15, International Rice Research Institute, Manila, Philippines, 1999a

Zhang WJ, Schoenly KG. IRRI Biodiversity Software Series. III. BOUNDARY: A Program for Detecting Boundaries in Ecological Landscapes. *IRRI Technical*

Bulletin No. 4. 1–17, International Rice Research Institute, Manila, Philippines, 1999b

Zhang WJ, Schoenly KG. 2001. A randomization test and software to compare ecological communities. *International Rice Research Notes,* 2: 48–49

Zhang WJ, Wopke van der Werf, Pang Y. 2011. A simulation model for vegetable-insect pest-insect nucleopolyhedrovirus epidemic system. *Journal of Environmental Entomology,* 33(3): 283–301

Zhang WJ, Wei W. 2009. Spatial succession modeling of biological communities: A multi-model approach. *Environmental Monitoring and Assessment* 158: 213–230

Zhang WJ, Zhang XY. 2008. Neural network modeling of survival dynamics of holometabolous insects: A case study. *Ecological Modelling,* 211: 433–443

Zhang WJ, Zhong XQ, Liu GH. 2008b. Recognizing spatial distribution patterns of grassland insects: Neural network approaches. *Stochastic Environmental Research and Risk Assessment,* 22(2): 207–216

Zhang WJ, Zheng H. 2012. A program for statistic test of community evenness. *Computational Ecology and Software,* 2(1): 80–82

Zhang XP, Wang JY, Yang AY, *et al.* 2013. Identifying interacting genetic variations by fish-swarm logic regression. *BioMed Research International* (doi: 10.1186/1471-2105-14-246 PMCID: PMC375134511)

Zhang YT, Fang KT. *Introduction to Multivariate Statistics.* 393–401, Science Press, Beijing, China, 1982

Zhao SN, Yu YX. *Catastrophe Theory and Its Application in Biomedical Science.* Science Press, Bejing, China, 1987

Zhao XC. 2014. Swarm intelligence optimization algorithms and life. http://blog.sciencenet.cn/blog-86581-772563.html. Accessed Mar 4, 2014

Zhao Y, Zhang WJ. 2013. Organizational theory: With its applications in biology and ecology. *Network Biology,* 3(1): 45–53

Zheng X. 2008. The adaptation of living creatures and adaptive education. *Journal of Qiongzhou University,* 15(3)

Zhou LK, Liu HZ. 2013. The application of three-dimensional path planning of wheeled robot based on adaptive ant colony system algorithm. *Mechanical Science and Technology for Aerospace Engineering,* 32(1): 54–58

Zhou Q, Li YD. 2003. Directed variation in evolution strategies. *IEEE Transactions on Evolutionary Computation,* 7(4): 356–366

Zipf GK. *Human Behavior and the Principle of Least Effort.* Addison-Wesley Press, New Jersey, USA, 1949

Zou XH, He P, Guo LP. 2001. From the evolution of the theory of evolution to that of science. *Journal of Chongqing Normal University (Natural Science Edition),* 18(3): 81–86

Index